Quantum Concepts
in Space and Time

QUANTUM CONCEPTS
IN
SPACE AND TIME

Edited by

R. PENROSE

Mathematical Institute, University of Oxford

and

C. J. ISHAM

The Blackett Laboratory,
Imperial College of Science and Technology

CLARENDON PRESS · OXFORD
1986

Oxford University Press, Walton Street, Oxford OX2 6DP

Oxford New York Toronto
Delhi Bombay Calcutta Madras Karachi
Petaling Jaya Singapore Hong Kong Tokyo
Nairobi Dar es Salaam Cape Town
Melbourne Auckland

and associated companies in
Beirut Berlin Ibadan Nicosia

Oxford is a trade mark of Oxford University Press

Published in the United States
by Oxford University Press, New York

British Library Cataloguing in Publication Data
Quantum concepts in space and time.
1. Quantum theory
I. Penrose, R.   II. Isham, C. J.
530.1'2   QC174.12
ISBN 0-19-851972-9

Library of Congress Cataloging in Publication Data
Main entry under title:
Quantum concepts in space and time.
"A Third Oxford Symposium"—P.
Bibliography: p.
Includes index.
1. Quantum theory—Congresses.   2. General relativity (Physics)—Congresses.   3. Quantum gravity—Congresses.   4. Space and time—Congresses.
I. Penrose, Roger.   II. Isham, C. J.
QC173.96.Q36   1986   530.1'2   85-18963
ISBN 0-19-851972-9.

Typeset and Printed in Northern Ireland by
The Universities Press, Belfast.

Preface

In 1975 and 1981 the Oxford University Press produced, under our editorship (together with that of our colleague Dennis Sciama), separate volumes with the title *Quantum Gravity*. These were the proceedings of the 1974 Conference at the Rutherford Laboratory and the 1980 Conference at the Mathematical Institute in Oxford.

In March 1984 we held a further conference, this time at Lincoln College, Oxford, and which, like the preceding one, was generously supported by the Nuffield Foundation. However, we felt that there had been insufficient progress in the intervening years to merit another meeting attempting once again to cover all aspects of work done in the subject. Instead, it seemed more opportune to re-examine certain foundational questions relevant to quantum gravity; in particular we wished to explore the possibility that the rules of quantum theory itself might need to be modified before a successful union with general relativity can be achieved. Our meeting was therefore more concerned with quantum theory and its foundations than with gravitational theory proper. Thus, while many of the talks did have direct relevance to general relativity, others were concerned with quantum mechanics, its experimental support, its strange and sometimes paradoxical features, its underlying philosophy, and its possible modifications. There was however one overriding and unifying theme: the conceptual problems of quantum physics in relation to space and time. This is reflected in our choice of title for the present volume which (as the more perceptive reader may observe) although actually meaningless, nevertheless encompasses the overall spirit of our endeavours!

We found some overlap with a conference held the previous year in Tokyo entitled 'International Symposium on Foundations of Quantum Mechanics in the Light of New Technology'. Three of our speakers, Aharonov, Aspect, and Leggett, considered that their talks at our meeting were essentially identical to the ones they had previously delivered at the Tokyo conference and felt that it would be redundant to provide a second write-up of the same material. Accordingly, we have obtained the generous permission of the organizer of that meeting and the publisher of their proceedings (Dr Y. Murayama and the Physical Society of Japan) to reprint the articles of these three talks from that volume. We express our grateful thanks for this permission. The remaining 24 articles appear here for the first time.

For the most part the articles are discursive rather than of a detailed technical nature since it was more our intention to explore foundations than to develop specific techniques. The problems of quantum non-locality, of state vector reduction, and of the possible links with gravity were questions closest to our minds, in addition to theory. But the discussions were free-ranging and many other topics were covered. We think that the accounts that we present here convey the flavour of what we believe was a very successful meeting. Even if the participants did not present the solutions to the major problems they provided much food for thought and, we hope, some promising directions for future development.

Oxford   R.P.
and London   C.J.I.
March 1985

Contents

Contributors

Y. AHARONOV, Department of Physics and Astronomy, Tel Aviv University, Tel Aviv, Israel.

J. ANANDAN, Max Planck Institut für Physik and Astrophysik, Werner Heisenberg Institut für Physik, Föhringer Ring 6, 8000 Munich 40, West Germany.

A. ASHTEKAR, Physics Department, Syracuse University, Syracuse, New York 13210, USA, and Physique Théoretique, Université de Paris VI, 75231 Paris, France.

A. ASPECT, Institut d'Optique Theorique et Appliquée, Université Paris-Sud, Centre d'Orsay B.P. 43, F 91406 Orsay Cédex, France

J. B. BARBOUR, College Farm, South Newington, Banbury, Oxon., UK.

I. BIALYNICKI-BIRULA, Institute for Theoretical Physics, Polish Academy of Sciences, Lotnikow 32/46, 02-668 Warsaw, Poland.

P. D. D'EATH, Department of Applied Mathematics and Theoretical Physics, University of Cambridge, Silver Street, Cambridge, UK.

D. DEUTSCH, Department of Astrophysics, University of Oxford, Oxford, UK, and Center for Theoretical Physics, University of Texas at Austin, Texas 78712, USA.

B. DeWITT, Department of Physics, University of Texas at Austin, Austin, Texas 78712, USA.

D. FINKELSTEIN, Georgia Institute of Technology, Atlanta, Georgia 30332, USA.

A. FRENKEL, Department of Theoretical Particle Physics, Central Research Institute for Physics, Budapest, Hungary.

C. GEFWERT, Stanford Linear Accelerator Center, Stanford University, Stanford, California 94035, USA.

P. GRANGIER, Institut d'Optique Theorique et Appliquée, Université Paris-Sud, Centre d'Orsay B.P. 43, F 91406 Orsay Cédex, France.

P. HAJICEK, Institute for Theoretical Physics, University of Bern, Sidlerstrasse 5, CH-3012 Bern, Switzerland.

L. P. HUGHSTON, Lincoln College, Oxford OX1 3DR, UK.

F. KÁROLYHÁZY, Department of Theoretical Physics, Eötvös Loránd University, Budapest, Hungary.

A. J. LEGGETT, Laboratory of Atomic and Solid State Physics, Cornell University, Ithaca, New York 14853, USA.

B. LUKÁCS, Department of Theoretical Particle Physics, Central Research Institute for Physics, Budapest, Hungary.

M. J. MANTHEY, Department of Computer Science, University of New Mexico, Albuquerque, New Mexico 87131, USA.

H. P. NOYES, Stanford Linear Accelerator Center, Stanford University, Stanford, California 94035, USA.

D. N. PAGE, Department of Physics, The Pennsylvania State University, University Park, Pennsylvania 16802, USA.

P. PEARLE, Hamilton College, Clinton, New York 13323, USA.

R. PENROSE, Mathematical Institute, University of Oxford, Oxford, UK.

E. RODRIGUEZ, Georgia Institute of Technology, Atlanta, Georgia 30332, USA.

N. SANCHEZ, ER 176 (CNRS), D.A.F. Observatoire de Meudon, 92190-Meudon, France.

M. SCHWARTZ, School of Physics and Astronomy, Tel Aviv University, Tel Aviv 69978, Israel.

A. SHIMONY, Departments of Philosophy and Physics, Boston University, Boston, Massachusetts 02215, USA.

L. SMOLIN, Department of Physics, Yale University, New Haven, Connecticut 06511, USA.

A. SUDBERY, Department of Mathematics, University of York, Heslington, York YO1 5DD, UK.

F. J. TIPLER, Department of Mathematics and Department of Physics, Tulane University, New Orleans, Louisiana 70118, USA.

TSOU SHEUNG TSUN, Mathematical Institute, University of Oxford, Oxford, UK.

R. M. WALD, Enrico Fermi Institute and Department of Physics, University of Chicago, 5640 South Ellis Avenue, Chicago, Illinois 60637, USA.

B. F. WHITING, ER 176 (CNRS), D.A.F. Observatoire de Meudon, 92190-Meudon, France.

A. ZEILINGER, Atominstitut der Österreichischen Universitäten, Schüttelstrasse 115, A-1020 Wien, Austria.

1

Experiments on Einstein–Podolsky–Rosen-type correlations with pairs of visible photons

Alain Aspect and Philippe Grangier

Our subject is related to a general question, raised early in the development of quantum mechanics (Von Neumann 1955): is it possible (is it necessary) to understand the probabilistic nature of quantum mechanical predictions by invoking a more precise description of the world at a deeper level? Such a description would complete quantum mechanics as statistical mechanics completes thermodynamics.

In a famous paper, Einstein *et al.* (1935) used a reasoning based on a thought experiment to conclude that quantum mechanics must be completed. Since Bohr disagreed with this conclusion, most physicists thought that the commitment to either position was just a matter of taste (Bohr 1935). A great advance in this question was attained with the discovery by Bell (1964) that the two points of view led to different numerical predictions when applied to Bohm's version (Bohm 1951) of the EPR *Gedankenexperiment.* Bell's paper then opened the route to real experiments which were designed at the beginning of the 1970s. We will be concerned here with experiments using pairs of low energy photons, inspired by Clauser *et al.* (1969). Since there are good reviews of previous experiments (Clauser and Shimony 1978; Pipkin 1978; Selleri and Tarrozzi 1981) we will report in particular on the three last experiments of this type that we carried out at the Institut d'Optique d'Orsay. We will see that great progress was possible thanks to the wonderful tool in atomic physics which is the laser.

1.1. The Einstein–Podolsky–Rosen–Bohm thought experiment with photons

In D. Bohm (1951) a simpler version of the EPR thought experiment was given using measurements of spin components of spin-$\frac{1}{2}$ particles. There is

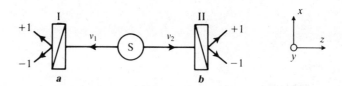

Fig. 1.1. EPRB *Gedankenexperiment* with photons. The two photons v_1 and v_2 counter-propagate along Oz and impinge on the linear polarization analysers I and II. The results $+1$ and -1 are assigned to linear polarizations parallel or perpendicular to the orientation of the polarizer (this orientation is characterized by a unit vector a or b). For a suitable state vector $|\psi(1, 2)\rangle$, quantum mechanics predicts strong correlations between the results of measurements on both sides.

a straightforward correspondence with measurements of the linear polarization of photons, and we will rather consider this latter case, which is closer to our experiments.

Let us consider pairs of photons of different energies, v_1 and v_2, counter-propagating along $+Oz$ and $-Oz$ (Fig. 1.1). We suppose that they are in a state

$$|\psi(v_1, v_2)\rangle = \frac{1}{\sqrt{2}}(|x_1, x_2\rangle + |y_1, y_2\rangle), \tag{1.1}$$

where $|x_1\rangle$ refers to a photon v_1 propagating along $-Oz$ linearly polarized along Ox, etc.

The apparatuses I and II are linear polarization analysers (for instance Wollaston prisms). They perform dichotomic measurements, i.e. a photon can be found in one of the two exit channels, labelled $+1$ or -1. This is similar to a Stern–Gerlach filter acting on spin-$\frac{1}{2}$ particles.

It is an elementary exercise in quantum mechanics to derive the probabilities of the results of the various measurements. For single measurements, with analysers I and II in orientations a and b, we obtain

$$\begin{aligned} P_+(a) &= P_-(a) = \tfrac{1}{2} \\ P_+(b) &= P_-(b) = \tfrac{1}{2}. \end{aligned} \tag{1.2}$$

For joint measurements, the quantum mechanical predictions are

$$\begin{aligned} P_{++}(a, b) &= P_{--}(a, b) = \tfrac{1}{2}\cos^2(a \cdot b) \\ P_{+-}(a, b) &= P_{-+}(a, b) = \tfrac{1}{2}\sin^2(a \cdot b). \end{aligned} \tag{1.3}$$

If we consider the special situation $a \cdot b = 0$ (same directions of analysis for both photons) we find

$$\begin{aligned} P_{++}(0) &= P_{--}(0) = \tfrac{1}{2} \\ P_{+-}(0) &= P_{-+}(0) = 0. \end{aligned}$$

We can then conclude that there is a strong correlation between the results of measurements on both photons. As a matter of fact, if we find v_1 in the +1 channel (the probability of which is 50 per cent), we are then sure to find v_2 in the +1 channel. But if we had found v_1 in the −1 channel we would have found v_2 in the −1 channel. The results are thus strongly correlated.

It will be convenient for the following to introduce the coefficient of correlation of polarization

$$E(a, b) = P_{++}(a, b) + P_{--}(a, b) - P_{+-}(a, b) - P_{-+}(a, b). \quad (1.4)$$

Using eqn (1.3), we obtain the quantum mechanical value for this coefficient:

$$E_{QM}(a, b) = \cos^2(a \cdot b) \quad (1.5)$$

This function can assume the values +1 or −1, which is another way of showing a complete correlation.

It is difficult to understand these correlations with the standard interpretation of quantum mechanics. According to this interpretation, photon v_1 has a 50 per cent chance of going into channel +1 and a 50 per cent chance of going into channel −1 until the moment when the measurement takes place, and similarly for photon v_2. But if v_1 goes into channel +1, then v_2 goes into channel +1, and conversely. One can wonder how v_2 'knows' which channel was chosen (at the last moment) for v_1.

On the other hand, it is easy to understand strong correlations between distant measurements on two systems that have previously interacted, by a classical picture involving common properties of the two members of a pair. The existence of such properties can be derived from a reasoning similar to EPR. Let us return to the special case $(a \cdot b) = 0$ and consider a result +1 for v_1. We are then sure to find +1 for v_2, and we are thus led to admit that there is some property (Einstein spoke of 'an element of physical reality') determining this result. For another pair, yielding the results $(-1, -1)$, this property would be different.

We have thus been led to introduce properties differing for the various pairs, while the quantum mechanical state vector $|\psi(1, 2)\rangle$ is the same for all pairs. This is why one can conclude at this stage—following EPR—that quantum mechanics is not complete, since we had to introduce supplementary parameters (often called 'hidden variables').

1.2. Bell's theorem

J. Bell has drawn the consequences from the preceding discussion, and has written some mathematics in agreement with its conclusions. He has introduced explicit supplementary parameters, denoted by λ, distributed over the ensemble of emitted pairs according to the probability distribution $\rho(\lambda)$

$$\rho(\lambda) \geqq 0 \quad \text{and} \quad \int d\lambda \rho(\lambda) = 1. \tag{1.6}$$

For a pair characterized by a supplementary parameter λ, the results of measurements will be

$$A(\lambda, \boldsymbol{a}) = +1 \text{ or } -1 \text{ at analyser I}$$
$$B(\lambda, \boldsymbol{b}) = +1 \text{ or } -1 \text{ at analyser II.} \tag{1.6'}$$

This formalism holds for a whole class of theories. A specific theory will yield the functions $A(\lambda, \boldsymbol{a})$, $B(\lambda, \boldsymbol{b})$ and $\rho(\lambda)$. It will then be easy to express the probabilities of the results of the various measurements. For our purpose, it is sufficient to notice that the correlation function can be written

$$E(\boldsymbol{a}, \boldsymbol{b}) = \int d\lambda \rho(\lambda) A(\lambda, \boldsymbol{a}) B(\lambda, \boldsymbol{b}). \tag{1.6''}$$

Assuming only that eqn (1.6) holds, it is straightforward to demonstrate Bell's inequalities. A convenient form of these inequalities is the one found in Clauser *et al.* (1969)

$$-2 \leqq S \leqq 2 \tag{1.7}$$

with

$$S = E(\boldsymbol{a}, \boldsymbol{b}) - E(\boldsymbol{a}, \boldsymbol{b}') + E(\boldsymbol{a}', \boldsymbol{b}) + E(\boldsymbol{a}', \boldsymbol{b}'). \tag{1.7'}$$

These inequalities appear as a constraint on a combination of polarization correlation functions, measured in various orientations of the polarizers.

On the other hand, we can find situations in which the quantum mechanical calculations lead to a value of S that does not obey Bell's inequalities. For instance, for the EPR situation with photons described by the state vector (1.1), the quantum mechanical predictions are given by eqn (1.5). If we consider the particular set of orientations of Fig. 1.2,

Fig. 1.2. A set of orientations leading to a strong conflict between quantum mechanics and Bell's inequalities.

the quantity S then assumes the value

$$S_{QM} = 2\sqrt{2}$$

in conflict with inequality (1.7).

It is thus clear that no theory following Bell's formalism (eqn (1.6)) can reproduce all the quantum mechanical predictions, since some of these predictions violate a consequence of eqn (1.6). This result is the essence of Bell's theorem.

There has been a great deal of discussion, to elucidate the significance of this result. Obviously, it is important to point out the hypotheses assumed by eqn (1.6). We can first remark that this formalism is deterministic, since the results of measurements $A(\lambda, a)$ and $B(\lambda, b)$ are certain, given the supplementary parameter λ and the orientations a and b. At first sight, one could believe that this is the reason for the conflict. But further generalizations by Bell (1971), and by Clauser and Horne (1974), have shown that there exist non-deterministic supplementary parameter theories that lead to the same conflict with quantum mechanics. Although there is some controversy on this point (Fine 1982), it thus seems that determinism is not the reason for the conflict.

On the contrary, there is an assumption which is essential in the derivation of Bell's inequalities—i.e. of a conflict with quantum mechanics—that is, Bell's locality condition. This condition is involved in eqn (1.6) since the results of the measurements by I—$A(\lambda, a)$—do not depend on the orientation b of the remote polarizer II (and vice versa). Similarly, the distribution $\rho(\lambda)$ that specifies the way in which the pairs are emitted does not depend on the orientations a and b.

Since his first paper, Bell has insisted upon the necessity of this locality condition for a conflict with quantum mechanics. It can be considered a natural hypothesis, according to our knowledge of the working of the polarizers and of the source. But it is not necessitated by any fundamental law. Bell thus insisted upon the importance of 'experiments of the type proposed by Bohm and Aharonov, in which the settings are changed during the time of flight of the particles'.† In such a time-dependent experiment, the locality condition would then become a consequence of Einstein's causality, which precludes faster than light influences.

In summary, Bell's theorem has established the impossibility of reproducing all quantum mechanical predictions with local supplementary parameter theories. Generalizing somewhat, one can say that Bell's theorem claims that there is no classical-looking picture, in the spirit of Einstein's idea, able to mimic all of the predictions of quantum mechanics. The answer to the question in the introduction is thus that it

† This idea was already expressed in Bohm (1951).

is *not* possible to consider quantum mechanics as an average of a deeper level theory, at least if we impose certain classical-looking features at this level.

1.3. Experiments—generalities

When Bell's theorem was published in 1965, quantum mechanics was a very well established theory, supported experimentally by a very large number of situations. One could thus consider Bell's theorem as a proof of the impossibility of supplementary parameters. However, it was soon realized that situations in which a conflict arises (sensitive situations) are so rare that none had occurred up to that time.

Obviously, the whole of classical physics obeys Bell's inequalities, since Bell's formalism (1.6) (or its generalizations) applies to classical mechanics and classical electrodynamics (in this latter case, just take the currents and charges of the sources as λ). Moreover, even in a situation describable only by quantum mechanics (and similar to that of Fig. 1.1, i.e. involving correlated measurements on two separated subsystems) there is seldom a predicted conflict with Bell's inequalities. We can point out two important necessary conditions for a sensitive situation:

(i) the two subsystems must not be in a mixture of factorizing states;

(ii) for each subsystem, it must be possible to choose at will a measurable quantity between two non-commuting observables (such as polarization measurements along two different directions *a* and *a'*).

These two conditions clearly show that the conflict only arises when the quantum mechanical calculations involves an interference between terms where each subsystem has a definite state. Then the result is different from that obtained with a mixture of such factorizing terms (for instance, state (1.1) leads to predictions different from a mixture of states $|x_1, x_2\rangle$ and $|y_1, y_2\rangle$).

When people realized that there were no experimental data available for a test of Bell's inequalities versus quantum mechanics, real experiments were devised for this specific purpose. We will not describe these experiments in detail because of the excellent reviews already cited. Let us just mention two kinds of experiments. The first one used pairs of γ photons produced by annihilation of positronium in its ground (singlet) state. Since there are no good polarizers available for such γ photons, the test was actually quite indirect and is questionable in the context of Bell's theorem. Anyway, these experiments (most of them) have shown a good agreement with quantum mechanics. We can put in the same class an experiment with protons scattered in a singlet state.

Closer to the thought experiment of Fig. 1.1 were the experiments using visible photons produced in atomic cascades. Four of these experiments (Freedman and Clauser 1972; Holt 1973; Clauser 1976; Fry and Thomson 1976) were carried out during the seventies, prompted by the paper of Clauser *et al.* (1969). These experiments gave conflicting results, which can easily be understood when one knows the difficulty and the poor signal in the first three experiments. However, a majority of them yielded results in agreement with quantum mechanics, especially that of Fry and Thomson, which used a laser for a better signal.

We must mention here that these experiments differ from the thought experiment of Fig. 1.1 in several respects. A major difference is the use of single channel polarizers instead of true polarization analysers with two channels (+1 and −1). These experiments therefore require indirect reasoning and auxiliary calibrations for a test of Bell's inequalities. Let us also remark that none of these experiments offered the possibility of rapidly changing the settings of the polarizers.

This is why we thought that some experimental progress could be achieved in this kind of experiment. With the technological progress in atomic physics—mostly related to advances in laser technology—it seemed possible, at the end of the 1970s, to build a better source of pairs of correlated photons. With such a source it would be possible to perform more accurate experiments involving various auxiliary checks. Moreover, new experimental schemes closer to the thought experiment would become feasible (Aspect 1975, 1976).

1.4. Orsay experiments

Much of our work was first devoted to building a good source of pairs of photons correlated in polarization. Clauser *et al.* (1969) had shown that some atomic cascades could yield suitable pairs of photons. For instance, a $J = 0 \rightarrow J = 1 \rightarrow J = 0$ cascade (J is the atomic angular momentum) yields photons in the state (1.1), if these photons are filtered in energy and in direction (Fig. 1.3).

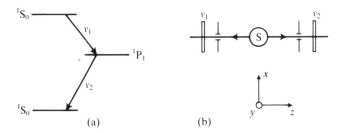

Fig. 1.3. Production of pairs of photons in an EPR-type state. (a) Convenient atomic radiative cascade; (b) ideal geometry (infinitely small solid angles).

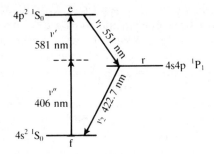

Fig. 1.4. Two-photon excitation of the chosen cascade in calcium.

Moreover, when one considers a more realistic experiment with light collected in finite solid angles, such a cascade is particularly convenient (Fry 1973) since the correlation of polarization predicted by quantum mechanics decreases only slightly (by 1 per cent) for an angle as large as 60°.

Such a cascade, namely the $4p^2\,^1S_0$–$4s\,4p\,^1P_1$–$4s^2\,^1S_0$ cascade in calcium, had been used in the first experiments of this kind (Freedman and Clauser 1972). However, at the time of those experiments it was not possible to produce an efficient excitation, and the source was weak. We have since been able to build a far better source by using a direct two-photon excitation of the upper level of this cascade (Fig. 1.4). This two photon process could be achieved with use of two lasers—a krypton laser at 406 nm, and a tunable dye laser at 580 nm focused onto a calcium atomic beam (Aspect *et al.* 1980). Both lasers are single mode operated, and the dye laser is tuned at resonance for the two-photon process. Several feedback loops control the lasers, so providing the required stability of the source (better than 1 per cent for several hours).

With a few tens of milliwatts from each laser, focused on less than $0.01\,\text{mm}^2$ on to an atomic beam with a density about 10^{10} atoms cm^{-3}, we could achieve a cascade rate $\mathcal{N}$ higher than 10^7 cascades s^{-1}. An increase over this rate would not significantly improve the signal-to-noise ratio for a coincidence counting experiment. Indeed the accidental background (Fig. 1.5) increases as $\mathcal{N}^2$, while the true coincidence rate increases as $\mathcal{N}$, and it is easy to show that it is not worth increasing $\mathcal{N}$ when this rate gets close to the inverse of the lifetime of the intermediate state of the cascade (5 ns in our case).

At such rates, and with our detection efficiencies (over 10^{-3} on each side) we could achieve coincidence rates of a few tens per second, yielding an accuracy close to 1 per cent for only 100s of coincidence counting (in the best of the previous experiments (Fry and Thomson 1976), such an accuracy was achieved in 80 min).

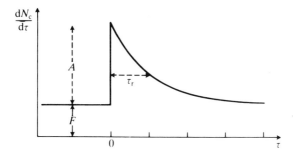

Fig. 1.5. Time-delay spectrum. Number of detected pairs as a function of the delay between the detections of the two-photons. The signal is the peak, of height A, exhibiting a decay time-constant τ_r (lifetime of the intermediate state of the cascade). The flat background is due to accidental coincidences between photons emitted by different atoms.

1.4.1. *Experiment with one-channel polarizers* (Aspect *et al.* 1981)

Our first experiment was of the same kind as the previous ones, since it used single channel polarizers. Our home-built, 'pile-of-plates' polarizers had an excellent optical grade, and no displacement of the beams was observed when rotating or withdrawing these polarizers. Consequently, it was not necessary to average over the various orientations as in previous experiments. Our experimental data have shown a violation of Bell's inequalities (adapted to this type of experiment (Clauser *et al.* 1969)) by 9 standard deviations. Simultaneously, as shown in Fig. 1.6, we have

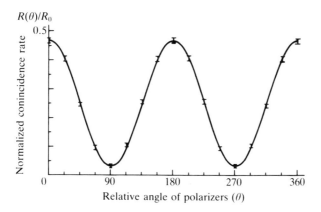

Fig. 1.6. Experiment with one-channel polarizers. The indicated errors on experimental results are ±1 standard deviation. The solid curve is not a fit to the data but the prediction by quantum mechanics for the real experiment.

found an excellent agreement with the predictions of quantum mechanics for the actual experiment, taking into account the efficiencies of the polarizers and the finite solid angles.

In order to test the suggested possibility that the correlations decrease at large distance (de Broglie 1974; this point had been much discussed in the context of γ photon experiments (Bohm and Hiley 1976)) we have repeated these measurements with the polarizers at 6.5 m from the source, i.e. at four coherence-lengths associated with the lifetime of the intermediate state of the cascade. In such an experiment, the detection events on both sides are space-like separated. We have not observed any modification of the polarization correlation in such a situation.

1.4.2. *Experiment with two-channel polarizers* (Aspect *et al.* 1982*b*)

With single-channel polarizers, the polarization measurements are inherently incomplete. When a pair has been emitted, if no count is got at one of the photomultipliers, there is no way to know if it has been 'missed' by the detector (due to the poor detection efficiency) or if it has been blocked by the polarizer (only the latter case corresponds to a -1 result for the measurements). This is why one has to resort to auxiliary experiments, and indirect reasoning, in order to test Bell's inequalities (Clauser *et al.* 1969).

With the use of two-channel polarizers, it is possible to follow much more closely the ideal experimental scheme of Fig. 1.1. We have been able to perform such an experiment (Fig. 1.7) with polarization splitters made of thin dielectric layers deposited on glass. Each splitter is followed

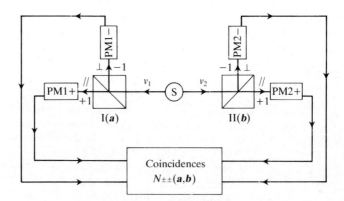

Fig. 1.7. Experiment with two-channel polarizers. Each polarization splitter is followed by two photomultipliers, feeding a four-fold coincidence circuit, which simultaneously monitors the four coincidence rates $N_{\pm\pm}(\boldsymbol{a}, \boldsymbol{b})$. Each analyser can rotate around the beam axis for the choice of the orientation.

by two photomultipliers, and the whole device is mounted in a mechanism rotatable around the axis of the analysed beam. Such a device is the optical analogue of a Stern–Gerlach filter.†

The two polarizers, analysing photons v_1 and v_2, feed a four-fold coincidence system. In a single run we monitor the four coincidence rates $N_{\pm\pm}(a, b)$, from which we can directly infer the polarization correlation coefficient for orientations a and b:

$$E(a, b) = \frac{N_{++}(a, b) + N_{--}(a, b) - N_{+-}(a, b) - N_{-+}(a, b)}{N_{++}(a, b) + N_{--}(a, b) + N_{+-}(a, b) + N_{-+}(a, b)} \quad (1.8)$$

We have just to repeat this measurement in three other orientations in order to test the inequalities (1.7).

As in previous experiments, we have the problem of the low efficiency of the photomultipliers, since only a small number of pairs are actually detected. In order to have a significant test, we must therefore assume that the measured value (eqn (1.8)) of $E(a, b)$ is a good estimator of the value defined over all emitted pairs (eqn (1.4)). That is to say, we have to assume that the ensemble of the pairs actually detected is a faithful sample of all emitted pairs. This assumption is very reasonable in our very symmetrical scheme, where the two result channels ($+1$ and -1) are treated in a similar way. Moreover, we have made various checks which justify this assumption. For instance, we have verified that the sum of the single rates on the $+1$ and -1 channels are constant when the analyser is rotated with a polarized (linearly or circularly) impinging light beam. We have also checked, during the correlation measurements, that the sum of the four coincidence rates $N_{\pm\pm}(a, b)$ is constant when the orientations are changed (the source being driven at constant intensity). This latter test shows that the size of the selected sample is constant, which is in agreement with our assumption.

Measurements have been made at the orientations in Fig. 1.2 (in which the greatest conflict is predicted). We have found

$$S_{\text{exp}} = 2.70 \pm 0.015. \quad (1.9)$$

This result violates the inequality (1.7) by more than 40 standard deviations, and it is in excellent agreement with the quantum mechanical predictions for our polarizers and solid angles:

$$S_{\text{QM}} = 2.70.$$

For a direct comparison with the quantum mechanical calculations, we

---

† A similar experiment with calcite two-channel polarizers has been undertaken at the University of Catania, Italy.

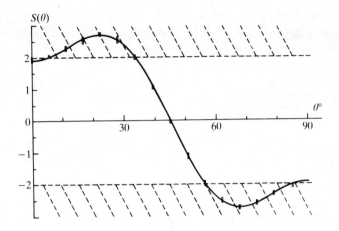

Fig. 1.8. Relevant combination of correlation coefficients for various sets of orientations. According to Bell's inequalities, S would not be in the hatched region. The errors are ± 2 standard deviations. The solid curve is the prediction of quantum mechanics.

have measured the polarization correlation coefficient at various orientations, and found an excellent agreement (Aspect *et al.* 1982). In order to achieve a more expressive presentation of the results, we have carried out new measurements yielding the quantity $S_{\exp}(\theta)$ (cf. eqn (1.7')) for sets of orientations such that

$$\boldsymbol{a}\cdot\boldsymbol{b}=\boldsymbol{b}\cdot\boldsymbol{a}'=\boldsymbol{a}'\cdot\boldsymbol{b}'=\theta \quad \text{and} \quad \boldsymbol{a}\cdot\boldsymbol{b}'=3\theta. \tag{1.10}$$

(Figure 1.2 shows such a set of orientations with $\theta = 22.5°$.) The results are displayed in Fig. 1.8, with the predictions of quantum mechanics, and the limit corresponding to Bell's inequalities (1.7). The agreement of the experimental data with quantum mechanics, and the violation of Bell's inequalities, are clearly shown.

1.4.3. *Experiment with time-varying polarizers* (Aspect *et al.* 1982a)

As emphasized in the first part, it would be worth doing an experiment in which the orientations of the polarizers 'are changed during the flight of the photons'. More precisely, such a thought experiment would need random changes with an auto-correlation time shorter than L/c (Aspect 1975, 1976) (L is the distance between the two polarizers, and c the speed of light).

Following our proposal in Aspect (1975, 1976), we have made a step towards such a thought experiment by replacing each polarizer by an optical switch followed by two polarizers in different orientations (Fig.

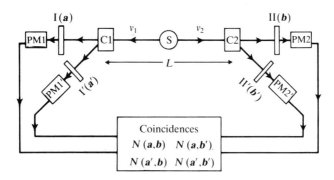

Fig. 1.9. Experiment with time varying analysers. Each optical switch, followed by two polarizers, is equivalent to a single polarizer switched rapidly between two orientations. A switching occurs each 10 ns, while L/c is 40 ns.

1.9). Each set-up is therefore equivalent to a single polarizer, the orientation of which is switched from one direction to another.

The switching of the light is effected by acousto-optical interaction of the light with an ultrasonic standing wave at 25 MHz, providing a commutation at 50 MHz, i.e. a change of orientation each 10 ns. This time is short compared to L/c (40 ns), but unfortunately it is not possible with these devices to achieve a random switching. In this respect, the experiment is far from the thought experiment. Nevertheless, let us mention that the two switches on both sides were driven by independent generators drifting separately.

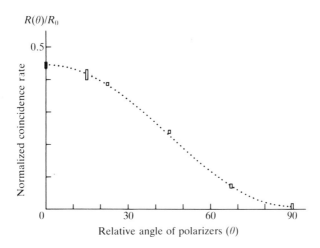

Fig. 1.10. Experiment with optical switches. Indicated errors are ±1 standard deviation. The broken curve is the prediction of quantum mechanics.

In this experiment we had to reduce the size of the beams, simultaneously weakening the coincidence signal by a factor 40. We had thus to collect the data for longer periods, and there were drift problems. This is why our results are far less accurate than in the two first experiments. In about 10 hours of experiment, we obtained data violating the Bell's inequalities (in an appropriate form for this experiment) by 5 standard deviations. These data also allow a comparison to be made with quantum mechanics, with which they show a reasonable agreement in the limit of our accuracy (Fig. 1.10).

1.5. Conclusion

As in previous ones, our experiments still have some imperfections that leaves open some loopholes for the advocates of hidden variable theories obeying Einstein's causality. Improved experiments can be devised (Lo and Shimony 1981) but the agreement of existing data with quantum mechanics is already impressive.

We are thus led to the conclusion that the predictions of quantum mechanics in EPR-type situations are vindicated by the experiments. Bell's theorem shows us that it .is not possible to understand these correlations in a 'classical-looking' fashion, and thus why we can consider these correlations surprising. As often, the situation is difficult to understand classically when we have interferences between terms to which we assign a macroscopic scale (each term $|x_1x_2\rangle$ and $|y_1y_2\rangle$ in eqn (1.1), describes a pair of photons separated by 12 m in our experiments).

On the other hand, let us remark that quantum mechanics gives definite predictions for these experiments, and these predictions are in excellent agreement with the experimental data. We can thus ask the question: 'Is there a real problem?'. Rather than my opinion, I will give a more authoritative one, namely that of R. Feynman, who wrote recently about this point: 'It has not yet become obvious to me that there is no real problem. I cannot define the real problem, therefore I suspect there is no real problem, but I am not sure there is no real problem. So that is why I like to investigate things' (Feynman 1982).

References

Aspect, A. (1975). *Phys. Lett.* **54A,** 117.
—— (1976). *Phys. Rev.* D **14,** 1944.
——, Imbert, C. and Roger, G. (1980). *Opt. Comm.* **34,** 46.
——, Grangier, P., and Roger, G. (1981). *Phys. Rev. Lett.* **47,** 460.
——, Dalibard, J., and Roger, G. (1982*a*). *Phys. Rev. Lett.* **49,** 1804.

——, Grangier, P., and Roger, G. (1982*b*). *Phys. Rev. Lett.* **49,** 91.

Bell, J. S. (1964). *Physics* **1,** 195.

—— (1971). In *Foundations of quantum mechanics* (ed. B. d'Espagnat). Academic Press, New York.

Bohm, D. (1951). *Quantum theory*. Prentice Hall, Englewood Cliffs, N.J.

——, and Aharonov, Y. (1957). *Phys. Rev.* **108,** 1070.

——, and Hiley, B. J. (1976). *Nuovo Cim.* B **35,** 137.

Bohr, N. (1935). *Phys. Rev.* **48,** 696.

de Broglie, L. (1974). *C. R. Acad. Sci., Paris* **278B,** 721.

Clauser, J. F. (1976). *Phys. Rev. Lett.* **36,** 1223.

——, and Horne, M. A. (1974). *Phys. Rev.* D **10,** 526.

——, and Shimony, A. (1978). *Rep. Prog. Phys.* **41,** 1881.

——, Horne, M. A., Shimony, A., and Holt, R. A. (1969). *Phys. Rev. Lett.* **23,** 880.

Einstein, A., Podolsky, B., and Rosen, N. (1935). *Phys. Rev.* **47,** 777.

d'Espagnat, B. (1975). *Phys. Rev.* D **11,** 1454; (1978) D **18,** 349.

Feynman, R. P. (1982). *Internatl. J. Theor. Phys.* **21,** 467.

Fine, A. (1982). *Phys. Rev. Lett.* **48,** 291.

Freedman, S. J., and Clauser, J. F. (1972). *Phys. Rev. Lett.* **28,** 938.

Fry, E. S. (1973). *Phys. Rev.* A **8,** 1319.

—— and Thomson, R. C. (1976). *Phys. Rev. Lett.* **37,** 465.

Holt, R. A. (1973). *Ph.D. Thesis*. Harvard University.

Lo, T. K., and Shimony, A. (1981). *Phys. Rev.* A **23,** 3003.

Pipkin, F. M. (1978). In *Advances in atomic and molecular physics* (ed. D. R. Bates, and B. Bederson). Academic Press, New York.

Selleri, F., and Tarrozzi, G. (1981). *Riv. Nuovo Cim.* **4,** 1.

Stapp, H. P. (1977). *Found Phys.* **7,** 313; (1979) **9,** 1.

Von Neumann, J. (1955). *Mathematical foundation of quantum mechanics*. Princeton University Press, Princeton.

Wigner, E. P. (1970). *Am. J. Phys.* **38,** 1005.

2

Testing quantum superposition with cold neutrons

Anton Zeilinger

2.1. Introduction

The historical development of quantum mechanics, particularly its interpretation, has been accompanied and signposted by an extended discussion of various brilliant *Gedankenexperiments*. These were of particular interest whenever a strange or seemingly counter-intuitive prediction of quantum mechanics for the results of observations were at the focus of scientific discussion. A high point in that respect was certainly the famed Bohr–Einstein dialogue (Bohr 1949) where Einstein, through a series of ever more sophisticated *Gedankenexperiments*, purported to demonstrate a supposed internal consistency of quantum mechanics, a position which, in all cases, could be refuted elegantly by Bohr. Despite the fact that quantum mechanics today is probably the single most successful physical theory in terms of breadth, correctness and variety of its predictions, its foundations and the epistemological questions raised by it are increasingly attracting interest.

It is in this context of an increase in interest in the foundations of quantum mechanics that attention is again focusing on conceptually simple and basic experiments. A significant difference, as compared to the earlier situation, lies in the fact that due to various technological advances in the meantime many of these experiments, or basically very similar ones, could actually be performed i.e. be moved out of the domain of 'mere' *Gedankenexperiments* to that of real ones (see e.g. the series of experiments presented and discussed in Kamefuchi *et al.* (1984)). In a sense, these experiments therefore serve to demonstrate some of the strange features of quantum mechanics in a rather direct way. In addition, if performed in a sufficiently precise and controlled way, some of these experiments provide evidence against or put upper limits on other theories being alternative to or extensions of quantum mechanics.

This chapter deals with some of these experiments performed in recent years with cold neutrons. Also, further experiments will be proposed aimed at elucidating further basic points. The reasons why neutrons are particularly well suited for some investigations of this kind stem from various facts. For example, compared with electrons, the larger mass of the neutron may be important if experiments aimed at possible deviations from standard quantum mechanics due to gravitational effects are performed. The larger mass also implies a slower speed at a given wavelength as compared with electrons. The property of the neutron of having no electric charge, which has been tested experimentally to extremely low limits (Gähler *et al.* 1982), on the one hand implies some disadvantages as compared, again, with electrons in beam handling, yet it also implies a significant reduction of the sensitivity of an experiment to the ever-present stray electric and magnetic field, a feature not insignificant in precision experiments. Such a disturbance due to stray fields in the neutron case can only be due to an interaction with the neutron's magnetic moment which again is significantly smaller than that of the electron. Neutron experiments have been particularly facilitated by the development of cold neutron sources, which provide beams of slow neutrons of appreciable intensity.

2.2. Linearity versus nonlinearity

The linearity of quantum mechanics may be viewed as one of its basic axioms (d'Espagnat 1976). It is reflected in the linearity of the Schrödinger equation and the superposition principle. Yet it is just that linearity which leads to some of the epistemologically most complex issues, like the spreading of wave packets beyond limit, and to the questions relating to the problem of the reduction of the wave packet upon measurement (Wigner 1963). Therefore it is not surprising, that nonlinear generalizations of the Schrödinger equation have been proposed (Bialynicki-Birula and Mycielski 1976; cf. also Chapter 9) with the aim at eliminating these issues. It is also interesting that many physicists expect a successful quantum formulation of gravity theory to lead to nonlinear quantum mechanics (Chapter 9). It is quite evident that, in view of the success of linear quantum mechanics, any nonlinear deviations must be very small.

Starting from the observation that in physics many linear equations are only the limiting cases of more general nonlinear equations, Bialynicki-Birula and Mycielski (1976) investigated the class of equations obtained by adding to the standard linear Schrödinger equation a term which is a function of the probability density

$$\left[-\frac{\hbar^2}{2m} \nabla^2 + V(\boldsymbol{r}, t) + F(|\psi|^2) \right] \psi(\boldsymbol{r}, t) = i\hbar \frac{\partial}{\partial t} \psi(\boldsymbol{r}, t). \qquad (2.1)$$

These authors then find that many of the customary features of the solutions of the Schrödinger equation are still obeyed by their nonlinear generalization. Of particular interest is a variant of eq (2.1), where the nonlinearity is logarithmic:

$$F(|\psi|^2) = -b \ln(|\psi|^2 a^n). \tag{2.2}$$

Here, b has the dimension of energy and is a measure of the strength of the nonlinear term, a is an unimportant constant and n is the dimension of the definition space of ψ. The reason for a logarithmic nonlinearity of the kind of eqn (2.2) is that this specific form provides for the separability of non-interacting subsystems. Based on the impressive agreement of experimental results for the Lamb shift with theoretical predictions, Bialynicki-Birula and Mycielski were obliged to place an upper limit of 10^{-10} eV on the magnitude of b.

Considerations of the nonlinear equation as represented in eqns (1.1) and (1.2) led Shimony (1979) to predict the existence of a purely amplitude-dependent phase shift in a neutron interferometer experiment. The experiment proposed by Shimony was subsequently performed by Shull *et al.* (1980) using the MIT two-crystal interferometer (Zeilinger *et al.* 1979). The experiment consisted of searching for a phase difference between an attenuated beam and an unattenuated one. No such phase difference was found within experimental accuracy, which observation permitted Shull *et al.* (1980) to lower the upper limit of b to 3.4×10^{-13} eV.

As already pointed out by Bialynicki-Birula and Mycielski, a non-vanishing value for the quantity b would prevent wave packets from spreading beyond limit. In particular, it can be shown that there exist soliton-like solutions of the logarithmic nonlinear equation of width (Bialynicki and Mycielski 1976, 1979)

$$l = \hbar/(2mb)^{1/2}. \tag{2.3}$$

It is therefore reasonable to search for changes in the free-space propagation of neutrons effected by the nonlinear term in Schrödinger's equation. Since the nonlinearity of eqns (2.1) and (2.2) is a function of $|\psi|^2$, the largest effect is to be expected for the most abrupt change in $|\psi|^2$ possible. This is just the case in the diffraction at an absorbing edge. In detail, one can argue that a gradient of $|\psi|^2$ in a direction normal to the wavefront leads to a related gradient of the phase of ψ. Yet, as can easily be seen, such a gradient in the phase leads to a bending of the wavefront, i.e. to a deflection. Assuming that any deviation due to a nonlinear term is very small, a more rigorous derivation leads to the

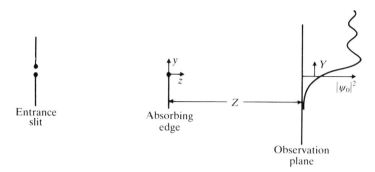

Fig. 2.1. Diffraction of cold neutrons at an absorbing straight edge: principle of the arrangement and co-ordinates used.

following expression for the deflection (Gähler *et al.* 1981):

$$Y = \frac{b}{E} \int_0^Z \frac{1}{|\psi_0|} \frac{d|\psi_0|}{dy} (Z - z)\, dz. \tag{2.4}$$

Here, y designates a direction orthogonal to the absorbing edge, z is the neutron propagation direction, Y denotes a given position in the diffraction pattern, and Z is the distance between the absorbing edge and the observation plane (Fig. 2.1). E is the kinetic energy of the neutrons and ψ_0 is the solution of the linear Schrödinger equation. The integral is to be taken along the whole path of the neutron from absorbing edge to observation plane. It is immediately obvious from eqn (2.4) that any effect would be the smaller the larger the neutron wavelength λ i.e. the smaller its kinetic energy is.

The experiment was performed at a cold neutron beam at the high-flux reactor of the Institute Laue–Langevin in Grenoble. The wavelength used was 20 Å with a bandwidth of ±0.5 Å. The experimental set-up was arranged on a large optical bench with a distance between absorbing edge and detector of 5 m. The observed diffraction pattern did not show any evidence for a statistically significant deviation from the predictions of the linear Schrödinger equation (Fig. 2.2). Based on the experimental resolution one therefore arrives at a new upper limit for the nonlinear term (Gähler *et al.* 1981):

$$b < 3.3 \times 10^{-15}\,\text{eV}. \tag{2.5}$$

We note that, using this limit, one can calculate the maximum to which an electron wave packet would spread, since b should be a universal constant (Bialynicki-Birula and Mycielski 1976). One thus obtains for the width of an electron gausson, which is the free-space soliton-like solution of the nonlinear Schrödinger equation, the value of 3 mm. It is

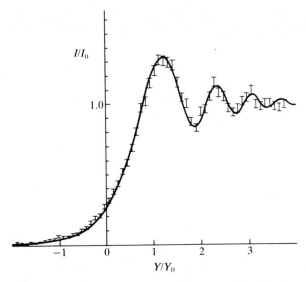

I/I_0

1.0

-1 0   1   2   3

Y/Y_0

Fig. 2.2. Measured edge diffraction pattern compared with the prediction of standard linear quantum mechanics.

remarkable, that therefore even the nonlinear theory has to admit the existence of macroscopic quantum objects.

At present we plan to improve the lower limit for b obtained so far by exploiting a neutron beam of about 80–100 Å wavelength available at the ILL in the future. The resulting decrease in neutron energy by a factor of 20 together with improved counting statistics should permit a limit of the order of 10^{-17} eV to be achievable.

2.3. Two-slit diffraction

As a *Gedankenexperiment*, two-slit diffraction has and is continuing to serve as one of the basic paradigmatic examples for demonstrating peculiar features of quantum mechanics (see e.g. Feynman *et al.* 1965). This is due to the fact that in that conceptually simple experiment some of the most fundamental features of the interpretation of quantum mechanics can be demonstrated directly. These are (1) the superposition of probability amplitudes, (2) the complementarity between different kinds of information, and (3) the reduction of the wave packet.

For massive particles, two-slit diffraction was first successfully shown by Jönnson (1961, 1974) with a lengthy series of experiments on electron diffraction at various slit assemblies. The results observed are in excellent

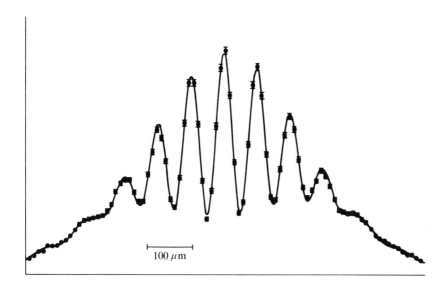

100 μm

Fig. 2.3. Measured two-slit diffraction pattern compared with the prediction of standard linear quantum mechanics.

agreement with theoretical prediction as regards the spacing of the interference fringes. Yet, to my knowledge, there exists no detailed comparison of the intensity distribution seen in the electron interference pattern with detailed theoretical calculation, although qualitative agreement has certainly been established.

Having had the optical bench set-up available for cold neutrons as mentioned in the last paragraph, the classic two-slit experiment was also performed for neutrons of wavelength 18.45 Å with a bandwidth of ±1.4 Å (Zeilinger *et al.* 1982). In that experiment, the centre-to-centre separation between the two slit openings was 126 μm and the combined width of the slits was 44 μm, making each slit about 22 μm wide. Due to experimental difficulties there was a slight asymmetry between the two slits which showed up as a corresponding asymmetry in the interference pattern (Fig. 2.3). It may be noted that, because of the fact that the wavelength of the neutrons used was so many orders of magnitude smaller than the size of the diffracting two-slit assembly, the angles of diffraction were only of the order of seconds of arc. Hence, a separation distance between object and observation plane of 5 m had to be employed.

In order to compare the experimentally obtained interference pattern with theoretical expectation, a detailed calculation based on scalar

Fresnel–Kirchhoff diffraction theory was performed (see e.g. Born and Wolf 1975). In that approach one has to include the various experimental details very carefully. First, the source has to be defined, which, for precision calculations, is by no means a trivial procedure. It was found, that it suffices to regard the entrance slit as the effective source. This is due to the fact that the width of that slit is very narrow (20 μm) and thus the slit effectively operated as the source of a beam coherent over both slits in the two-slit diaphragm. An interesting point is the treatment of the radiation right at the source, i.e. inside the source slit, since the radiation incident on that slit consists of both different direction and different wavelength components—in principle treatable as being either coherent or incoherent with respect to each other. In extensive numerical calculations it was found that the difference between treating different incident directions as being coherent or incoherent to each other did lead to differences in the interference pattern too small to be detectable in the experiment performed. Also, since the experiment was not performed in a time-dependent mode, any coherence between different wavelength, i.e. energy, components would be undetectable.

Therefore, the amplitude at an observation point P is

$$u_P = \text{const} \int\int e^{ik(r+s)} \, dS_1 \, dS_2 \qquad (2.6)$$

where r is distance from a source point to a point in the diffraction screen and s is the distance from that latter point to P. The integrations are to be performed over the source and over the two-slit diaphragm. The intensity is then found by incoherent integration over the directions and wavelengths incident on the entrance slit

$$I_P = \int\int |u_P|^2 \, d\alpha \, d\lambda. \qquad (2.7)$$

The smooth curve in Fig. 2.3 is the result of numerical calculations performed according to this equation.

The excellent agreement thus obtained between experiment and standard Schrödinger wave mechanics can be interpreted as implying upper limits on deviations from that linear theory. In particular, the fact that the interference contrast is less than 100 per cent can be fully accounted for by the wavelength spread of the incident radiation, which implies a coherence length of

$$L_c = \lambda^2 / \Delta\lambda = 243 \text{ Å} \qquad (2.8)$$

for the present experiment. Any deviation of the experimentally observed interference contrast from the quantum mechanically predicted one therefore has to be smaller than the experimental accuracy. This implies that, if such a deviation actually existed, it could not be larger than $6 \cdot 10^{-3} = \Delta C/C$, where C is the interference contrast defined as

$$C = (I_{max} - I_{min})/(I_{max} + I_{min}) \tag{2.9}$$

with I_{max} and I_{min} being the maximum and minimum intensities of the interference pattern at the innermost maxima and minima respectively.

We note that this agreement indicates that any presently unknown mechanism for a further reduction of the interference contrast is restricted to bounds which are given by the present experiment. Such additional mechanisms have been proposed explicitly and independently by Pearle (1976, 1979, 1982) and Hawking (1982). In both cases it is proposed that there may exist intrinsic mechanisms breaking the unitary evolution of the state vector. Such an evolution would then lead to a dynamically describable reduction mechanism which, for the two-slit experiment, could result in a loss of interference contrast. Such a loss of interference contrast should then depend on the time the state vector spends evolving.

Following Pearle (1984) we parametrize such a dynamic reduction mechanism by an exponential law as

$$\Delta C/C = e^{-t/\tau}, \tag{2.10}$$

where t is the time available for evolution of the state vector and τ is a characteristic reduction time, the properties of which depend on the particular theory chosen. Using the neutron speed of $217 \, \mathrm{m \, s^{-1}}$ we find the flight time for the distance of 5 m between the slits and the detector to be $2.3 \times 10^{-2} \, \mathrm{s}$. Hence, with the maximum possible interference contrast reduction of 6×10^{-3} which would not be in disagreement with experiment we obtain as a lower limit for the spontaneous reduction time (Zeilinger *et al.* 1984)

$$\tau > 4\mathrm{s}. \tag{2.11}$$

Looking again into the future, it can be estimated that with a neutron beam of a wavelength of 100 Å, as may be available in the near future at the high-flux reactor of the ILL in Grenoble, an improvement of the lower limit of the reduction time by possibly as much as two orders of magnitude could be achievable. This assumes that an experiment dedicated explicitly to the search for an unknown reduction mechanism would be carried out. In such an experiment one could think of using a flight path of about 10 m length and of counting neutrons in the central portion of the interference pattern only in order to improve the counting statistics in the determination of the interference contrast. We hope that such an experiment may be performed in the near future.

2.4. Two-slit complementarity experiments

The two-slit experiment is usually regarded as providing a beautiful explicit example for complementarity in quantum mechanics. In that experiment it is the complementarity between the interference pattern and the information which slit the particle passed through (Wooters and Zurek 1979). Yet, to our knowledge, no experiment exists which shows explicitly this complementarity feature in a two-slit set-up. Here we will propose rather simple extensions of the experiment reported in the previous paragraph which would permit an explicit demonstration of two-slit complementarity.

First, it is evident, yet still important, that the width of the scanning slit in front of the detector is crucial. In the conventional way of analysing the two-slit experiment, it is usually said that the scanning slit has to be narrower than the characteristic width of the features in the interference pattern, i.e. the minimum-maximum distance. This often is connected to the implied, yet incorrect, mental picture of the interference pattern as having some kind of *a priori* reality independent of whether a detector is actually placed there to detect it. Analysing in detail the operation of the detector slit—and for that of a narrow detector too—we have to investigate the diffraction taking place at that slit. Doing that we assume that the slit width is just equal to the distance between the interference minima and maxima

$$\delta = \frac{\lambda L}{2s}. \tag{2.12}$$

Here, L is the distance between the two-slit diaphragm and the detector slit and s is the centre–centre distance between the two slits in the two-slit diaphragm itself. Diffraction at a slit of that width of incident unidirectional radiation results in a single slit pattern of angular full width at half maximum of

$$\Delta\theta = \lambda/\delta = 2s/L. \tag{2.13}$$

Yet, we note that this is just twice the angular separation between the two slits as seen from the detector slit. Hence, observation of the interference pattern using a narrow enough detector (slit) results in destruction of the information about the path the particle took when passing through the two-slit assembly. It is, we submit, just this latter property which is responsible for the creation of the interference pattern in the first place.

If, on the other hand, the detector slit is much wider than the width given in eqn (2.13), diffraction at that slit is negligible and it is possible to determine the particle path even after passage through the detector slit.

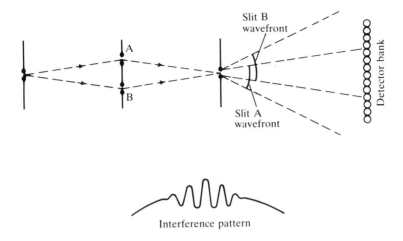

Fig. 2.4. Proposed two-slit complementarity experiment: diffraction at the narrow exit slit destroys the information about passage through slit A or B and hence permits the two-slit diffraction pattern to be observed upon variation of exit slit position.

Therefore, an experimental arrangement is proposed with a variable-width detector slit and a detector bank behind that slit (Figs. 2.4 and 2.5). If the slit is narrow (Fig. 2.4), diffraction at that slit creates the two-slit pattern in the detector bank due to overlap between the two wavefronts originating from the two different openings in the two-slit diaphragm. Such an overlap will no longer occur if the detector slit is

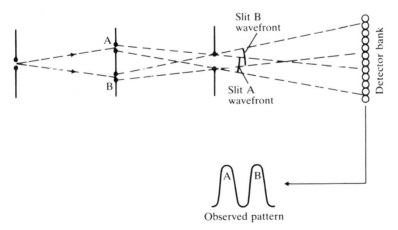

Fig. 2.5. Proposed two-slit complementarity experiment: wide exit slit. Here, diffraction at that slit is negligible and hence no two-slit pattern appears, yet slit passage through A or B may still be discriminated.

made much wider (Fig. 2.5). Then, the firing of a given detector would allow determination of the slit passed. Evidently this latter observation can be interpreted as a particle property, while the first one is a wave property. Also, we note that any intermediate type of result is possible.

An interesting extension of these experiments, which are readily performable with both neutrons and photons, results if we consider changing the detector slit width at a time after the particle has already passed the two-slit diaphragm. Then we can still decide to observe the interference pattern or to determine the particle's path. This clearly is a very explicit variant of a delayed choice experiment as proposed by Wheeler (1978). A most simple variant of these experiments results by replacing the detector bank by just two properly positioned detectors (Zeilinger 1984).

2.5. Concluding comments

The experiments presented here provide a nice example of technological progress—here the development of intense sources of cold neutrons—leading to fundamental experiments of hitherto unachieved precision. Whether these experiments will ever show a deviation from physics as known today is open. Yet, we note that increases in the precision of experimentation repeatedly in the history of physics have led to the observation of unexpected phenomena. An interesting field of work in that spirit will also be provided by the possibility of doing experiments, in the near future, aimed at testing the time-dependent Schrödinger equation prediction in great detail with precision far surpassing present capabilities. This again will be done with intense cold neutron beams.

Acknowledgements

I wish to thank the Institute Laue–Langevin for making the intense cold neutron beams available for the experiments reported here. I also wish to acknowledge the stimulating cooperation with Dr. R. Gähler (Munich), Professor A. G. Klein (Melbourne), Professor C. G. Shull (M.I.T.), and Dr. W. Treimer (Berlin). I also appreciate fruitful discussions with Professor P. Pearle and Professor H. Rauch (Vienna). This work was supported by the Fonds zur Förderung der wissenschaftlichen Forschung (Austria), project S34-01 and by the U.S. National Science Foundation.

References

Bialynicki-Birula, I. and Mycielski, J. (1976). *Ann. Phys. N.Y.* **100,** 62.
—— and Mycielski, J. (1979). *Phys. Scripta* **20,** 539.
Bohr, N. (1949). In *Albert Einstein: Philosopher–Scientist* (ed. P. A. Schilpp). The Library of Living Philosophers, Open Court Publ. Co., La Salle, Illinois. pp. 200–41.

Born, M. and Wolf, E. (1975). *Principles of optics,* 5th edn. Pergamon, Oxford. p. 428.

d'Espagnat, B. (1976). *Conceptual foundations of quantum mechanics,* 2nd edn. Benjamin, Reading. p. 16.

Feynman, R. P., Leighton, R. B., and Sands, M. (1965). *The Feynman lectures on physics,* Vol. II. Addison-Wesley, Reading. pp. 1–4.

Gähler, R., Klein, A. G., and Zeilinger, A. (1981). *Phys. Rev.* A**23,** 1611.

——, Kalus, J., and Mampe, W. (1982). *Phys. Rev.* D**25,** 2887.

Hawking, S. (1982). *Commun. Math. Phys.* **87,** 395.

Jönsson, C. (1961). *Z. Phys.* **161,** 454.

—— (1974). *Am. J. Phys.* **42,** 4.

Kamefuchi, S., Ezawa, H., Murayama, Y., Namiki, M., Nomura, S., Ohnuki, Y., and Yajima, T. (Eds.) (1984). *Proc. Int. Symp. on Foundations of Quantum Mechanics in the Light of New Technology.* Physical Society of Japan, Tokyo.

Pearle, P. (1976). *Phys. Rev.* D**13,** 857.

—— (1979). *Int. J. Theor. Phys.* **18,** 489.

—— (1982). *Found. Phys.* **12,** 249.

—— (1984). *Phys. Rev.* D**29,** 235.

Shimony, A. (1979). *Phys. Rev.* A**20,** 394.

Shull, C. G., Atwood, D. K., Arthur, J., and Horne, M. A. (1980). *Phys. Rev. Lett.* **44,** 765.

Wheeler, J. A. (1978). In *Mathematical foundations of quantum theory* (ed. A. P. Marlow). Academic Press, New York, p. 9.

Wigner, E. P. (1963). *Am. J. Phys.* **31,** 6.

Wooters, W. K., and Zurek, W. H. (1979). *Phys. Rev.* D**19,** 473.

Zeilinger, A. (1984). *J. Physique, Colloque C3, Suppl.* **45,** C3–213.

——, Horne, M. A., and Shull, C. G. (1984). In Kamefuchi *et al.* (1984). p. 289.

——, Shull, C. G., Horne, M. A., and Squires, G. L. (1979). In *Neutron Interferometry* (ed. U. Bonse and H. Rauch) Oxford University Press, Oxford, p. 48.

——, Gähler, R., Shull, C. G., and Treimer, W. (1984). In *Neutron scattering* (ed. J. Faber Jr.). Am. Inst. Phys. Conf. Proc. Ser. 89, American Institute of Physics, New York, p. 93.

3

The superposition principle in macroscopic systems

Anthony J. Leggett

3.1. Introduction

It is well known that the extrapolation of the conventional formalism of quantum mechanics to the macroscopic level leads to highly paradoxical consequences. For example, consider a microscopic system (e.g. an atom) which can be in either of two orthogonal states ψ_1 and ψ_2 (e.g. the states of spin projection $\pm\frac{1}{2}$ on some direction $\hat{n}$). Suppose we connect the microsystem up to a macroscopic apparatus so as to 'measure' which of the two states it is in; for example, we pass the atom through a Stern–Gerlach magnet with the field gradient oriented (in spin space) along $\hat{n}$, and arrange that it triggers one or other of two Geiger counters according to which beam it comes out in. Then, obviously, if the atom was originally in state ψ_1 only one counter, say 1, will click, while if it was in ψ_2 only counter 2 will respond: the final state of the macroscopic world is recognizably different in the two cases. Suppose now that the original state of the microsystem was a *linear superposition*:

$$\psi = a\psi_1 + b\psi_2, \qquad |a|^2 + |b|^2 = 1. \tag{3.1}$$

Then, of course, application of the standard measurement axioms of quantum mechanics tells us that the probability of counter 1(2) clicking is simply $|a|^2 (|b|^2)$. But what if we were to describe not only the state of the atom but that of the counters (and, if necessary, the magnet) in quantum mechanical language, as we are *prima facie* entitled to do, and moreover assume that the general structure of quantum mechanics persists for bodies of arbitrary size, complexity etc. Then, if the initial state of the microsystem had been ψ_1, the final state of the universe (that is, the atom plus the counters plus anything else with which interactions may have taken place) would have been described by some state† Ψ_1, while if the microsystem had started in ψ_2, the final state of the universe would have been some state Ψ_2 which is not only orthogonal to Ψ_1, but possesses macroscopically different properties. It then follows rigorously from the

† In fact this will not be so if the initial state of the universe was not a pure state. This does not affect the essence of the paradox: see Wigner (1963).

strict linearity of the formalism of quantum theory that the final state of the universe corresponding to an initial microsystem state of the form (3.1) is

$$\Psi = a\Psi_1 + b\Psi_2, \tag{3.2}$$

that is, a superposition of *macroscopically different* states. On the other hand, we know perfectly well that a 'measurement' (such as direct observation with the naked eye) will always find the universe in one macroscopic state or the other. So even the macro-world appears not to 'be' in a definite state until it is observed! This uncomfortable conclusion is of course the essence of the famous 'Schrödinger's cat' paradox, and various exotic and non-exotic 'resolutions' of it have appeared in the literature.

Now, it is probably safe to say that the overwhelming majority of physicists who have considered this problem at all subscribe to two opinions:

(i) that under the conditions specified the description (3.2) of the final state of the universe is indeed the physically correct one, but

(ii) that this has no observable consequences, because in view of the complexity of any macroscopic system one will never be able to distinguish the state (3.2) from an incoherent mixture of the states Ψ_1, Ψ_2 (which can then be treated by purely classical probability ideas). If these two statements are correct, then it appears that physics as such has nothing to contribute to the solution of the cat paradox but must leave it to the mercy of the philosophers.

In this chapter I shall (a) give strong reasons to believe that opinion (ii) is not necessarily correct, and (b) point out that if this is so, it is possible to test opinion (i) (which is by no means self-evident). In other words, I shall suggest that *if* quantum mechanics can indeed be extrapolated to bodies of macroscopic size, complexity, etc., then it should not be totally impossible to obtain at least circumstantial evidence for the existence of macroscopic superpositions of the type (3.2); so that any failure to find such evidence when it is expected would cast doubt on the legitimacy of the extrapolation. Some experiments of this type have indeed already been done, and others are in the pipeline. Since I have recently written two fairly extended reviews (Leggett 1984a,b) of this area, I shall just give the main lines of the argument here and refer to these papers for the technical details (see also Leggett 1980, Caldeira and Leggett 1983).

3.2. Where to look for macroscopic superpositions

Let me start by asking: Why is it the general belief that 'macroscopic superpositions' will be unobservable? The reason lies in the *complexity* of macroscopic bodies: generally speaking, states of (say) a Geiger counter which are macroscopically different (e.g., the states corresponding to having been triggered or not) will be different in the behaviour of a macroscopically large number of degrees of freedom. Let us for example label the degrees of freedom ξ_i $(i = 1, 2, \ldots, M)$ and suppose that $N - M$ of these behave in the same way in the two states, but M in totally different ways, where M is a large number. Then a schematic representation of the two different macrostates Ψ_1, Ψ_2 would be

$$
\begin{aligned}
\Psi_1 &= X_{N-M} \prod_{i=1}^{M} \phi_i(\xi_i), \\
\Psi_2 &= X_{N-M} \prod_{i=1}^{M} \psi_i(\xi_i),
\end{aligned}
\tag{3.3}
$$

when each ϕ_i is orthogonal to the corresponding ψ_i, and X_{N-M} is the wave function of the last $N - M$ degrees of freedom. It is obvious that the linear superposition

$$
\Psi = a\Psi_1 + b\Psi_2
\tag{3.4}
$$

cannot be distinguished from an incoherent mixture of Ψ_1 and Ψ_2 by measuring any operator which involves less than M-fold correlations. (This is a generalization of the well-known argument that to learn anything about the EPR paradox you have to measure two-photon correlations—measurements of one-photon properties tell you nothing.) Since for large M such measurements are in practice impossible, it is often concluded that the detection of macroscopic superpositions is unfeasible in real life.

However, this argument is a bit too simple. Indeed, if it were really true as it stands it would imply, for example, that we should never be able to see interference phenomena of the 'two-slit' type with heavy atoms, since the states which are interfering differ in the behaviour of (say) 50 nuclei and electrons, and we cannot measure 50-particle correlations. This conclusion is of course absurd: indeed, although to the best of my knowledge no two-slit diffraction experiments have been done for heavy atoms, a *single-slit* interference experiment has been done with potassium (Leavitt and Bills 1969) $(A + Z \sim 60)$ with the results predicted by quantum mechanics. The point we have missed, of course, is that we do not have to detect the simultaneous existence of the two states (corresponding to passage through the different slits) directly: rather, we let nature do the work for us by applying an operator, namely the time-dependent operator $\hat{U}(t) = \exp - i\hat{H}t/h$ which does have 50-particle (and higher) correlations built into it. (Although $\hat{H}$ contains only one and two-particle operators, arbitrarily high powers of $\hat{H}$ occur in $\hat{U}(t)$.) The result of applying this operator is that the 50-odd microscopic (electronic and nuclear) co-ordinates are all locked adiabatically to a single degree of freedom, namely the centre-of-mass co-ordinates; since this is a *sum* of one-particle operators, not a product, there is no problem about measuring it. Of course, we still cannot detect the existence of the superposition at the diffracting screen directly, any more than we could for a single electron or photon: but just as in that case, we can detect its effects later, when the quantum-mechanical time development operator has recombined the two waves at the detecting screen (see Leggett, (1980), §3 for details).

Why then should we not extrapolate this argument to the genuinely macroscopic scale, and look for the interference of (say) macroscopically different states of the centre-of-mass co-ordinate of a billiard ball? In the standard textbook discussions the reason often given is that the de Broglie wavelength associated with the billiard ball under any reasonable conditions is so small that the interference effects would be totally unobservable. A second, related, reason sometimes given is that the energy level spacing of macroscopic objects is so tiny that it would be very much less than the thermal energy kT at all conceivably attainable temperatures, so that the thermal noise will totally obscure any quantum interference behaviour. Both arguments have some validity as applied to the billiard ball, but both fail as soon as we consider more general types of macroscopic (collective) co-ordinate. For example, consider a simple LC-circuit, for which the macroscopic variable is the flux threading the inductance. This system behaves like a simple harmonic oscillator with classical frequency $\omega_0 = (LC)^{-1/2}$, and hence the correct quantum mechanical description presumably implies a level spacing $\hbar\omega_0$. With modern fabrication techniques and cryogenics this need by no means be small compared to kT, even for a circuit of dimensions ~ 1 cm. Moreover, the predicted r.m.s. flux uncertainty in the ground state $(\hbar/\omega_0 C)^{1/2}$, can easily be of the order of 10^{-17} Wb, a level easily measurable by modern magnetometers. Hence, at first sight at least, there is no obstacle to seeing quantum mechanical interference of states differing in flux value by an amount of this order—states, that is, which by any normal criterion would reasonably be called macroscopically different.

There are, however, two very serious difficulties to this proposal. The first is that under most circumstances the quantum mechanical description of the system, interference terms and all, will lead to precisely the same dynamics, and hence the same observable effects, as the application of classical mechanics. This is always true for a simple harmonic oscillator such as the LC-circuit considered above, and it will be true more generally to the extent that we are in the correspondence limit. That is, to see characteristically quantum-mechanical effects a minimum condition

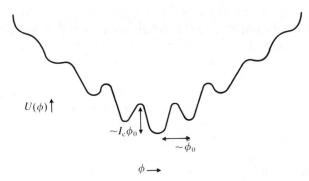

$U(\phi)\uparrow$

$\sim I_c\phi_0\downarrow$

$\sim\phi_0$

$\phi\longrightarrow$

Fig. 3.1. The form of the potential energy $U(\phi)$ of an r.f. SQUID as a function of the trapped flux ϕ. The external d.c. flux is taken to be zero.

is that we evade this limit: crudely speaking, the characteristic energy scale V_0 of the potential in which the system moves must be comparable to (or not much larger than) the level spacing, which is of order $\hbar\omega_c$ where ω_c is the classical frequency of the motion. Now, generally speaking, when we are talking about a macroscopic variable the associated potential is also macroscopic, while the quantity $\hbar\omega_c$ is of course very small indeed; thus we are always willy-nilly in the correspondence limit. (This can be checked by considering, e.g. a body of macroscopic size moving in any ordinary electric, magnetic or gravitational potential: cf. Leggett 1984a.) Thus, in reality it is the fact that the *energy* scale, rather than the ordinary geometrical scale, is macroscopic which prevents us seeing interference effects with macroscopic variables.

There is, however, at least one† very striking case in which we do have a genuinely macroscopic variable whose motion is controlled by a 'potential' which itself implicitly contains $\hbar$ and hence is of a *microscopic* order of magnitude. This is the case of phenomena associated with the Josephson effect, in which the Cooper pairs formed in a bulk superconductor tunnel across an insulating barrier (Josephson junction). The energy scale associated with this phenomenon is of order $I_c\phi_0$, where the junction critical current I_c may be in the range, say, 1 nA–1 mA, and the flux quantum $\phi_0 \equiv h/2e$ is $\sim 2 \times 10^{-15}$ Wb; the 'flux scale' is ϕ_0 itself (which is in fact what permits the accurate magnetometry mentioned above). In particular, if the junction is inserted in a closed bulk superconducting ring ('r.f. SQUID'), the motion of the flux trapped in the ring is controlled by a potential of the general form shown in Fig. 3.1; the distance between the various minima along the flux axis is $\sim\phi_0$ and

† A second possible candidate is the system of charge density waves which occur in some quasi-one-dimensional metals, where the pinning potential for the macroscopic variable (the phase of the wave) is also microscopic (see Bardeen 1979, 1980).

the height of the barriers between them is $\sim I_c \phi_0$. The scale of the parabolic envelope of the curve, and hence the number of minima, is determined by the ring self inductance. The 'mass' (inertia) associated with the motion of the flux turns out to be the effective capacitance shunting the junction. Since the characteristic frequency of classical motion of a system of mass m in a potential of magnitude $\sim V_0$ which varies over a region $\sim l$ is $\sim (V_0/ml^2)^{1/2}$, in this case it is $\sim (I_c/C\phi_0)^{1/2}$. Thus at first sight the condition $V_0 \gg \hbar\omega_c$ to see characteristically quantum effects becomes approximately

$$\lambda \equiv (8CI_c\phi_0^3/\pi^3\hbar^2)^{1/2} \gg 1 \qquad (3.5)$$

(where the symbol $\gg$ means 'not too large compared with'). The true condition actually turns out to be somewhat weaker than this, because by biasing the ring with a d.c. external flux we can change the shape of the potential $u(\Phi)$ (Fig. 3.1) in such a way as to reduce both the barrier height and the effective scale: see Leggett (1984) for the details. In any case, with attainable junction parameters ($C \sim 10^{-13}$ F, $I_c \sim 1$ nA $- 1\,\mu$A) the condition (3.5) can be reasonably well fulfilled. Of course, if any quantum effects we are looking for are not to be blurred out by thermal disorder, it is *prima facie* also necessary to ensure the condition $k_B T \ll \hbar\omega_c$ (though see below): but it turns out that with modern cryogenic techniques this poses no special problem. Thus, if we ignore for the moment the second of the two difficulties mentioned above, all the conditions necessary to see quantum mechanical effects on a macroscopic scale should be satisfied in a system of the type described, that is a bulk superconducting ring interrupted by a Josephson junction ('r.f. SQUID').

It is necessary to emphasize at this point, that although the system most suited to our purpose involves superconductivity, the effects we are about to consider are quite different from the 'macroscopic quantum effects' conventionally associated with superconductors. The latter have to do with the fact that in a superconductor a macroscopic number of Cooper pairs are condensed into the same pair state and hence behave coherently; however, once we have defined the relevant macroscopic variable (the current carried by the pairs or a related quantity, e.g. the trapped flux), these latter phenomena can be described by treating it purely classically. These phenomena do however provide the basis for the 'second-level' effects to be considered below. (For a discussion of this point, see Likharev 1983.)

It should also be mentioned at this point that although from the theorist's point of view the r.f. SQUID forms the ideal system, for practical reasons some of the relevant experiments have been done on a related system, namely an isolated Josephson junction biased by a fixed external current. With regard to the 'MQT' experiment to be discussed below, I believe the two systems are essentially equivalent; however, there is no analog of the 'MQC' experiment in a current-biased junction. This point is fully discussed in Leggett (1984*b*). There are also related experiments on charge-density waves in quasi-one-dimensional solids, a system analogous in many ways to a current-biased junction (see Grüner 1983). At the time of writing there is some controversy over the interpretation of these (Bardeen 1979, 1980; Sneddon *et al.* 1982).

3.3. Macroscopic quantum tunnelling and coherence

Assuming, then, that we have attained the condition (3.5) and also the condition $kT \ll \hbar\omega_c$, what characteristically quantum-mechanical effects shall we look for in order to establish that the motion of a macroscopic variable (the flux) does indeed show the characteristic quantum mechanical superposition of amplitudes? The form of the potential curve of Fig. 3.1 suggests two possible experiments, one indirect but certainly possible, the other direct but problematic, both associated with the classically totally forbidden phenomenon of penetration through an energy barrier. The first, which I shall call simply 'macroscopic quantum tunnelling' (MQT) or, better, macroscopic quantum decay, simply involves the escape of the system from a metastable state into the quasi-continuum formed by the potential outside: we assume implicitly that there is sufficient dissipation in the latter region that once the system has escaped, any coherence with the part of the wave function remaining inside the well can be neglected. This phenomenon is the precise analogue at the macroscopic level of the decay of a heavy nucleus by spontaneous fission or by emission of an alpha-particle. While the conditions for observing it are not impossibly stringent (see below), the question of precisely what aspects of the quantum-mechanical structure its observation would verify is somewhat problematic: see the discussions in Leggett (1980, §4; 1984*a*, §2).

The second type of experiment, which if feasible is a much more direct test of the superposition principle, is what I shall call 'macroscopic quantum coherence' (MQC). It requires that we bias the external d.c. flux at precisely half a flux quantum; it then turns out that (with suitable choice of ring parameters) the potential $U(\Phi)$ has the form shown in Fig. 3.2, with two degenerate wells. Then, in analogy to what happens in a microscopic system such as the NH_3 molecule, the ground state splits into an even and an odd-parity state, with separation $\hbar\Gamma$, and if we initially place the system, in, say, the left-hand well it will perform a coherent oscillation, so that the probability of finding it subsequently in that well is given by the expression

$$P(t) = \tfrac{1}{2}(1 + \cos \Gamma t). \tag{3.6}$$

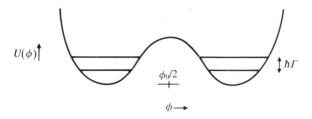

Fig. 3.2. The form of $U(\varphi)$ for a weakly hysteric r.f. SQUID with external d.c. flux bias of $1/2\phi_0$.

Although it would obviously not be possible to follow this oscillation directly (since any attempt to determine which well the system is in will of course destroy the interference necessary to maintain the oscillation) it is possible to do an experiment analogous to those used by particle physicists to detect the strangeness oscillations in the neutral K-meson system: that is, effectively start the system in one well, turn off one's measuring apparatus for a time t, turn it on again and 'observe' which well the system is in, and plot the statistical data obtained as a function of t. Precisely such an experiment was in fact conducted by Bol *et al.* (see also de Bruyn Ouboter 1983), though probably in a region of parameters where one would not expect to see coherent oscillation (cf. below).

It is clear that the occurrence of MQT is a necessary but by no means sufficient condition for that of MQC. If we simply follow through the standard WKB theory for tunnelling out of a metastable well, it turns out that the conditions to see MQT in a realistic experiment are (a) $V_0 \gg \hbar\omega_0$ and (b) $k_{\rm B}T \ll \hbar\omega_0$ where ω_0 is the small-oscillation frequency in the metastable well and V_0 is the barrier height. (The first condition ensures that the lifetimes involved are not astronomical, the second that quantum tunnelling is not overwhelmed by purely classical thermal escape.) These conditions arc just the exemplification, for the special case of tunnelling, of the general conditions derived above to see characteristically quantum effects. Now by varying the external d.c. bias and hence the shape of the potential $u(\Phi)$ (Fig. 3.1) we can make $V_0/\hbar\omega_0$ as small as we choose; unfortunately we then also reduce the absolute value of ω_0 and hence the ratio $\hbar\omega_0/k_{\rm B}T$. A detailed calculation (see Leggett 1984b) shows that the condition for (a) and (b) to be satisfied simultaneously is roughly

$$k_{\rm B}T/\hbar\omega_{\rm J} \ll \lambda^{-0.2}, \tag{3.7}$$

where $\omega_{\rm J}$ is the plasma frequency $(2\pi I_{\rm c}/c\phi_0)^{1/2}$ associated with the junction (this is the typical frequency of small oscillations of the flux in a 'deep' well) and is typically of the order of a few tens of GHz, and λ is defined by eqn (3.5) above. Because of the very weak dependence on λ, this condition can be well satisfied at attainable temperatures for rather large values of λ. However, it turns out (Leggett 1984b) that the width of the barrier through which the tunnelling takes place is of order $\lambda^{-0.4}\phi_0$, so that one might perhaps argue that for λ too large the tunnelling is not a 'macroscopic' event in the fullest sense of the word.

We turn now to the phenomenon of macroscopic quantum coherence. Here, we must bear in mind that a realistic experiment is unlikely to work if the tunnelling frequency Γ is greater than, say, a few MHz, in view of the need to switch the electronics on and off in a time short compared to Γ^{-1}. Since Γ depends exponentially on the height of the barrier, which in turn is a function of the ring parameters, there is at best a very small 'window' of these parameters in which Γ is $\lesssim$ a few MHz but not astronomically low. We simultaneously need $k_B T \ll V_0$, so that incoherent thermal transitions will not blur out any coherence effects. It turns out that a necessary, though by no means sufficient, condition is (Leggett 1984b)

$$k_B T / I_c \phi_0 \ll \lambda^{-4/3}, \tag{3.8}$$

and that this may just be satisfied if the capacitance C can be made as small as 10^{-14} F. (A detailed discussion is given in Leggett (1984b)). This condition is even more stringent than it looks: whereas for purposes of calculating the MQT behaviour, which takes place at frequencies of the order of the 'attempt' frequency ω_0, which is usually of the order of a few GHz, the effective capacitance shunting the junction may well be that associated with the junction itself and hence perhaps 10^{-13}–10^{-15} F, the MQC experiment, as we have seen, must be done with tunnelling frequency $\Gamma \lesssim$ a few MHz, and at these low frequencies the effective capacitance is almost certainly the much larger geometrical capacitance of the circuit. Thus, even at this stage of the argument, the observability of MQC in practice looks somewhat problematical.

3.4. The watched-pot effect: experimental feasibility

We still have to discuss the second of the two difficulties, which brings ut full circle back to the considerations mentioned at the beginning of this discussion and is, to my mind, the whole crux of the problem. Namely, we have behaved so far as if the macroscopic degree of freedom were completely decoupled from its environment and subject to its own Schrödinger equation. This was justified, implicitly, by the argument that all the other, microscopic degrees of freedom would follow it adiabatically. If this is really the case, we are home and dry. However, the whole point of much of the extensive literature on the quantum measurement problem in recent years has been that macroscopic systems not only are very complex but show *highly irreversible* behaviour. If this is so, then the energy and, more importantly, the phase information stored in the macroscopic degree of freedom will be rapidly dissipated into its environment (microscopic degrees of freedom) and the superposition of macroscopically different states will be unobservable, in accordance with the general conclusions of many authors who have considered this problem. To put it another way, the system is being continually 'observed' by its environment (Zurek 1981; Cini 1983) and if this 'observation' is too effective the 'watched-pot' effect will come into play and we will never see the characteristic quantum behaviour. Since the interactions which destroy the phase information are precisely those which fail to satisfy the adiabatic principle, and these in turn are the ones which dissipate the energy of the macroscopic system, the problem is to determine the *effect of dissipation on macroscopic quantum tunnelling and coherence*. Until this question is resolved, the simple atomic analogies on which we have relied so far are essentially worthless. The last three or four years have seen considerable progress in this area, and I now summarize our current understanding.

The effect of dissipation on quantum tunnelling was considered by Caldeira and Leggett (1981). They assumed that the quasi-classical motion of the system (described here by a coordinate q and an associated potential $V(q)$) in the region near the metastable minimum was given by a damped equation of motion of the form

$$M\ddot{q} + \eta\dot{q} + \partial V/\partial q = F_{ext}(t) \qquad (3.9)$$

where the phenomonological friction coefficient η (which in the SQUID problem would correspond to the normal conductance shunting the junction) is independent of amplitude and frequency. They were then able to show that, as compared to the system with the same $V(q)$, but $\eta = 0$, the system tunnels more slowly: crudely speaking, the effect of the dissipation is to multiply the effective WKB exponent by a factor of order $(1 + \gamma/\omega_0)$, where ω_0 is the small-oscillation ('attempt') frequency in the metastable well and $\gamma \equiv \eta/2M$ is its damping factor. Thus, tunnelling will be strongly suppressed if the small oscillations in the metastable well are overdamped (as is usually the case for practical SQUIDs) but in the opposite limit will be relatively little affected. Subsequent work (Caldeira and Leggett 1983; Ambegaokar *et al.* 1982; Zwerger 1982; Wonneberger 1983) has generalized this result considerably, but the qualitative conclusion remains the same: provided that the *classical* motion is only weakly damped, the tunnelling behaviour is not much affected. It should be mentioned that there does exist a class of problems (which I have elsewhere (Caldeira and Leggett 1983) called the 'anomalous' case) for which the classical equation of motion is of a different nature from eqn (3.9) and in which the introduction of dissipation actually *increases* the tunnelling rate (Widom and Clark 1980); however, it does so, in effect, only by providing an impedance mechanism which blocks out a large inertia which would otherwise have hindered the tunnelling. Such cases seem rather unlikely to be relevant in real life, least of all in the SQUID case. This question is discussed in detail in Leggett (1984c, §§3, 4). Generally speaking, one can say that, at least in practical cases such as SQUIDs dissipation will tend to suppress the tunnelling rate, and the effect is measured in order of magnitude by the relative damping of the small oscillations in the metastable well.

We finally turn to the delicate problem of the effect of dissipation on macroscopic quantum coherence. In a practical situation 'bare' tunnelling frequency in the absence of coupling to the environment, say Γ_0, will be small compared to the small-oscillation frequency ω_0 in each well and we can therefore plausibly truncate the problem to the two-dimensional space spanned by the lowest levels in each well. With a reasonable description of the coupling to the environment which is responsible for the dissipation, the problem then reduces to the familiar 'spin-boson problem' which arises in many different areas of physics: that is, the problem of a particle of spin $\frac{1}{2}$ subjected to a constant 'magnetic field' (the 'bare' level splitting $h\Gamma$) in the x-direction and a fluctuating field in the z-direction proportional to the displacement of a set of simple harmonic oscillators. The relevant Hamiltonian is

$$\hat{H} = -\Gamma_0 S_x + \sum_\alpha \tfrac{1}{2} m_\alpha (\dot{x}_\alpha^2 + \omega_\alpha^2 x_\alpha^2) + S_z q_0 \sum_\alpha C_\alpha x_\alpha, \qquad (3.10)$$

where q_0 is the distance between the minima. The most important case for our purpose is that in which the original (untruncated) classical motion was described by eqn (3.9), which corresponds to the widely used 'resistively shunted junction' model of a SQUID. In that case one can show (Caldeira and Leggett 1983) that the parameters of (3.10) are related to the phenomenological viscosity η (or shunting conductance) by

$$\frac{\pi}{2} \sum_\alpha (C_\alpha^2/m_\alpha \omega_\alpha) \delta(\omega - \omega_\alpha) \equiv J(\omega) = \eta\omega \quad \text{for} \quad \omega \ll \omega_c, \qquad (3.11)$$
$$\rightarrow 0 \quad \text{for} \quad \omega \gtrsim \omega_c.$$

We then have two dimensionless parameters in the problem, the ratio (Γ_0/ω_c) of the 'bare' tunnelling frequency to the environment cutoff frequency ω_c, and the dimensionless coupling constant $\alpha \equiv \eta q^2/2\pi\hbar$. In the SQUID case the latter becomes $(\Delta\phi)^2/2\pi\hbar R_n \equiv (\Delta\phi/\phi_0)^2 (R_0/R_n)$, where $\Delta\phi$ is the spacing between minima in flux space, ϕ_0 is the flux quantum, R_n is the shunting resistance of the junction, and R_0 is the characteristic quantum unit of resistance, $h/4e^2 \sim 7 \text{ k}\Omega$. In any practical case the ratio Δ_0/ω_c will be much less than 1 (typically $\sim 10^{-5}$) but α can be quite comparable to 1.

Very recently, S. Chakravarty and Leggett (1984) have investigated the dynamics of the hamiltonian (3.10) in the limit of large ω_c/Γ_0, with the following conclusions, which we believe to be accurate to within corrections of higher order in Γ_0/ω_c: In this limit the only relevant frequency scale, apart from $kT/\hbar$, is the renormalized tunnelling rate $\Gamma_r \equiv \Gamma_0(\Gamma_0/\omega_c)^{\alpha/(1-\alpha)}$, which except for very small α will be much less than the 'bare' tunnelling rate Γ_0. We then find that if the dynamics of the two-level system is plotted as a function of α and of $(k_B T/\hbar\Gamma_r)$, the *only* region of the parameter space in which we see anything resembling the sinusoidal oscillations of the original undamped system is the low-damping, low-temperature region $\alpha < \frac{1}{2}$, $\alpha k_B T/\hbar\Gamma_r \lesssim 1$, where we indeed get damped harmonic oscillations with frequency $\sim\Gamma_r$ and a Q-factor which for small α is proportional to α, as we should expect. Over all other regions of the parameter space the behaviour is an incoherent exponential relaxation which gives little direct evidence for the quantum nature of the system. This conclusion agrees qualitatively with the conclusions of a phenomenological argument put forward previously by Leggett (1984*b*).

Thus, to see quantum coherence in a SQUID we would need, in addition to the various other conditions specified above, the conditions

$$\alpha \equiv (\Delta\phi/\phi_0)^2(R_0/R_n) \ll \tfrac{1}{2}, \qquad \alpha k_B T/\hbar\Gamma_r \lesssim 1. \qquad (3.12)$$

If we assume that we can work with a *renormalized* tunnelling frequency Γ_r of the order of say 10 MHz, and at a temperature of ~ 1 mK, then satisfaction of the first condition will automatically imply satisfaction of the second. Whether, given the current constraints on fabrication of Josephson junctions, the conditions (3.12) can in practice be satisfied simultaneously with all the others mentioned above remains an open question.

It should be emphasized that the above conditions are in any case *minimum* conditions. It may well be that in practice, at the fairly low frequencies necessary for the projected experiment, other complications such as $1/f$ noise (whose nature and origins are still only poorly understood) come into play and obscure the expected oscillation behaviour. A theory of MQC which incorporates such complications is at the moment nonexistent.

3.5. Conclusion

Thus, while the conditions to see macroscopic quantum tunnelling are not too severe, and indeed various groups have already reported its detection (see in particular Jackel *et al.* 1981; Voss and Webb 1981), the observability of macroscopic quantum coherence is at the time of writing an open question: if it can be seen at all, it will certainly put extreme demands not only on the fabrication technique but on the cryogenics and noise control. Since MQC would be a much more direct and spectacular demonstration of the validity of the superposition principle at the macroscopic level than MQT, it is certainly a goal worth aiming for.

Acknowledgements

I would like to acknowledge conversations some years ago with A. Widom, whose enthusiasm for the idea of macroscopic quantum coherence induced me to undertake a detailed investigation of how severe would be the difficulties of seeing it in practice. I also thank Vinay Ambegaokar, R. de Bruyn Ouboter, Sudip Chakravarty, David Waxman, and Willi Zwerger for helpful discussions.

References

Ambegaokar, V., Eckern, U., and Schön, G. (1982). *Phys. Rev. Lett.* **48,** 1795.
Bardeen, J. (1979). *Phys. Rev. Lett.* **42,** 1498.
—— (1980). *Phys. Rev. Lett.* **45,** 1978.
Bol, D., van Weelderen, R., and de Bruyn Ouboter, R. (1983). *Physica* B **122,** 1.
de Bruyn Ouboter, R. (1983). In *Proc. Int. Symp. on Foundations of Quantum Mechanics in the Light of New Technology* (ed. S. Kamefuchi, H. Ezawa, Y. Murayama, M. Namiki, S. Nomura, Y. Ohnuki, and T. Yajima). Physical Society of Japan.
Caldeira, A. O. and Leggett, A. J. (1981). *Phys. Rev. Lett.* 46, 211.
—— —— (1983). *Ann. Phys. N.Y.* **149,** 374.
Chakravarty, S. and Leggett, A. J. (1984). *Phys. Rev. Lett.* **52,** 5.
Grüner, G. (1983). *Comments on Solid State Phys.* (6B) **10,** 183.
Jackel, L. D., Gordon, J. P., Hu, E. L., Howard, R. E., Fetter, L. A., Tennant, D. M., Epworth, R. W., and Kurkijärvi, J. (1981). *Phys. Rev. Lett.* **47,** 265.
Leavitt, L. A. and Bills, F. A. (1969). *Am. J. Phys.* **37,** 905.
Leggett, A. J. (1980). *Progr. Theor. Phys. Suppl.* **69,** 80.
—— (1984a). In *Essays in theoretical physics in honor of Dirk ter Haar* (ed. W. E. Parry). Pergamon, Oxford.
—— (1984b). In *Proceedings of the* 1983 *NATO ASI on percolation, localization and superconductivity* (ed. A. M. Goldman and S. A. Wolf). Plenum, New York.
—— (1984c). *Phys. Rev.* B **30,** 1208.
Likharev, K. K. (1983). *Usp. Fiz. Nauk.* **139,** 170.
Sneddon, L., Cross, M. C., and Fisher, D. S. (1982). *Phys. Rev. Lett.* **49,** 292.
Voss, R. F. and Webb, R. A. (1981). *Phys. Rev. Lett.* **47,** 265.
Widom, A. and Clark, T. D. 1984. *Phys. Rev.* B **30,** 1205.
Wigner, E. P. (1963). *Am. J. Phys.* **31,** 6.
Wonneberger, W. (1983). *Z. Phys.* B **50,** 23.
Zurek, W. H. (1981). *Phys. Rev.* D **24,** 1516.
Zwerger, W. (1982). *Z. Phys.* B **47,** 129.

4

Non-local phenomena and the Aharonov–Bohm effect

Yakir Aharonov

4.1. Introduction

In recent years it has become clear that non-local phenomena and topological effects play an important role in many areas of physics. In this chapter we will review the Aharonov–Bohm (A–B) effect, which is common to all gauge theories, and provides a particularly clear example of non-local phenomena. We will show that effects analogous to the A–B effect exist in classical theories as well, but that in quantum theories they acquire dynamical significance which has no classical analogue. We will discuss these dynamical aspects and show that they provide a general characterization of non-local phenomena. Finally we will show how this approach can guide us in finding other families of non-local effects which have no classical analogue.

Let us first review the A–B effect.† Imagine a thin solenoid confining a magnetic flux Φ surrounded by an impenetrable barrier (Fig. 4.1). The region outside the barrier is free of fields and therefore all local experiments confined to this region will be insensitive to the flux. The region seems (locally, at any rate) empty. Indeed, a classical charged particle moving in this region will behave as if nothing whatever is there. On the other hand, the wave function of a quantum mechanical charged particle will, in these circumstances include interference terms which depend on the flux, Φ modulo Φ_0 (where Φ_0 is a unit of flux ch/e, called a fluxon). Thus the behaviour of charged, quantum mechanical particles is sensitive to the non-vanishing line integral of the vector potential around the barrier. This is an example of non-local phenomena, since the effect of the vector potential in any local region is not gauge invariant and therefore not measurable. We may speak, if we choose to, about a non-local effect of the enclosed flux on the charged particle, since the final effect depends only on the inaccessible flux.

---

† A fairly complete and up-to-date list of references concerning the A–B effect is given in Ruijsenaars (1983).

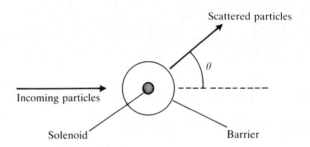

Fig. 4.1. Scattering from a solenoid surrounded by a barrier.

A further insight into the nature of this non-local effect is provided by the scattering of the particle from the barrier surrounding the flux. The scattering cross-section of the particle depends both on R, the radius of the barrier, and on Φ, the flux enclosed therein. If the flux is quantized ($\Phi = n\Phi_0$), the scattering will vanish as $R \to 0$. For non-quantized flux values the scattering does not vanish in that limit, but, rather, is proportional to the wavelength λ of the scattered particle [more particularly: $\sigma \sim \lambda \sin(\pi\Phi/\Phi_0)$]. This periodicity of the scattering cross-section as a function of the external 'disturbance' (i.e., the external flux) is characteristic of non-local phenomena in general and offers a clue to the dynamical view of such phenomena to be developed below. A second clue is connected with the vanishing of the cross-section in the classical limit.

Before continuing, it is perhaps worth disposing of attempts to explain the effect via local means. Certain authors have claimed that under ideal conditions, when the particle is completely confined to the field-free region, the effect will disappear.

That this is not so can be shown as follows: First, the solution of the Schrödinger equation for this ideal case predicts an effect, provided that we use single-valued wave functions. That the wave functions *must* be single-valued, even in this ideal case, can be seen from the following argument. Suppose that initially the flux in the solenoid is zero, and is switched on later. The time-dependent Hamiltonian appropriate to these circumstances is manifestly single-valued; the initial state, too, is single-valued, and therefore the final state must be single-valued as well.

As another argument (following closely the arguments of Furry and Ramsey 1960), consider a set-up in which a charged particle q_1 is prepared in a coherent superposition of two states confined to the interior of two perfectly conducting spheres (Fig. 4.2). A second charge q_2 is placed in the region between the two spheres. By observing the momentum transfer to q_2, we can tell in which of the two spheres q_1 is

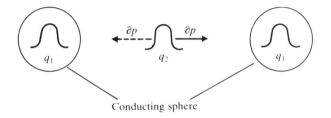

Fig. 4.2. Interaction between a shielded charge and an external charge.

located, and thus destroy the coherency of the packets. Obviously this can be done even in the ideal case when the state of q_1 vanishes exactly outside the spheres. Thus the relative phase of the state describing q_1 must be sensitive to the fields produced by q_2 even when q_1 is entirely confined to the field-free region.

Another attempt to account for the A–B effect locally is based upon the so-called hydrodynamical form of Schrödinger equations (Casati 1979). To obtain the hydrodynamical equations, we replace Ψ by ρ and v, which are defined as follows:

$$\Psi = Re^{i\Phi}, \tag{4.1}$$

$$\rho = R^2, \tag{4.2}$$

$$v = \nabla\Phi - \frac{e}{c}A \tag{4.3}$$

The equations for ρ and v are:

$$\partial\rho/\partial t + \operatorname{div}\rho v = 0, \tag{4.4}$$

$$\partial v/\partial t + (\nabla \cdot v)v = eE - \nabla(\Delta^2 R/2R), \tag{4.5}$$

$$\nabla \times v = \frac{e}{c}\nabla \times B, \tag{4.6}$$

where E and B are the external electric and magnetic fields respectively. (We have assumed for simplicity $m = 1$ and taken $\hbar = 1$).

Since these equations are manifestly local and gauge invariant, we may assume that they account locally for the A–B effect. The situation, however, is not so simple. Consider an initial condition where we have finite quantities of fluid in two non-overlapping regions, while in the region separating them there is only infinitesimal quantity (Fig. 4.3). Let ρ_1, v_1 and ρ_2, v_2 represent the properties of the fluid in the two non-overlapping regions, and ρ_{in}, v_{in} those of the fluid which occupies the region in between. An electric field is then switched on and off again in

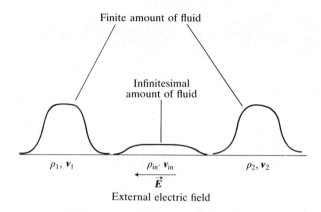

Fig. 4.3. A–B effect in the hydrodynamical picture.

the intermediate region and changes v_{in} in eqn (4.5). Later, when ρ_1 and ρ_2 are allowed to meet, the resulting distribution will depend on $\int_1^2 v_{in} \cdot dl \bmod 2\pi$ (since this integral represents the relative phase in the Schrödinger representation). In the hydrodynamical approach we have to assume that the resulting change in the final ρ is due to the nonlinear interaction between ρ_{in} and ρ_1 and ρ_2. This interpretation is unacceptable because the final effect on ρ is independent of the size of ρ_{in}. Thus, even if $\rho_{in} \to 0$, i.e. there is initially no fluid at all in the intermediate region, the resulting interaction must nevertheless occur. We are therefore led to the conclusion that $\int_1^2 v \cdot dl \bmod 2\pi$ is a non-local property of ρ_1 and ρ_2 rather than a sum of local properties of the fluid in the intermediate region. We have here a very important lesson. Nonlinear equations, though local in appearance, may nevertheless conceal non-local effects.

4.2. Non-local phenomena of the first type

One hopes that the arguments in the preceding section have convinced the reader that the A–B effect is indeed non-local. I would like now to outline some considerations touching upon the idea of non-locality in general.

Let us, first of all, say quite generally what we mean by a 'non-local' property of a physical system. Suppose, we have a system which occupies two separate regions of space (the system might consist, for example, of two objects, one in each region; or, if it is a quantum system, it may consist of a single object whose wave function is non-zero in these regions, but zero elsewhere). The essential difference between local and non-local properties of the system is that in the former case all possible information can be obtained by independent measurements made in the two regions, while in the latter case this is not true.

Let us start with a classical example. Suppose that in one region there exists a constant (constant in space, that is, but time-dependent) gravitational potential, and that in the other the potential is zero (Wisnivesky and Aharonov 1967), and that our system consists of two clocks, one in each of the regions. Despite the fact that there is no gravitational force in either region, it is a well-known consequence of general relativity that the clock in the region where the potential is finite will run more slowly than the one in the region where the potential is zero. We should like to know whether an observer confined to one of the regions can, by means of experiments conducted within that region alone, determine whether it is the clock in *his* region that is running slowly (that is, we want to know whether the rate difference between these two clocks is a local property of the system or a non-local one). Since the effect of the gravitational potential is universal, that is, the potential will affect any experimental apparatus of which such an observer makes use in precisely the same way as it does the clock itself, no local determination will be possible. Let the observer compare the clock in question with any 'standard' clock in his own region; that latter clock will be slowed by the gravitational potential (or not slowed, if the potential is absent) in precisely the same way as the former. Such a standard clock, then, will necessarily fail to detect any slowing of the original clock relative to itself. Thus the effect of the potential here is a purely non-local one, which can be ascertained only later when a comparison of the times on the two clocks becomes possible.

The quantum mechanical situation has much in common with this. Suppose that we prepare a charged particle in the following superposition of two localized wave packets,

$$\Psi_+ = \Psi_1 + \Psi_2, \tag{4.7}$$

where Ψ_1, Ψ_2 are small packets centred at x_1 and x_2. By means of the application of localized electric potentials in the vicinities of x_1 or x_2 (potentials, that is, which are constant in space through the neighbourhood of x_1 or x_2 and thus will produce *no* forces on the particle), the phase difference between the two components can manifestly be altered so as to produce, for example,

$$\Psi_- = \Psi_1 - \Psi_2. \tag{4.8}$$

However, Ψ_- cannot be distinguished in any of its *local* properties from Ψ_+. It differs from Ψ_+ only in certain of its observable *non*-local properties.

We thus see that both classical and quantum gauge theories lead to non-local phenomena. In both theories the local description includes non-gauge invariant quantities, i.e., potentials, phases, time, etc. These gauge-dependent quantities cannot be observed locally. Only later, when comparison between the variables in one region of space and those in another, becomes possible, do gauge-invariant observable effects arise. So far, the classical and quantum non-locality seem to be on an equal footing conceptually. There is, nevertheless, an important distinction between the two, as we proceed to show in the following section.

4.3. Dynamical considerations

In order to understand the dynamical issues involved in our problem it is best to switch from the Schrödinger picture (which we have been using thus far) to the Heisenberg picture. Consider again the ideal case where an external electric field E affects the behaviour of a charged particle which is confined to the field-free region. In the Heisenberg picture it would seem that such an electric field can have no effect on the particle because the Heisenberg equations of motion for x and p are identical, in this case, to the classical ones, i.e.

$$\frac{\mathrm{d}x}{\mathrm{d}t} = \frac{p}{m},$$

$$\frac{\mathrm{d}p}{\mathrm{d}t} = eE. \tag{4.9}$$

Since E vanishes where the particle is located, these equations of motion seem not to depend on E at all. However, this first impression is misleading (Aharonov *et al.* 1969). While, for classical theory, the vanishing of $\mathrm{d}p/\mathrm{d}t$ implies the vanishing of $\mathrm{d}f(p)/\mathrm{d}t$, where $f(p)$ is any arbitrary function of p, this is not necessarily true quantum mechanically. Indeed, consider the function

$$f(p) = \exp(\mathrm{i}p \cdot L/\hbar) \tag{4.10}$$

(i.e. the translation operator). Then

$$\frac{\mathrm{d}f}{\mathrm{d}t} = [f(p_1), H] = \left(\int_x^{x+L} E \cdot \mathrm{d}l \right) f, \tag{4.11}$$

where x is the position of the particle. The Heisenberg equations of motion are then manifestly non-local for $f(p)$ of the type considered above.

In order to relate this result to the Schrödinger picture consider the state:

$$\Psi_\alpha = \Psi_1 + \mathrm{e}^{\mathrm{i}\alpha}\Psi_2, \tag{4.12}$$

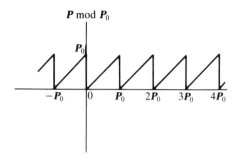

Fig. 4.4. Modular momentum.

where again Ψ_1 and Ψ_2 occupy non-overlapping regions (Fig. 4.3). Let us also assume that $\Psi_2(x) = \Psi_1(x + L)$. Now the effect of the electric or magnetic fields in the A–B example is to produce a change in the relative phase α without disturbing Ψ_1 and Ψ_2 in any way. The creation of such a shift in the relative phase will change neither $\langle p \rangle$ (the expectation value of the momentum of the particle) nor $\langle x \rangle$ nor the expectation value of any polynomial† in x and p. The expectation value of the translation operator $e^{iP \cdot L/\hbar}$, though, depends on α since for this case‡

$$\langle e^{ip \cdot L/\hbar} \rangle = e^{-i\alpha}/2 \qquad (4.13)$$

Let us proceed by noticing that the eigenstates of the translation operator $\exp(iP \cdot L/\hbar)$ are also eigenstates of the modular momentum P modulo P_0 (Fig. 4.4) defined by:

$$P \bmod P_0 = P - nP_0. \qquad (4.14)$$

Here n is an integer vector satisfying

$$0 \leqq P - nP_0 \leqq P_0, \qquad (4.15)$$

componentwise and

$$P_0 = h/L. \qquad (4.16)$$

---

† The proof is straightforward: In evaluating $\langle \Psi_\alpha | ax^n + bp^n | \Psi_\alpha \rangle$ it will emerge that only the cross terms, of the form $\langle \Psi_1 | ax^n + bp^n | \Psi_2 \rangle$ will depend on α. Now, since Ψ_1 and Ψ_2, by hypothesis, do not overlap in space, and the action of any finite polynomial in x and p on any of them does not change that fact, these terms will vanish. Therefore, the expectation value of any polynomial in x and p will not depend on α.

‡ Note that in our case $e^{iP \cdot L/\hbar}\Psi_1 = \Psi_2$. Therefore $e^{-i\alpha}\langle \Psi_2 e^{iP \cdot L/\hbar}\Psi_1 \rangle = e^{-i\alpha}\langle \Psi_1 \Psi_1 \rangle = e^{-i\alpha}/2$.

Note also that Ψ_α is a non-analytic function of x and therefore

$$\langle e^{iP \cdot L/\hbar} \rangle = \left\langle \sum_n (iP \cdot L/\hbar)^n / n! \right\rangle \neq \sum_n 1/n! \langle (iP \cdot L/\hbar)^n \rangle.$$

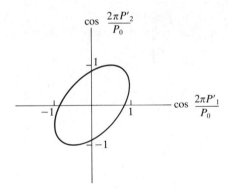

Fig. 4.5. Conservation law for modular momentum.

We see then that the non-local effect of the fields is to produce a shift in the modular momentum of the charged particle while leaving the expectation values of moments of its momentum unaltered.

We next show that modular variables satisfy conservation laws of their own. Consider a collision between two systems, 1 and 2. Momentum conservation implies that for components in any direction:

$$P_1 + P_2 = P_1' + P_2'. \tag{4.17}$$

Consider now the quantities

$$\pi_1(P_1) = \cos(2\pi P_1/P_0) \tag{4.18}$$

and

$$\pi_2(P_2) = \cos(2\pi P_2/P_0). \tag{4.19}$$

As indicated above, measuring π_i is the same as measuring $P_i \bmod P_0$. Momentum conservation implies that

$$\pi_1\pi_2 - (1 - \pi_1^2)^{1/2}(1 - \pi_2^2)^{1/2} = \pi_1'\pi_2' - [1 - (\pi_1')^2]^{1/2}[1 - (\pi_2')^2]^{1/2}. \tag{4.21}$$

We therefore get, after simple manipulation,

$$(\pi_1')^2 + (\pi_2')^2 - 2C\pi_1'\pi_2' = 1 - C^2, \tag{4.22}$$

where

$$C = \cos[2\pi(P_1 + P_2)/P_0].$$

We see, then, that modular variables satisfy conservation laws of their own. Instead of the conserved unbounded line $P_1 + P_2 = \text{const}$, we have a conserved ellipse, as shown in Fig. 4.5.

If we know the initial values of the modular momentum of the two interacting systems, we may represent their initial state by a point on the conserved ellipse of Fig. 4.5. As the interaction between the two systems proceeds, the point representing the system will move along the ellipse and eventually come back to its original position. We see then how the periodicity of the non-local phenomena is reflected in the conservation laws for the relevant modular variables. We also note that in the classical limit $P_0 \rightarrow 0$, so that $P \bmod P_0$ changes so rapidly as a function of P as to become entirely unobservable. We see then that it is possible to think about the A–B effect as a non-local exchange of conserved modular variables, something which is obviously completely quantum in origin.

In the rest of this section, and in the next section, we shall consider a number of examples which demonstrate the usefulness of the dynamical approach advocated above.

4.3.1. Example: Scattering by electric flux lines

Consider the effect of a 'thin' electric flux line on a charged particle (Fig. 4.6). Let the electric field producing the flux be given by

$$E_x(x, t) = E_0 \quad \text{for} \quad |x| \leq \Delta X/2 \quad \text{and} \quad |t| \leq \Delta t/2$$

and $E_x(x, t) = 0$ elsewhere.

The electric flux in this case is given by

$$\Phi = E_x c \, \Delta x \, \Delta t. \tag{4.23}$$

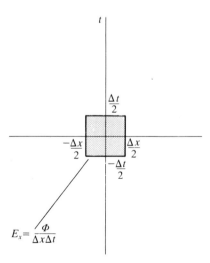

Fig. 4.6. Electric flux confined in a small space–time region.

Let us calculate the scattering cross-section for this case, assuming for simplicity that the particle has been prepared initially in an eigenstate of momentum $P_x = P_y = P_z = 0$. It is straightforward then to show that, provided Δt is sufficiently small, so that the impulse approximation is valid, the probability amplitude for the momentum of the particle after the flux is turned off is given by[†]

$$f(k) = \frac{\sin \pi \alpha}{k} + \Delta x \frac{\sin[(k \, \Delta x - \Phi)/2]}{k \, \Delta x - \Phi}, \qquad (4.24)$$

where

$$\alpha = \Phi/\Phi_0, \qquad k = p/\hbar, \qquad \Phi_0 = c\hbar/e. \qquad (4.25)$$

This result may be understood using the dynamical insight we have gained. We notice that $f(k)$ is made up of two terms. The first term is periodic in the flux and is independent of Δx. This term evidently represents the non-local exchanges of the modular momentum. The second term, which is small in magnitude (of the order of Δx), accounts for a large exchange of momentum (of the order of $E_0 c \, \Delta t = \Phi/\Delta x$). This term evidently represents the rare occasions when the particle interacts locally with the field. Consider now the scattering of the particle from a periodic array of identical flux 'lines'. If we choose

$$\Phi = n\Phi_0 + \varepsilon\Phi_0, \qquad (4.26)$$

where $\varepsilon \ll 1$, we find that the resulting momentum transfer equals $\varepsilon\hbar/L$, where L is the distance between neighbouring flux lines.[‡]

Thus for negative ε the particle will be scattered in a direction *opposite* to that of the force. Of course, the Ehrenfest theorem tells us that the average momentum shift must be in the same direction as that of the force. Indeed, very rarely, with probability proportional to $\Delta x/L$, a local collision will occur and the large momentum exchange that takes place (this time in the 'right' direction) compensates for all the rest of the collisions that went in the opposite direction.

This example demonstrates a general situation that prevails in quantum interactions which vary sharply in space and in time. The effect of such interactions is generally composed of two parts. The first part represents the non-local exchange of modular conserved quantities, and is therefore periodic as a function of the strength of the interaction. This periodicity allows us to choose the interaction strength so that the resulting effect will go in the opposite direction to that predicted by classical theory. The

---

[†] In the impulse case, $\psi' = e^{-iV(X)\Delta t/\hbar}\psi$, where ψ' is the wave function immediately after the flux is turned off, Ψ is the wave function just before it was turned on, and $V(x)$ is the potential, i.e., $eE_x = -\partial V/\partial x$.

[‡] This can be seen by taking the Fourier transform of $e^{-iV(x)/\hbar}$, where $V(x)$ is the potential produced by the array of electric flux lines.

second part corresponds to the local exchanges. By choosing interactions that are sufficiently sharp, i.e., reducing the regions in which the local forces act (and at the same time increasing their strength so that the non-local exchanges, which are proportional to the appropriate flux, remain unaltered) we may reduce the probability for local exchanges at will. Note, though, that in this case in the rare events where the local interaction does manifest itself, the resulting exchanges will be sufficiently large to compensate for all the other events that went in the 'wrong' direction.

4.3.2. Non-local phenomena of the second type

So far, we have considered non-local effects unique to gauge theories. These effects were characterized by the property of being mediated locally by gauge-dependent potentials. We want to show now that more general effects of a non-local nature exist which are mediated directly by local forces, and which have no classical analogue. We begin with an example. Consider a spin-$\frac{1}{2}$ particle confined to a given region of space. The spin of the particle is pointing in the z-direction. An external spatially constant magnetic field in the same direction acts on the particle for some finite time. As is well known, this field will cause the components of the spin in the $x-y$ plane to precess around the z-direction. If this were a classical magnetic moment such a precession could certainly be observed. But for our quantum system, since the direction of the spin in the $x-y$ plane is completely uncertain, there is no local change that is observable, and the only effect of the magnetic field is to produce a change in the phase of the state. Consider now a more elaborate case where our particle is initially prepared in the coherent superposition of two non-overlapping regions of space. Assume that we apply the same magnetic field in one of the regions only. The resulting effect will be to produce a change in the relative phase of the two packets. This, as we have seen, will also cause an exchange of $P \bmod P_0$ where $P_0 = h/L$, L being the distance between the two packets.

Classically, the magnetic field will exchange momentum only with particles which are located in regions where the magnetic field varies in space. But quantum mechanically this is not the case, and the particle can experience the effect of the spatial dependence of the magnetic field even when, as in our example, the variation exists in a region from which the particle is excluded. We see in this example the two ingredients essential for non-local phenomena of the second type:

(1) The local effects of the interaction are completely masked by the quantum uncertainties of the system, so that the system has no local 'memory' of the interaction after it is over.

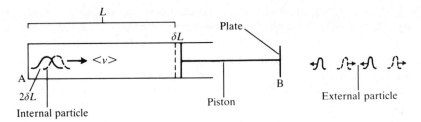

Fig. 4.7. An example of a non-local exchange of modular energy.

(2) The interaction varies in space in the region separating the two regions in which the particle is located, and this variation is responsible for the non-local exchanges of the relevant modular variables.

Let us consider one more example of such phenomena. Suppose (as shown in Fig. 4.7) a particle is confined within a box, one wall of which is actually a moveable piston. Suppose, also that at $t = 0$ the wave function of the particle is negligible everywhere save within a packet whose width Δx is small compared to the length L of the box, and which is located at the end of the box opposite to the piston. Finally, suppose that the expectation value of the velocity of the particle in the x-direction (see Fig. 4.7) is sufficiently large that the spread of the packet can be neglected over times of the order of the period T of oscillation of the particle within the box.

Obviously, the energy of such a particle will be rather poorly defined; but, because the particle, after a period T, will return nearly to its original state, its *modular* energy, $e^{iET/\hbar}$, will be quite sharp.

Imagine now that at $t = 0$, when the particle in the box is far from the piston, the piston is pushed in a short distance $\delta L \ll \Delta x$ and then brought once again to rest. This can be affected by an elastic collision with an external particle. The modular energy of the particle in the box will now be changed, and it follows from the conservation of modular energy that it must necessarily have exchanged that modular energy with the particle outside. This exchange is obviously another example of Type 2 non-locality. Let's understand the nature of this exchange in greater detail. Suppose that the initial state of the particle in the box satisfies

$$e^{-iET/\hbar}\Psi(0) = \Psi(0), \tag{4.27}$$

i.e., it is an eigenstate of $E \bmod h/T$ with eigenvalue zero. Because of the displacement of the piston, we now have, after one period T,

$$\Psi(T, x) = \Psi(0, x - 2\delta L) = e^{-i2p\delta L/\hbar}\Psi(0, x). \tag{4.28}$$

In our case, where $\Delta x \gg \delta L$, we may write

$$\Psi(T, x) = e^{-i2\langle p \rangle \delta L/\hbar}\Psi(0, x) \tag{4.29}$$

where $\langle p \rangle$ is the initial expectation value of the momentum.

For $\delta L = nh/2\langle p \rangle$, with n any integer, the modular energy will be entirely unaffected. This periodicity reminds us of what we have encountered before in connection with quantized fluxes in the A–B effect.

It should be noted that, in the process just described, the modular energy, $E \bmod h/T$, of the external particle is completely uncertain because the time at which it strikes the piston is specified with an uncertainty much less than T. (This is an essential feature of non-local interactions of this type, the absence of which can be shown to imply acausality.) The effect of this uncertainty is to prevent detection of the exchange of modular energy by observations on the external particle. However, reduction of the uncertainty makes this detection possible. This reduction is achieved by preparing the external particle in a state consisting of two wave packets. If the separation of these packets is given by $L' = \langle v_{\text{ext}} \rangle T$, where $\langle v_{\text{ext}} \rangle$ is the mean velocity of the external particle, there will again be non-local interaction with the internal particle. The result of the collision, then, will be to introduce the relative phase shift whose magnitude is exactly that encountered in eqn (4.28) and opposite in sign. This exemplifies the exchange of modular variables (in this case energy) subject to an overall conservation law, as discussed above, and is the analogue of local exchanges of conserved quantities.

If the internal particle were initially in a pure energy eigenstate, the interaction with the external particle would no longer be purely non-local, but rather partly local and partly non-local. For the non-local part to be predominant, however, the energy must be large compared to the ground state energy of the box. Just as in the electric flux example of the previous section, there will be a periodic non-local exchange of energy (periodic in $\delta L(2mE)^{1/2}/2h$) plus a term representing the usual local exchange. This latter term will be vanishingly small as E becomes large. Again, the periodicity implies that, with probability arbitrarily close to unity, the exchange can go the 'wrong way', i.e., can result in the lowering of the internal particle's energy as the piston is pushed in. (The expectation value of its energy will of course rise as a result of those rare, but highly energetic, local exchanges.)

4.3.3. Semi-classical considerations

Further insight into non-local phenomena may be gained by considering the semi-classical approximation. As we know, in the semi-classical approximation the phase of the wave function in the Schrödinger representation corresponds to the classical action. Thus we have to find a method of changing the classical action of the set of orbits in a given region of space–time without affecting the orbits themselves. The simplest way of doing this is to introduce a potential which is purely a function of time in the region under consideration since that will augment $S \equiv \int L \, dt$ by an amount $\delta S = \int V(t) \, dt$. (Of course, we can achieve a similar result with the aid of a vector potential A.) Thus the A–B effect corresponds to an interaction which produces a change in the action of the classical orbits in a given region of space–time without changing the orbits themselves. (To complete the effect we need to introduce a different change in a second region, which of course implies that we will have non-vanishing fields in the intervening region.)

We call such phenomena non-local phenomena of the first kind. We can now generalize our discussion in the following way. Suppose we introduce in a given region of space a non-zero force field which will modify each orbit in the region for a finite time, but return it afterwards to what it would have been without the field. If we do so for all the orbits in the region, the end result will be to produce no observable change in the final behaviour of the classical system. So although we have changed the behaviour of the system while the interaction took place, after the interaction is over no 'memory' of such changes will persist. What is important for the quantum case, though, is whether any change of *action* occurred due to the interaction, and we shall now show by a simple example that this indeed may be the case.

Consider a situation where we have switched on, and then off again, a given force $F(t)$ which is constant in the whole spatial region. This single impulse will produce a change in both the velocity and the position of each orbit. By introducing a second impulse equal and opposite to the first we may compensate for the change of the velocity but not for the change of positions along each orbit. Let us call this double impulse a 'di-pulse'. If we now consider adding subsequently an opposite 'di-pulse', we will correct also for the change in the position of the orbit. (This consideration is only approximately true if the particle is not free, but we can make this approximation as good as we like by approaching the limit of finite impulses which are sufficiently dense in time.) 'Quadra-pulses' of this kind will cause no change in the orbits, but as we shall now see they *will* produce a change in the action of the orbits, and *will* produce a phase change in the corresponding quantum state. For simplicity we shall consider here only a free case, although the result is valid for the more general case if we again make our impulses sufficiently dense in time.

The action of the particle is

$$S = (m/2) \int (v_0 + \delta v)^2 \, dt$$

where δv is the change of the velocity due to the effect of the impulses.

$$\delta S = m \int v_0 \, \delta \psi \, \mathrm{d}t + (m/2) \int (\delta v)^2 \, \mathrm{d}t,$$

wherein the first term vanishes since the impulses are chosen so that $\int \delta n \, \mathrm{d}t = \delta X = 0$ where δX is the resulting change in the position of the particle. Thus there is a change in the action which is independent of the orbit, and which will result in a change of phase for the state of the particle if it is confined to the considered region. If we confine our particle to two separate regions of space–time in one of which we introduce these impulses, and in the other of which we do not, we will end up with a result identical to the pure A–B effect. This constitutes an example of non-local phenomena of the *second* kind (those mediated by *local* interactions which nonetheless produce only non-local changes in the system.) Indeed, if the world were full of such 'quadra-pulses', they would produce (at least in the limits of singular impulses) no observable effect classically (just like singular non-quantized lines of flux); but they *would* produce observable effects in the quantum mechanical case.

The final example of non-local phenomena mediated by local means is of a more general kind and is related in a non-trivial way to the quantum uncertainty principle. Consider again our classical orbits and imagine an interaction that modifies not only the action of the orbit but also the orbit itself. Classically, the effect of the interaction will be observable locally. In the quantum case, though, we can have states for which the resulting changes of the orbit will be completely masked by the uncertainties of the orbit and the only effect of the added interaction will be to produce a pure phase change of the state. In this case we will end up with non-local phenomena for a family of states but not for every state. This situation corresponds of course to the two examples of the last section.

4.4. Conclusion

In this disussion, I have tried to give a general characterization of non-local phenomena and to outline how one might understand such phenomena in a dynamical way. We have encountered two distinct types of non-local phenomena:

(1) The A–B effect, which is common to all gauge theories, and is characterized by phenomena associated with systems confined to field-free, non-simply connected regions. These were classified as non-local phenomena of type 1.

(2) Effects which arise through local interactions with fields (or other forces), but in such a way as to produce only non-local changes in the

systems in question. These were classified as non-local phenomena of type 2.

New types of variables, the modular variables, were introduced, and were shown to be the natural tools for describing non-local dynamics.

References

Aharonov, Y., Pendleton, H., and Petersen, A. (1969). *Int. J. Theor. Phys.* **2,** 213.
Casati, G. (1979). *Phys. Rev. Lett.* **42,** 1579.
Furry, W. H. and Ramsey, N. F. (1960). *Phys. Rev.* **118,** 623.
Ruijsenaars, S. N. M. (1983). *Ann. Phys. N.Y.* **146,** 1.
Wisnivesky, D. and Aharonov, Y. (1967). *Ann. Phys. N.Y.* **45,** 479.

5

Gravitational effects on charged quantum systems

J. Anandan

5.1. Introduction

The importance of the operational procedure for measuring the quantities used in physics has been emphasized by many physicists. The usual probes used for measuring the gravitational field are neutral and 'classical' in the sense that their response to the gravitational field can be explained using just the laws of classical physics. I shall consider here, however, probes of the gravitational field that differ from the usual probes in two respects: they are electrically charged and quantum mechanical. Since they are charged, their unavoidable interaction with the electromagnetic field must also be taken into account.

Thus I am concerned with three physical systems: the charged test particle (treated quantum mechanically), the gravitational field, and the electromagnetic field. Each of these systems interacts with the other two. The interaction between the gravitational and electromagnetic fields that will be considered here can be regarded in a quantum theory of gravity and electromagnetism as interaction between gravitons and photons, but I think that it is fair to say that we still do not have a quantum theory of gravity; in any case, both fields will be treated here classically. Also, the effect of the electromagnetic field on the gravitational field and the effect of the test particle on either field are very small for all the configurations that we will consider and therefore will be neglected.

In Section 5.2, I consider the modification of the phase shifts in the quantum interference of a charged particle due to the electric and magnetic fields, in the presence of a gravitational field. This generalizes the celebrated Aharonov–Bohm effects (Ehrenberg and Siday 1949; Aharonov and Bohm 1959) when the gravitational field is present. In Section 5.3, the influence of the gravitational field on the electrons in a conductor is considered. Two new effects, which involve the interplay of thermodynamics, general relativity and electromagnetism, are mentioned. The influence of the gravitational field on the quantum mechanical wave function of the Cooper pair is considered in Section 5.4, from a general relativistic point of view. Some superconducting devices that can, in principle, measure the gravitational field by its effect on the wave function of the Cooper pairs are described.

5.2. Gravitationally coupled electromagnetic fields and quantum interference

I shall assume that space–time is pseudo-Riemannian with metric $g_{\mu\nu}$ of signature $(+ - - -)$. The electromagnetic field $F_{\mu\gamma}$ satisfies the general relativistic Maxwell equations

$$F^{\nu\mu}_{;\nu} = j^{\mu}, \qquad \partial_{[\mu}F_{\nu\rho]} = 0. \tag{5.1}$$

Consider now an apparatus, which may be a conductor, whose linear dimensions are small compared with the radius of curvature and which is non-rotating relative to the local inertial frames. The gravitational field will be assumed throughout to be stationary with respect to the apparatus, i.e. there exists a Killing vector field ξ^{μ} having the same direction as the four-velocity field t^{μ} of the apparatus. Also any current density j^{μ} flowing in the apparatus and the electromagnetic field in a region containing the apparatus are assumed to be stationary, i.e. $\mathcal{L}_{\xi}j = 0$, and $\mathcal{L}_{\xi}F = 0$, where $\mathcal{L}_{\xi}$ denotes the Lie derivative with respect to ξ^{μ}. The electric and magnetic fields as seen by observers who are at rest with respect to the apparatus are $E^{\mu} = F^{\mu}_{\nu}t^{\nu}$ and $B^{\mu} = (\frac{1}{2})\eta^{\mu\rho\nu\sigma}t_{\nu}F_{\rho\sigma}$.

On defining $\Lambda = \xi^{\mu}\xi_{\mu}$, Killing's equation $\xi_{(\mu;\nu)} = 0$ implies that $\partial_{\mu}\Lambda^{1/2} = -\Lambda^{1/2}a_{\mu}$, where $a^{\mu} = t^{\nu}t^{\mu}_{;\nu}$ is the acceleration of the apparatus relative to the local inertial frames. Under the above assumptions, a local co-ordinate system can be chosen such that $\xi^{\mu} = (1, 0, 0, 0)$ and $g_{\mu\nu} = \mathrm{diag}(\Lambda, -1, -1, -1)$, to a good approximation, where $\Lambda = \xi^{\mu}\xi_{\mu} = g_{00}$. Then $\mathbf{g} \equiv -c^2 \nabla\Lambda^{1/2}/\Lambda^{1/2}$ is the 'acceleration due to gravity'. Also, on defining $\mathbf{E} = (E^1, E^2, E^3)$, $\mathbf{B} = (B^1, B^2, B^3)$, the charge density $\rho = \Lambda^{1/2}j^0$ and the current density $\mathbf{j} = \Lambda^{1/2}(j^1, j^2, j^3)$, eqn (5.1) may be written in this co-ordinate system as (Anandan 1984c),

$$\mathrm{div}\,\mathbf{E} = \rho, \qquad \mathrm{curl}(\Lambda^{1/2}\mathbf{B}) = \tilde{\mathbf{j}}$$
$$\mathrm{div}\,\mathbf{B} = 0, \qquad \mathrm{curl}(\Lambda^{1/2}\mathbf{E}) = 0. \tag{5.2}$$

Now write $\mathbf{E} = \mathbf{E}_0 + \mathbf{E}_1$ and $\mathbf{B} = \mathbf{B}_0 + \mathbf{B}_1$, where $\mathbf{E}_0$ and $\mathbf{B}_0$ are the unperturbed electric and magnetic fields (the solutions of the usual Maxwell's equations in the absence of the gravitational field for the same charge and current densities ρ and $\tilde{\mathbf{j}}$) and $\mathbf{E}_1$, $\mathbf{B}_1$ are the perturbations of order g. Then, on neglecting $O(g^2)$ terms, from (5.2)

$$\operatorname{div} \boldsymbol{E}_1 = 0, \qquad \operatorname{curl} \boldsymbol{B}_1 = \frac{\boldsymbol{g} \cdot \boldsymbol{r}}{c^2} \tilde{\boldsymbol{j}} + \frac{1}{c^2} \boldsymbol{g} \times \boldsymbol{B}_0$$

$$\operatorname{div} \boldsymbol{B}_1 = 0, \qquad \operatorname{curl} \boldsymbol{E}_1 = \frac{1}{c^2} \boldsymbol{g} \times \boldsymbol{E}_0.$$

(5.3)

We shall now determine the modification of the electric field of a point charge due to its coupling to the gravitational field. Since the eqns (5.3) are linear, for a fixed background gravitational field, the electric field due to an arbitrary charge distribution can be obtained by superposition from the solution for a point charge. For a charge q at rest at the origin of the above co-ordinate system, $\boldsymbol{j} = 0$, $\boldsymbol{E}_0 = (q/4\pi r^3)\boldsymbol{r}$ and $\boldsymbol{B}_0 = 0$, where $r = (x^1, x^2, x^3)$ and $r = |\boldsymbol{r}|$. Since $\boldsymbol{E}_1$ is of order g, on dimensional grounds

$$\boldsymbol{E}_1 = a \frac{q\boldsymbol{g}}{c^2 r} + b \frac{q\boldsymbol{g} \cdot \boldsymbol{r}}{c^2 r^3} \boldsymbol{r},$$

where a and b are dimensionless constants. Substitution into eqn (5.3) yields $a = b = 1/8\pi$. Hence the required solution is

$$\boldsymbol{E}_1 = \frac{q}{8\pi c^2 r} \left(\boldsymbol{g} + \frac{\boldsymbol{g} \cdot \boldsymbol{r}}{r^2} \boldsymbol{r} \right), \qquad \boldsymbol{B}_1 = 0.$$

(5.4)

Thus, we see explicitly that a charge at rest in a stationary gravitational field does not radiate even though it is accelerating relative to the local inertial frames. In the chosen co-ordinate system, the total electric field $E^i = (g_{00})^{-1/2} F_{0i} = -(g_{00})^{-1/2} \partial_i A_0$, where A_0 is the electrostatic potential in a gauge in which the vector potential $A_i = 0$. Hence,

$$\nabla A_0 = -\left(1 - \frac{\boldsymbol{g} \cdot \boldsymbol{r}}{c^2} \right) \boldsymbol{E} = \frac{-q}{4\pi r^3} \boldsymbol{r} - \frac{q}{8\pi c^2 r} \left(\boldsymbol{g} - \frac{\boldsymbol{g} \cdot \boldsymbol{r}}{r^2} \boldsymbol{r} \right),$$

and, therefore,

$$A_0 = \frac{q}{4\pi r} - \frac{q}{8\pi c^2} \frac{\boldsymbol{g} \cdot \boldsymbol{r}}{r}.$$

(5.5)

The phase shift in electron interference due to the electric field of two identical parallel-plate capacitors, back-to-back, in the presence of a gravitational field was obtained previously in Anandan and Stodolsky (1983) and Anandan (1984b). We consider this problem now for the electric field of a spherically symmetric charge distribution with total charge q at the centre of the interferometer. Suppose that the interferometer is in the x^1–x^3 plane, with $\boldsymbol{g} = -(0, 0, g)$ and the charge centered at the origin. A beam of particles with charge e and constant energy is split coherently at $A(-X/2, 0, -Z/2)$ and the resulting two beams are reflected at B $(X/2, 0, -Z/2)$ and $C(-X/2, 0, Z/2)$ to be recombined and detected at $D(X/2, 0, Z/2)$. To lowest order, the beams AC, BD travel under identical conditions and, therefore, do not contribute to the phase shift. The first term in eqn (5.5) does not contribute to the total phase shift due to the electric field owing to the symmetry of the configuration, to lowest order. The phase shift due to the second term in eqn (5.5) is

$$\Delta \theta_E = -\frac{e}{\hbar c} \oint A_\mu \, dx^\mu = -\frac{e}{\hbar} \oint \frac{qgz}{8\pi c^2 (x^2 + z^2)^{1/2}} \, dt$$

$$= -\frac{eqgZ}{8\pi \hbar c^2 v} \int_{-X/2}^{X/2} \frac{dx}{(x^2 + z^2)^{1/2}} = -\frac{eqgZ}{4\pi \hbar c^2 v} \sinh^{-1}\left(\frac{X}{Z}\right), \quad (5.6)$$

where $x = x^1$, $z = x^3$ and the integration is along the unperturbed trajectory so that v is the velocity of the unperturbed beams AB and CD.

There is also a phase shift due to the gravitational field alone, namely $\Delta \phi = -\varepsilon g Z X / \hbar c^2 v$, where $\varepsilon = mc^2/(1 - v^2/c^2)^{1/2}$ is the relativistic kinetic energy of the incoming beam. This $\Delta \phi$ is the relativistic generalization of the phase shift measured for neutrons by Colella *et al.* (1975). If the interferometer is rotated by an angle θ about a horizontal axis then the factor g in both phase shifts must be replaced by $g \cos \theta$. The phase shift due to gravity can, therefore, be varied by varying θ. It should be noted that unlike the configuration considered previously (Anandan and Stodolsky 1983), the electric field is non-zero along the 'horizontal' beams. It has been shown that the phase shift due to the bending of trajectories due to an external 'force', which in the present case is $eE + mg$, is second order in this force (Greenberger and Overhauser 1979, Chiu and Stodolsky 1980). The contribution to this phase shift from the beams AB and CD is the sum of two terms of $O(g^2)$ and $O(E^2)$, since the $O(g \cdot E)$ contributions cancel. However, the momenta of the beams AC and BD may differ by a term of $O(g \cdot E)$ due to this bending effect, which may contribute an $O(gq)$ term that may mask $\Delta \theta_E$. We shall not compute this contribution because $\Delta \theta_E$ is too small, in any case, to be measurable at present. Nevertheless, the ratio $\Delta \theta_E/\Delta \phi \sim eq/\varepsilon X$, which can be made an appreciable fraction of unity. This suggests that the new effect may be detectable if a proton interferometer is developed.

We consider now the modification of the magnetic field due to a wire carrying a steady current. Let I be the proper current at an arbitrary point on the wire, i.e. I is the current as measured by a local observer who is at rest at this point with respect to the wire, using his proper time. It can be shown using the charge conservation law that $\tilde{I} \equiv \Lambda^{1/2} I = (1 - g \cdot r/c^2)I$ is constant along the wire (Anandan 1984*a,b*). Let $I = I_0$ at a fixed point O on the wire and let B_0 be the magnetic field due to this current if there were no gravitational field. Then, solving eqn (5.3), the total magnetic field in the presence of gravity is (Anandan 1981b,c)

$$B(r) = B_0(r) + B_1(r)$$
$$= \left(1 - \frac{gz}{c^2}\right)B_0(r) - \int d^3r' \frac{g \cdot B_0(r')}{c^2 |r - r'|^3}(r - r'), \tag{5.7}$$

where z is the height of the point r above O. A special case is a long horizontal solenoid for which $g \cdot B_0 = 0$ inside the solenoid and the contribution from $g \cdot B_0$ outside the solenoid is negligible. Then $B(r) = (1 - gz/c^2)B_0(r)$. Hence the phase shift due to the magnetic flux inside this solenoid in an Aharonov–Bohm type of interference experiment is

$$\Delta\theta_B = \left(1 - \frac{g\bar{z}}{c^2}\right)\frac{e}{\hbar c}\int_S B_0 \cdot dS, \tag{5.8}$$

where S is an area of cross-section of the solenoid and $\bar{z}$ is the height of the centroid of the projection of this area on a plane normal to B_0.

In this experiment, the current I_0 at O can be kept fixed, when the gravitational field is 'turned on', by having a battery of constant e.m.f. at O, assuming that the total resistance does not change. Then (5.8) predicts a fractional change in phase shift of $g\bar{z}/c^2$. But, of course, the fluctuations in the e.m.f. of the battery alone would completely mask this effect. It is, therefore, better to produce the magnetic flux by means of a superconducting circuit in which the current would be constant to a very good approximation. Such circuits and the effect on them due to the gravitational field will be considered in Section 5.4.

5.3. Thermoelectric–gravitational effects

Let us consider now charge and heat flow in a conductor in a stationary gravitational field. Suppose that the time-like vector w^μ represents the density of flow of total kinetic energy, including rest mass energy, as measured by observers whose four-velocity is the same as the four-velocity t^μ of the conductor. Using the projection tensor $P_{\mu\nu} = g_{\mu\nu} - t_\mu t_\nu$, we define also the conduction current density $j_c^\mu = P^{\mu\nu}j_\nu$ and the flux of total kinetic energy (including heat energy) $w_c^\mu = P^{\mu\nu}w_\nu$. Then the known equations for charge and heat conduction for an isotropic conductor (Sommerfeld 1964), in the absence of a magnetic field, can be covariantly generalized to

$$w_c^\nu = KP^{\nu\rho}(\partial_\rho T - a_\rho T) - (\Pi - \zeta/e)j_c^\nu, \tag{5.9}$$

$$j_c^\nu = \sigma E^\nu + \frac{\sigma}{e}P^{\nu\rho}(\partial_\rho \zeta - a_\rho \zeta) - \sigma\varepsilon P^{\nu\rho}(\partial_\rho T - a_\rho T), \tag{5.10}$$

where K, σ are the thermal and electrical conductivities, ζ is the chemical potential (including the rest mass energy), T is the temperature, Π is the Peltier coefficient, e is the charge of the current carrier, which in a metal is an electron, and $a^\mu = t^\nu t^\mu_{;\nu}$. In a stationary electromagnetic field, the electrochemical potential is $\mu = \zeta + A_\mu t^\mu$, where the electromagnetic potential A_μ is in a gauge in which the Lie derivative $\mathcal{L}_g A = 0$. Then eqn (5.10) may be rewritten as

$$j^\nu_c = \frac{\sigma}{e} \boldsymbol{P}^{\nu\rho}(\partial_\rho \mu - a_\rho \mu) - \sigma\varepsilon \boldsymbol{P}^{\nu\rho}(\partial_\rho T - a_\rho T). \qquad (5.11)$$

It is interesting that eqns (5.9) and (5.10) imply the following general relativistic generalizations of the Peltier and Seebeck effects (Anandan 1984d): if a current is driven through a circuit formed out of two different wires, then the heats U_1 and U_2 given out per unit proper time at the upper and lower junctions are related by $U_2 = -(1 + gH/c^2)U_1$, where H is the height between the junctions. Also if the temperatures at the upper and lower junctions are maintained at T_1 and T_2 respectively, and there is an open switch in the circuit, then the e.m.f. across this switch is $V = (\varepsilon_I - \varepsilon_{II})\{T_1(1 + gH_1/c^2) - T_2(1 + gH_2/c^2)\}$, where ε_I, ε_{II} are the values of ε for the two metals and H_1, H_2 are the heights of the junctions above the switch.

5.4. Quantum gravimeters

We consider now the effect of the gravitational field on superconductors, from a general relativistic point of view. Let $\psi = \chi e^{i\phi}$ be the wave function of the centre of mass of each Cooper pair, which is spread out over the entire superconductor, where χ and ϕ are real functions. It is known that current can flow only on the surface of a superconductor. Hence, in the interior, the current density is given by

$$j_\mu = (i\hbar/2m)\left\{\psi^*\left(\partial_\mu + i\frac{2e}{\hbar c}A_\mu\right)\psi - \psi\left(\partial_\mu - i\frac{2e}{\hbar c}A_\mu\right)\psi^*\right\}$$

$$= -\frac{\chi^2}{m}\left(\hbar\,\partial_\mu\phi + \frac{2e}{c}A_\mu\right) \propto t^\mu,$$

where t^μ is the four-velocity field of the superconductor. Hence

$$-\left(\hbar\,\partial_\mu\phi + \frac{2e}{c}A_\mu\right) = \frac{2}{c}\zeta t_\mu. \qquad (5.12)$$

The function ζ in eqn (5.12) must be the chemical potential in order that eqn (5.12) is in agreement with the Josephson equation

$$-\hbar t^\nu \partial_\nu \phi = 2\mu/c,$$

where μ is the electrochemical potential.†

If the superconductor has no rotation ($\eta^{\mu\nu\rho\sigma} t_\nu \nabla_\rho t_\sigma = 0$) then a space-like closed curve γ can be chosen inside the superconductor so that $\oint t_\mu \, dx^\mu = 0$. Then from eqn (5.12), since ψ must be single valued,

$$\int_S F_{\mu\nu} \frac{d\sigma^{\mu\nu}}{2} = n\Phi_0, \tag{5.13}$$

where $\Phi_0 = ch/2e$, $F_{\mu\nu} = \partial_\mu A_\nu - \partial_\nu A_\mu$, $n =$ integer and S is a surface spanned by γ. Equation (5.13) represents, in the present general relativistic context, the known result that the magnetic flux enclosed by a superconducting region is quantized. This is the macroscopic analogue of the quantization of angular momentum in an atom. More generally, if the superconductor has a rotation, eqn (5.12) implies generalizations of the fluxoid quantization and the London moment (Anandan 1984d).

It is possible to construct superconducting devices, which we call quantum gravimeters, that can in principle measure the gravitational field by its effect on the wave function ψ. I illustrate this by means of two examples; two more examples that use the d.c. and a.c. Josephson effects are given by Anandan (1984d). Consider first a superconducting circuit consisting of two horizontal superconducting solenoids, of inductances L_1 and L_2, in series, with L_1 at a height H above L_2. The diameters of the solenoids are much smaller than H. Suppose δI_1 and δI_2 are the changes in the current I in L_1 and L_2, respectively, when a gravitational field is 'turned on'. Then eqn (5.13) implies that the total magnetic flux should not change, assuming that there is no leakage of fluxoid. Hence,

$$L_1 \delta I_1 + L_2 \delta I_2 = 0. \tag{5.14}$$

Also, according to Section 5.2, charge conservation requires that $I + \delta I_1 = (I + \delta I_2)(1 - gH/c^2)$ or

$$\delta I_2 - \delta I_1 = IgH/c^2. \tag{5.15}$$

Solving eqns (5.14) and (5.15), $\delta I_1 = -[L_2/(L_1 + L_2)]IgH/c^2$ and $\delta I_2 = [L_1/(L_1 + L_2)]IgH/c^2$.

---

† If ψ satisfies the following general relativistic generalization of the time-dependent Ginzburg–Landau equation

$$g^{\mu\nu}\left(\nabla_\mu + \frac{2ie}{\hbar c}A_\mu\right)\left(\nabla_\nu + \frac{2ie}{\hbar c}A_\nu\right)\psi + \frac{4m^*\alpha}{\hbar^2}\psi + \frac{4m^*\beta}{\hbar^2}|\psi|^2\,\psi + \frac{4m^{*2}c^2}{\hbar^2}\psi = 0,$$

where $2m^*$ is the mass of the Cooper pair, then ζ has the following expression in terms of the Ginzburg–Landau parameters α and β, on using eqn (5.12) and neglecting $(\hbar^2/m^*c^2)\nabla^\mu \nabla_\mu \chi$: $\zeta = m^*c^2 + \alpha + \beta\chi^2 +$ higher-order terms in α/m^*c^2 and β/m^*c^2.

There would then be corresponding changes in the magnetic fluxes in the solenoids. These changes can, in principle, be detected by SQUIDs that measure magnetic field by the phase shift in the quantum interference of a Cooper pair across Josephson junctions, given by eqn (5.8). However, the fractional changes in the current and therefore the magnetic field are so small that it is practically impossible to measure them. An improvement on the above device would be to short out the circuit by means of an approximately horizontal superconducting wire so that, in the absence of gravity, it carries no current. When gravity is 'turned on', according to eqn (5.13), the proper current in each solenoid remains the same, assuming that L_1 and L_2 do not change. But a current $\delta I = 2IgH/c^2$ must then flow through this horizontal wire because of general relativistic charge conservation. If the circuit is rotated about this wire then an alternating current, with the same frequency as the rotational frequency, would flow through this wire. Even this experiment seems very difficult to do at present because moving the solenoids in the gravitational field so as to vary the effect of gravity may change L_1 and L_2. Indeed, all the ideas proposed here seem to be only of theoretical interest, at present.

References

Aharonov, Y. and Bohm, D. (1959). *Phys. Rev.* **115,** 485.

Anandan, J. (1984a). *Gen. Rel. Grav.* **16,** 33.

—— (1984b). In *Proc. Int. Symp. on Foundations of Quantum Mechanics in the Light of New Technology, Tokyo, 1983,* (ed. S. Kamefuchi, H. Ezawa, Y. Murayama, M. Namiki, S. Nomura, Y. Ohnuki and T. Yajima). Physical Society of Japan, Tokyo.

—— (1984c). *Class. Quantum Grav.* **I,** No. 5, L51.

—— (1984d). *Phys. Lett.* **105A,** 280.

—— and Stodolsky, L. (1983). *Phys. Rev. Lett.* **50,** 1730.

Chiu, C. and Stodolsky, L. (1980). *Phys. Rev.* D **22,** 1337.

Colella, R., Overhauser, A. W., and Werner, S. A. (1975). *Phys. Rev. Lett.* **34,** 1472.

Ehrenberg, W. and Siday, R. E. (1949). *Proc. Phys. Soc. Lond.* B **62,** 8.

Greenberger, D. M. and Overhauser, A. W. (1979). *Rev. Mod. Phys.* **51,** 43.

Sommerfeld, A. (1964). *Thermodynamics and statistical mechanics.* Academic Press, New York. Sec. 21, eqns (18a,b).

6

Continuous state reduction

Tony Sudbery

Change and decay in all around I see. (Hymn)

6.1. Introduction

'Quantum mechanics may be peculiar, but it works: it gives precise predictions of a wide range of physical phenomena, which have been verified with unprecedented accuracy.' So runs a defence of orthodoxy which, though it might not be expressed so pugnaciously today as was common twenty years ago, would still command widespread support among physicists. And with good reason. One might query whether a theory 'works' by making successful predictions—it is the function of a theory to give true understanding, and the point of making predictions is to test whether the theory is working successfully at this primary task— but one cannot deny that quantum physicists do make very good predictions. This certainly suggests that there is a good deal of truth in quantum mechanics, though we may have to work a bit harder for the understanding.

To be more precise, the inference is that there is a great deal of truth in the theory that quantum physicists use to make predictions. It is not clear that this is the same as 'quantum mechanics' as it is formulated in the textbooks. In this article I would like to look at a class of problems in physics which are handled very successfully by quantum physicists, but whose solutions do not seem to accord with the official principles of quantum mechanics. This situation puzzled me as a student, but since the grown-ups seemed to know what they were doing I thought it would be naive and pedantic to insist on the letter of the textbook formulation: clearly quantum mechanics was an art, not a science, and I ought to wait until I could do it too. Now I propose to take it seriously as a science by observing the practitioners of the art and precisely formulating the rules that govern their behaviour. These turn out to be slightly different from the usual postulates of quantum mechanics. I will argue that they are also rather easier to understand.

6.2. Decay in quantum mechanics

The process I want to discuss is the decay of an unstable system like a radioactive nucleus or an elementary particle which decays by weak interactions, say

$$A \rightarrow B + C. \tag{6.1}$$

The quantum mechanics which 'works' does so by describing this system by a time-dependent state vector

$$|\psi(t)\rangle = a(t)\,|A\rangle + b(t)\,|BC\rangle, \tag{6.2}$$

the solution of the Schrödinger equation for the system, and then interpreting this as meaning that the probability that the decay has happened by time t is $|b(t)|^2$. I want to support this interpretation, but I claim that in the conventional formulation of quantum mechanics it stands unsupported: there is no deductive path which takes one from the principles of quantum mechanics to a statement about the probability of the event (6.1) having happened by time t. According to the books, the interpretation of the state vector (6.2) is that if one performs an experiment on the system at time t to find out whether its state is A or B + C, the probability that the answer will be 'B + C' is $|b(t)|^2$. It is generally assumed that this is equivalent to the former interpretation, and one can see why: if the result of the measurement is 'B + C', then the decay has occurred, and once the decay has occurred the result of the measurement will be 'B + C'. But this only yields a decay probability if there is a definite act of experimental measurement at time t: if there is no such intervention the interpretation does not apply and quantum mechanics is silent about the meaning of the state vector (6.2). This is usually the case. In investigating a decay one does not bring one's apparatus to bear every now and then, asking the question 'A or B + C?'; the apparatus functions all the time, ready to record the decay as it happens. This situation is not covered by the usual formulation of quantum mechanics.

The general picture of physical change provided by the conventional principles of quantum mechanics is that the state of a system changes continuously, following the Schrödinger equation, as long as no measurements are made on the system: when a measurement is made (this being an instantaneous event), the state changes discontinuously, the state after the measurement being in correspondence with the actual outcome of the experiment. For simplicity, consider a system which has two states $|\psi_1\rangle$ and $|\psi_2\rangle$ which are eigenstates of the observable being measured. The system starts in a state $|\psi_0\rangle$ which evolves continuously to

$$|\psi(t)\rangle = e^{-iHt}\,|\psi_0\rangle = c_1(t)\,|\psi_1\rangle + c_2(t)\,|\psi_2\rangle \tag{6.3}$$

until the time of measurement; then the state becomes either $|\psi_1\rangle$ (with probability $|c_1|^2$) or $|\psi_2\rangle$ (with probability $|c_2|^2$). It is convenient to describe this by means of statistical operators. The statistical operator at $t = 0$ is

$$\rho_0 = |\psi_0\rangle\langle\psi_0|, \tag{6.4}$$

which evolves to

$$\rho(t) = |\psi(t)\rangle\langle\psi(t)|$$
$$= e^{-iHt}\rho_0 e^{iHt} \tag{6.5}$$

until the measurement, when it becomes

$$\rho'(t) = |c_1(t)|^2 |\psi_1\rangle\langle\psi_1| + |c_2(t)|^2 |\psi_2\rangle\langle\psi_2| \tag{6.6}$$
$$= P_1\rho(t)P_1 + P_2\rho(t)P_2, \tag{6.7}$$

where P_1 and P_2 are the projection operators on to the state vectors $|\psi_1\rangle$ and $|\psi_2\rangle$. In general, Dirac's† projection postulate (Dirac 1930) states that the effect of a measurement is to change the statistical operator from ρ to $\sum_i P_i\rho P_i$ where P_i are the projection operators onto the eigenspaces of the observable being measured.

The objection that ought to occur to every student who is asked to believe this story is that a measurement is a physical process just as much as the undisturbed evolution of the system being measured, and it ought also to be subject to a Schrödinger equation. If the student is very bright he or she will point out that the Schrödinger equation cannot give the statistical operator (6.7) from the initial condition (6.4), because the condition (6.4) has rank 1 while the operator (6.7) has rank 2, and the evolution given by eqn (6.5) preserves the rank of ρ. Within the conventional theory there are two answers to this. Von Neumann (1932) gave a general schematic description of the measurement process in which the apparatus is explicitly included in the quantum mechanical state: the state space is enlarged from $\mathcal{S}$, the state space of the system which was the original object of interest, to $\mathcal{S} \otimes \mathcal{A}$, where $\mathcal{A}$ is the state space of the apparatus. Thus the initial state is a product state $|\psi_0\rangle|\alpha_0\rangle$. While the system is developing and the apparatus is lying idle this evolves to $|\psi(t)\rangle|\alpha_0\rangle$. Then the measurement is set in train, which means that an interaction between the system and apparatus is switched on; the Hamiltonian acquires an extra term which causes the state $|\psi(t)\rangle|\alpha_0\rangle$ to evolve to

$$a(t)|\psi_1\rangle|\alpha_1\rangle + b(t)|\psi_2\rangle|\alpha_2\rangle, \tag{6.8}$$

† Why does everyone call it Von Neumann's projection postulate? Von Neumann stated a postulate about the results of measurement later than Dirac, and he got it wrong. Dirac's version is the one now generally used.

where $|\alpha_1\rangle$ and $|\alpha_2\rangle$ are the states of the apparatus which exhibit the measurement results corresponding to the system states $|\psi_1\rangle$ and $|\psi_2\rangle$. The state (6.8) now shows the correlation between the properties of the system and the state of the apparatus which it is the purpose of measurement to produce, and this has been achieved by means of a normal physical evolution according to the Schrödinger equation, of the sort that our rebellious student insisted on. We would like to say now that you can see from eqn (6.8) that if the state of the apparatus is $|\alpha_1\rangle$ (or $|\alpha_2\rangle$) then the state of the system is $|\psi_1\rangle$ (or $|\psi_2\rangle$), and this has all been achieved by the Schrödinger equation, without any appeal to the projection postulate; but of course this won't do. In order to say that the state of the apparatus is $|\alpha_1\rangle$, and therefore that the state of the collective apparatus and system is $|\psi_1\rangle\,|\alpha_1\rangle$ rather than the superposition (6.8), we have to observe the apparatus and then apply the projection postulate in accordance with the result of this observation. This analysis does not eliminate the need for a projection postulate, but it does show that there is consistency between two possible applications of it at different levels: whether it is applied to the measurement of the system made by the apparatus, or the measurement of the apparatus made by the experimenter, the results will be the same. (This, of course, leaves open the question of how far one can pursue this line of argument, and whether projection is ultimately a function of consciousness.)

There is another account of state reduction which does not require going to higher and higher levels and including more and more of the universe in one's state vector. The theory of Daneri *et al.* (1962) takes account of the fact that laboratory apparatus is necessarily macroscopic and that the measurement process must be irreversible (a measurement does not count unless it leaves a permanent record, if only in the brain of the experimenter). Using ergodicity conditions implied by these criteria, they prove that for all practical purposes the effect of the projection postulate is the same as the Schrödinger evolution of the apparatus: it is a very good approximation to say that

$$e^{-iHT}\rho e^{iHT} \simeq \sum_i P_i\rho P_i \tag{6.9}$$

where T is the time taken by the measurement, even though the two sides have different rank. The approximation here is intended in the sense that the probability of observing the difference between the two sides is very small. It can be argued that this does not affect the fact that they are different; that one *knows* (by consulting the apparatus) that the system is in one of the eigenstates contained in the mixture on the right-hand side, and not in the pure (superposition) state on the left-hand side; and that therefore, however small its observable consequences, there has actually been a change in the state of the system which did not obey the Schrödinger equation.

6.3. Continuous observation

Whether or not these accounts of the measurement process satisfactorily resolve the issue of state reduction, they share the feature that they describe an experimental intervention occupying a period of time which is short compared with the timescale of changes in the system being studied. They offer no account of an extended observation which is designed to monitor and record the natural evolution of the system. I shall distinguish these two types of experiment by calling them 'measurement' and 'observation' respectively. A *measurement* is a process during which the state of the system changes, if at all, only because of its interaction with the apparatus; it is described by a time-dependent Hamiltonian (the apparatus being switched on at a certain time). A measurement is typically the measurement of the value of an observable at a definite time. *Observation*, on the other hand, is a continuous process during which the state of the system, following its natural Schrödinger evolution, changes appreciably. This is described by a Hamiltonian which is constant in time, representing the continuous readiness of the apparatus. As a standard example of observation we can think of a radioactive nucleus surrounded by counters which will discharge if triggered by an α-particle. Clearly the Hamiltonian of this whole system does not depend explicitly on time.

The interpretative postulates of quantum mechanics have to relate the mathematical objects of the theory to experimental experience. They do so by referring only to measurement and not to observation; in order to apply quantum mechanics to observation, therefore, we have to analyse (continuous) observation in terms of (instantaneous) measurement. One possible way of doing this is to model a period of observation as a succession of repeated measurements, and then take the limit as the interval between the measurements tends to 0; indeed this would seem to be the only way of doing it. This, however, leads to an absurd result. Under very general conditions, it yields zero probability that the result of the last measurement is different from that of the first: a watched pot never boils. The precise statement is:

Theorem. Let X be a two-valued observable on a quantum system, with eigenvalues 0 and 1, and let P_0 and P_1 be the projection operators onto the eigenspaces of X. Suppose that the system is in a pure state at time $t = 0$ and that measurements of X are made in the time interval $[0, T]$ at times $0 = t_0, t_1, \ldots, t_N = T$, the projection postulate being applied after each measurement. Let p_n be the probability that the measurement at time t_n gives the value 0 for X, and let $\tau = \max(t_{n+1} - t_n)$. Then, provided that the Hamiltonian H of the system has a domain of definition which is invariant under P_0 and P_1,

$$p_N - p_0 \to 0 \quad \text{as} \quad \tau \to 0. \tag{6.10}$$

A simple proof of this theorem can be found in Sudbery (1984); see also Misra and Sudarshan (1976); Chiu *et al.* (1977); Fonda *et al.* (1973); Ghirardi *et al.* (1979). (The result is sometimes called Zeno's paradox, but this is a misnomer; Zeno was concerned by the fact that the equations of motion are second-order in time. There is no Zeno's paradox in quantum mechanics.)

Now suppose the system starts in an eigenstate of the observable X, like our unstable particle which starts in an eigenstate of the observable which distinguishes A from B + C; then $p_N = p_0 = 1$ and the system is still in the same eigenstate at the end of the period T. The successive measurements have kept projecting the system back into its original state (this being the most likely thing to happen in a short time interval because of the cosine-like time dependence) and it has never had a chance to build up a component orthogonal to this state.

Since watched pots do boil eventually, whatever it may feel like while you're waiting, there is clearly something wrong in the theoretical input into this argument. Either we must find a different description of continuous observation, or we will have to give up the projection postulate in its usual form.

At this point in the discussion Chris Isham asked why I unregenerately persisted in talking about a single system, when everyone knows that a careful treatment of quantum mechanics must be phrased in terms of ensembles: the state vector does not refer to an individual system. Indeed, Pearle (1980) has argued that the ensemble interpretation makes it possible to drop the projection postulate altogether, so that the basis for the watched pot theorem vanishes. In answer to Chris Isham's question, the reason for my delinquency is that I don't understand ensembles; but I would like to postpone consideration of ensembles, and assessment of Pearle's argument, until after a more careful investigation of the process of continuous observation. This will not assume any form of projection postulate, since the watched pot result renders this highly suspect, and though I will continue to have an individual system in my mind you can, I think, translate everything into the language of ensembles if that makes your mind more comfortable.

Clearly the projections that occur in the model of continuous observation as repeated measurement cannot be justified in the same way as the projection after a single measurement. The quantum mechanical description of the system plus apparatus in a process of repeated measurements would require a wildly time-varying Hamiltonian which is being switched on and off so fast that it is not surprising that all the fuses blow. Let us instead consider the quantum mechanics of a more realistic model in which the system and apparatus are permanently coupled by a time-independent Hamiltonian which causes the apparatus to respond to changes in the system.

6.4. The decay process

First we must consider the change of the system by itself. I will describe this by a Weisskopf–Wigner model, in which the system has a single undecayed state $|\psi_0\rangle$, but after decay there is a large space of states available to it. Thus the full state space is $= \mathcal{S}_0 \otimes \mathcal{S}'$, where $\mathcal{S}_0$ is the one-dimensional subspace containing $|\psi_0\rangle$. This is a reasonable description of the case of a single unstable particle if we restrict the system to a particular eigenvalue of total momentum. We suppose that the state $|\psi_0\rangle$ (the particle A in eqn (6.1)) decays to a state $|\psi_1\rangle$ of the two particles B and C, which then evolves in the BC subspace $\mathcal{S}'$; this can be described by a Hamiltonian $H_0 + \varepsilon V$, where H_0 describes the evolution of the particles separately and V describes the decay. Thus $|\psi_0\rangle$ is an eigenstate of H_0 (with eigenvalue E_0, say) and $\mathcal{S}'$ is invariant under H_0; V couples $|\psi_0\rangle$ to $|\psi_1\rangle$ but has no other matrix elements, so that as an operator it is given by

$$V|\psi_0\rangle = |\psi_1\rangle, \tag{6.14}$$

$$V|\psi'\rangle = \langle\psi_1|\psi'\rangle|\psi_0\rangle \quad \text{for any} \quad |\psi'\rangle \in \mathcal{S}'.$$

We can write the state at time t as

$$|\psi(t)\rangle = f(t)|\psi_0\rangle + |\psi'(t)\rangle, \tag{6.15}$$

with $|\psi'(t)\rangle \in \mathcal{S}'$; then the Schrödinger equation becomes

$$i\frac{df}{dt} = E_0 f(t) + \varepsilon\langle\psi_1|\psi'(t)\rangle \tag{6.16}$$

$$i\frac{d}{dt}|\psi'(t)\rangle = H_0|\psi'(t)\rangle + \varepsilon f(t)|\psi_1\rangle. \tag{6.17}$$

Integrating (6.17) gives

$$|\psi'(t)\rangle = -i\varepsilon\int_0^t dt' f(t')|\chi(t-t')\rangle, \tag{6.18}$$

where $|\chi(t)\rangle = e^{-iH_0 t}|\psi_1\rangle$ is the state to which the immediate decay state

$|\psi_1\rangle$ would evolve in time t in the absence of the decay interaction. This then gives an equation for $f(t)$ which is best expressed in terms of $F(t) = e^{-iE_0 t} f(t)$:

$$\frac{dF}{dt} = -\varepsilon^2 \int_0^t dt' F(t') \langle \psi_1 | \chi(t - t') \rangle. \tag{6.19}$$

If the decay products disperse quickly in the absence of the decay interaction, so that $|\chi(t)\rangle$ is orthogonal to $|\psi_1\rangle$ after some short time τ, then the integral can be replaced by a constant multiple of $F(t)$, so that eqn (6.19) has (approximately) an exponential solution $F(t) = e^{-\gamma t}$. However, this solution cannot be valid for short times, as eqn (6.19) shows that $dF/dt = 0$ when $t = 0$. It is a consequence of the positivity of the total energy (eigenvalue of H) that the exponential decay law must also break down at large times (Fonda *et al.* 1978).

We now have the full state of the system as

$$|\psi(t)\rangle = f(t) |\psi_0\rangle - i\varepsilon \int_0^t dt' f(t') |\chi(t - t')\rangle. \tag{6.20}$$

This can be very plausibly interpreted† in terms of physical processes. The state is a superposition of all the possible states at time t: it might have not decayed and still be in the state $|\psi_0\rangle$, or it might have decayed at some time t' and then evolved in the decay channel for a time $t - t'$. The amplitude that it was still undecayed at time t' is $f(t')$ and the amplitude for the decay to occur in time dt' is $-i\varepsilon\, dt'$; these factors multiply the appropriate term in the superposition.

The operational interpretation of the state (6.20) seems fairly clear, if one is not too legalistically worried about sticking closely to the letter of the principles of quantum mechanics. The probability that the decay has not occurred by time t is $|f(t)|^2$, the squared modulus of the coefficient of $|\psi_0\rangle$. This, we say, is 'the probability that the system is still in the state $|\psi_0\rangle$'. We can go further and discuss the time at which the decay occurred if $|\chi(t)\rangle$, the freely evolving state of the decay products, moves through a sequence of orthogonal states which can be taken as eigenstates of an observable 'time since decay'. Then by examining the state of the decay products at time t we find out how long they have been evolving since the decay (for example, we see how far apart they have moved). Thus the probability of finding that the decay occurred at time t_0

---

† The word 'interpret', as applied to mathematical expressions in a physical theory, is deeply ambiguous. I will distinguish the *pictorial* interpretation, which is an intuitive description of supposed underlying physical reality, prompted by the form of the mathematical expressions, from the *operational* interpretation, which is a prescription relating mathematical expressions to experimentally determined quantities. The two senses move in opposite directions: one goes towards theory, the other towards experiment.

is the squared modulus of the coefficient of $|\chi(t-t_0)\rangle$, times a constant to allow for the motion of $|\chi(t)\rangle$ between eigenstates. That is proportional to the interval δt between eigenvalues, i.e. the indeterminacy in the definition of t_0 (Sudbery 1984). From eqn (6.20) we deduce that the probability that the decay occurred between t_0 and $t_0 + \delta t$ is a constant $\times$ $|f(t_0)|^2$.

Putting these two conclusions together appears to give another argument that the decay law must be exponential: if the probability that the decay has not occurred by time t is $|f(t)|^2$, then the probability that it occurs between time t_0 and $t_0 + \delta t$ is $-(d/dt_0)|f(t_0)|^2 \delta t$. Since this is proportional to $|f(t_0)|^2$, this quantity must be exponential. However, the shift from t to t_0 in this argument is not allowed in a strict adherence to quantum mechanics. The first probability is the probability that the system will be found not to have decayed by time t if a measurement is made at t; the second is the probability that the decay will be found to have happened at t_0 if a measurement is made at t. There is no warrant for any talk of a probability of the decay actually happening at any time; we can only give the probability that we will make a deduction to this effect at a later time. The state (6.20) is the result of isolated evolution of the system, and it must be coupled to macroscopic apparatus before it will yield empirical consequences.

6.5. The observing apparatus

In order to see if there is any justification for the above operational interpretation of the time-dependent state (6.20) when the system is continuously observed (at least to the extent that von Neumann's analysis of measurement justifies the projection postulate), we must consider the quantum mechanics of a system which is permanently coupled to an observing apparatus. I have studied a simple model of this (Sudbery 1984), in which the response of the apparatus to a decay is a change of state which is modelled in the same way as the decay itself. Thus the apparatus state space is taken as $\mathscr{A}_0 \otimes \mathscr{A}'$, where $\mathscr{A}_0$ is a one-dimensional subspace containing the initial state $|\alpha_0\rangle$ of the apparatus, registering 'no decay', and $\mathscr{A}'$ is a large space of apparatus states which register that the decay has occurred. The full state space of the system plus apparatus is $\mathscr{S} \otimes \mathscr{A}$, where $\mathscr{S} = \mathscr{S}_0 \otimes \mathscr{S}'$, as before, is the state of the system being observed. The Hamiltonian is

$$H = H_S + H_A + \zeta W, \tag{6.21}$$

where $H_S = H_0 + \varepsilon V$ is the Hamiltonian of the system alone, which we have already discussed; W is the observation interaction which only has matrix elements between the apparatus state $|\alpha_0\rangle$ and a state $|\alpha_1\rangle \in \mathscr{A}'$ in the presence of system states from $\mathscr{S}'$ (i.e. it behaves like the decay interaction V, but only if the system has decayed); and H_A (corresponding to H_0 for the system) governs the behaviour of the apparatus in isolation. Thus W causes a transition from $|\alpha_0\rangle$ to $|\alpha_1\rangle$ after a decay of the system; H_A then causes $|\alpha_1\rangle$ to relax in $\mathscr{A}'$ so as to form an irreversible record.

If the apparatus is to monitor the change in the system accurately, its response must be fast compared with the timescale of the system; hence the apparatus state after a decay has registered must quickly evolve away from $|\alpha_1\rangle$. Under these conditions, as we saw in considering the system alone, the time development is exponential. Thus in the presence of decay products the apparatus state $|\alpha_0\rangle$ decays as $e^{-\beta t}$ into states in $\mathscr{A}'$, where β is much larger than the rate of decay of the system. With this assumption, the result of the analysis† is that the initial state $|\psi_0\rangle |\alpha_0\rangle$ evolves to

$$|\psi(t)\rangle = e^{-\beta t}\left(k(t)\,|\psi_0\rangle - i\varepsilon \int_0^t dt'k(t')\,|\chi(t-t')\rangle\right)|\alpha_0\rangle + |\Psi'(t)\rangle \quad (6.22)$$

where $|\Psi'(t)\rangle \in \mathscr{S} \otimes \mathscr{A}'$ is a state in which the apparatus registers 'decay', and the function $k(t)$ satisfies

$$k(t) = f(t) + \beta \int_0^t dt'k(t')f(t-t'), \quad (6.23)$$

$f(t)$ being the non-decay amplitude of the isolated system, defined by eqn (6.15).

From eqn (6.22) the probability that the apparatus registers 'no decay', calculated as the squared norm of the $\mathscr{S}$-vector coefficient of $|\alpha_0\rangle$, is

$$P(t) = e^{-2\beta t} + 2\beta e^{-2\beta t}\int_0^t |k(t')|^2\, dt'. \quad (6.24)$$

In the case where the decay of the system is exponential, so that $|f(t)| = e^{-\gamma t}$ with $\gamma \ll \beta$, this gives the very satisfactory result

$$P(t) = \frac{\beta e^{-2\gamma t} - \gamma e^{-2\beta t}}{\beta - \gamma}, \quad (6.25)$$

i.e. the probability that the apparatus registers 'no decay' is slightly greater than the probability that the system has not decayed, the difference being attributable to the possibility that the system has decayed but the apparatus has not yet responded.

However, this result is heavily dependent on the particular form assumed for the decay probability of the system. If, for example, instead

---

† Done in Sudbery (1984) with an unnecessary appeal to perturbation theory: the result does not depend on analyticity in the interaction parameters ε and ζ.

of an exponentially decaying system we assume an oscillating behaviour with $|f(t)| = \cos \omega t$ (which would result if $\mathcal{S}'$ was one-dimensional), then the probability that the system registers 'no decay' becomes

$$P(t) = A e^{-2\beta t} + B e^{2\omega^2 t/\beta} \tag{6.26}$$

which $\rightarrow 1$ as $\beta \rightarrow \infty$.

Thus the presence of the apparatus slows down the time development of the system; in the limit of instantaneous apparatus response the change of the system is inhibited altogether. (It is the change of the system that is inhibited, not just its registration by the apparatus: as $\beta \rightarrow \infty$ the state $|\psi(t)\rangle$ becomes $e^{-iE_0 t}|\psi_0\rangle|\alpha_0\rangle$, so that all change of state is frozen). This case was investigated by Kraus (1981), who called the inhibition of change the 'watchdog effect'. (The distinction between this watchdog effect and the watched-pot effect is that the former arises from continuous observation as a result of evolution according to the Schrödinger equation, the latter from repeated measurement as a result of repeated application of the projection postulate.)

This inhibition of change in the system occurs with other forms of $f(t)$, and I suspect that it is a general phenomenon; that whenever the isolated system has non-decay probability $|f(t)|^2$ with zero derivative at $t = 0$, then $P(t) \rightarrow 1$ as $\beta \rightarrow \infty$. This is worrying, since we know that the exponential form cannot be exactly correct and must be modified at small times so as to satisfy just this condition; indeed if one modifies it by taking

$$|f(t)| = \frac{\alpha e^{-\gamma t} - \gamma e^{-\alpha t}}{\alpha - \gamma}, \quad \text{with} \quad \alpha \gg \gamma, \tag{6.27}$$

then again one finds the watchdog effect. However, the limit $\beta \rightarrow \infty$ is unrealistic here; there is no need for the appaatus to be so fast that it can beat the modifying term $e^{-\alpha t}$. If one assumes that $e^{-\alpha t}$ is the fastest function in the problem, then one regains an apparatus probability $P(t)$ which approximately follows the exponential decay $e^{-2\gamma t}$ of the system.

6.6. Continuous projection postulates

This analysis of continuous observation does not, and cannot, resolve the problem posed by the watched-pot theorem. Let us return to the problem we started with and recapitulate the discussion in terms of the decay $A \rightarrow B + C$. We want to know the operational meaning of the state

$$|\psi(t)\rangle = a(t)|A\rangle + b(t)|BC\rangle \tag{6.2}$$

in the context of continuous observation. The textbooks give us a recipe for drawing empirical consequences from this state only in the context of a discrete measurement; when we try to adapt this to solve our problem, we get an absurd (or at least, empirically false) answer: a watched pot never boils. Taking the textbooks' prescription to be a summary of a more complicated calculation involving the apparatus, we repeat this calculation for continuous observation and find that in favourable cases the joint development of the system and apparatus can be approximated by the state vector

$$|\psi(t)\rangle = a(t)\,|A\rangle\,|\alpha_0\rangle + b(t)\,|BC\rangle\,|\alpha'\rangle. \tag{6.28}$$

We would like to use this to justify the conclusion that the probability that the particle has not decayed by time t is $|a(t)|^2$, the argument being that eqn (6.28) shows that the probability the apparatus registers 'no decay' at time t (by being in the state $|\alpha_0\rangle$) is $|a(t)|^2$.

But we still have no warrant for this interpretation. It is still true that the basic principles refer only to the context of a discrete measurement and leave us to make what shift we can in other contexts, and this applies when we try to interpret the state vector of the apparatus. This means that we still have no way of describing the experience of an observer who continuously observes the apparatus and becomes aware of it registering a decay as it happens. If we try to model this as the limit of a sequence of repeated measurements, then once again the watched pot theorem applies. As observers, it appears that we must restrict ourselves to one look at the apparatus, at a later time; it may bear marks which we interpret as meaning that certain things happened to it at certain times, but we are not allowed to know about those things at the time when they happened. The situation is the same for the apparatus as we found for the system in trying to interpret the state (6.20).

No doubt this interpretative scheme is fully consistent—one can imagine all records of experiments being placed in an archive which is described by a big state vector, giving probabilities for the various outcomes of the ultimate measurement of consulting the archive (which may include our own memories). But it does not seem satisfactory as an interpretation: it smacks uncomfortably of temporal solipsism ('only the present is real'), and it does not reflect the actual practice of those whose day-to-day job consists of interpreting the state vectors of quantum mechanics.

In their day-to-day practice, quantum physicists will invariably interpret the state vector (6.2) as meaning that

'the probability that the system is in the state $|A\rangle$ is $|a(t)|^2$'. (6.29)

On the face of it this is nonsense—the system is *not* in the state $|A\rangle$, it is in the state $a(t)\,|A\rangle + b(t)\,|BC\rangle$—but in fact eqn (6.28) shows that this is not true when the system is continuously coupled to an observing apparatus: in those circumstances the system does not have a definite state, and the statement (6.29) makes sense. It is not deducible from eqn (6.28), however, for that would require applying the statement to eqn (6.28) itself. It can only be assumed as an independent postulate.

In order to make this postulate precise, it is convenient to use the language of statistical operators again. The statistical operator appropriate to a system which is in the state $|A\rangle$ with probability $|a(t)|^2$ and state $|BC\rangle$ with probability $|b(t)|^2$ is

$$\rho(t) = |a(t)|^2 |A\rangle \langle A| + |b(t)|^2 |BC\rangle \langle BC|$$
$$= P_1 |\psi(t)\rangle \langle \psi(t)| P_1 + P_2 |\psi(t)\rangle \langle \psi(t)| P_2 \qquad (6.30)$$

where P_1 and P_2 are projection operators onto $|A\rangle$ and $|BC\rangle$. This suggests the following general form of the above interpretation of the state $|\psi(t)\rangle$:

Continuous projection postulate A. If an observable D is continuously observed, the statistical operator of the system at time t is

$$\rho(t) = \sum_i P_i e^{-iHt} \rho_0 e^{iHt} P_i \qquad (6.31)$$

where ρ_0 is the statistical operator at $t = 0$, H is the Hamiltonian of the system, and P_i are the projection operators on to the eigenspaces of D.

In other words, the system is always in an eigenstate of D, and makes spontaneous and unpredictable transitions between different eigenstates.

This postulate can be taken as a supplement to the conventional (discrete) projection postulate: it supplies empirical content to quantum mechanics in circumstances where a formal statement of empirical consequences has been lacking. Nevertheless, it is not a new addition or an alteration to generally accepted quantum mechanics; it merely formalizes an interpretation that has long been in general use.

One could make further claims for this continuous projection postulate, still without challenging the validity of orthodox quantum mechanics. First, it could be taken not as a supplement to but as a replacement for the discrete projection postulate. In the course of any measurement the macroscopic apparatus variables can be observed continuously: applying the continuous projection postulate gives the changes of state during the measurement, and reproduces the results of the discrete projection postulate when the measurement is over.

Second, in this form the continuous projection postulate, like the discrete one, depends on the particular experimental arrangement—and that, moreover, in an unspecified way. Its content depends on what observable is being observed, and it is not laid down what counts as observation of that observable. It can be freed from this particularity by assuming that the observables of which the system must be in an eigenstate are universally defined, as follows:

> *Continuous projection postulate B*. There are privileged observables, and the statistical operator of any system evolves according to eqn (6.30), where P_i are the projection operators onto the eigenspaces of the privileged observables.

As far as I can see, it is not possible to universalize the discrete projection postulate in this way.

6.7. The paradox score

Table 6.1. The paradox score

		Discrete projection	Continuous A	Projection B
	A watched pot never boils.	√	×	×
	Two laws of time development	√	×	×
	Two types of physical system	√	√	√
	Vagueness: definition of measurement	√	√	× (?)
Schrödinger's cat	{ Vagueness; time of reduction	√	×	×
	{ Reverse causality	√	×	×
Einstein–	{ Subjectivity	√	×	×
Podolsky–Rosen	{ Non-locality	√	√	√
	Semigroup law	×	√	√
	Loss of symmetry (canonical transformations)	×	×	√
	Loss of symmetry (time reversal)	×	√	√

Table 6.1 compares the three forms of projection postulate in respect of various paradoxical or otherwise undesirable features. A tick indicates the presence of the feature and is a Bad Thing. Discussion of these points can be found in Sudbery (1984); here I would like to comment briefly on a few of them only.

Reverse causality. The discrete projection postulate has the peculiar feature that the sudden and unpredictable changes which characterize the microscopic world are brought about by the act of measurement. Nature is passive, the observer active: observation causes the observed phenomena. The continuous projection postulates, on the other hand, show quantum changes happening spontaneously: nature is active, the observer passive, and phenomena are observed because they happened.

Subjectivity. This is not necessarily a feature of the discrete projection postulate, but can become one if the state vector is interpreted as a representation of the observer's knowledge in an attempt to circumvent some of the other undesirable features. The philosophy behind the continuous projection postulates is a thoroughgoing objectivism: the state vector (more accurately, the associated point in projective space) is taken to be an objective property of the system (Burgos 1984). This point will be discussed further in connection with the ensemble interpretation.

Failure of semigroup law. This is a very serious objection to the continuous projection postulate. Unless all decay probabilities are exponential, the transformations T_t which take the statistical operator ρ_0 to $\rho(t)$ according to eqn (6.30) do not satisfy the semigroup law

$$T_s T_t = T_{s+t}. \qquad (6.32)$$

In other words, letting the system develop for a time t and then letting it develop for a time s is not the same as letting it develop for a time $s + t$.

Loss of symmetry (time reversal). Here I regard a tick, indicating the absence of time reversal invariance, as a Good Thing. Others might disagree.

6.8. Reasons to be exponential

If one takes the view I have been advocating, that quantum systems make genuine transitions from one state to another, it becomes very puzzling that the decay law is not exponential. One would expect the transition rate to depend only on the state that the system is actually in, and this leads to an exponential time dependence. Any other time dependence causes a failure of the semigroup law, as indicated above. However, if the decay law is exponential then the watched pot theorem does not hold and the whole problem evaporates.

The analysis of decay described here has shown in several respects how perverse it is of quantum mechanics not to have an exponential decay law. First there was the argument following the interpretation of the time-dependent state (6.20) (this argument, however, has some play in it and will allow the decay law to be approximately exponential). Second, if one follows the development of the decayed part $|\psi(t)\rangle$ of the full state $f(t)|\psi_0\rangle + |\psi'((t)\rangle$, according to the Schrödinger equation, one finds

$$e^{-iHt_1}|\psi'(t_2)\rangle = [f(t_1 + t_2) - f(t_1)f(t_2)]|\psi_0\rangle + |\psi'(t_1 + t_2)\rangle - f(t_2)|\psi'(t_1)\rangle. \qquad (6.33)$$

Thus if the non-decay amplitude is exponential, and only in this case, there is no regeneration of the initial state $|\psi_0\rangle$: the decay is an irreversible event. Finally, an exponential decay law (or at least one with non-vanishing initial derivative) appears to be necessary to protect the decay against the watchdog effect brought about by an infinitely quickly responding apparatus. But, as we have seen, it is not clear that this protection is needed, since deviations from exponentiality do not bring about the watchdog effect if their timescale is smaller than that of the apparatus.

All these considerations carry more weight at long times than at short times: deviation from exponentiality at small t can probably be accommodated in the objective view of decay processes. Deviations at large times are more puzzling. The discrete projection postulate, on the other hand, runs foul of the short-time deviations, since it is these that are responsible for the watched-pot paradox.

A thorough investigation of non-exponential decay in a similar framework to that of this paper has been done by Peres (1980); see also Ghirardi *et al.* (1979). The experimental position (Butt and Wilson 1972) is that there are possible indications of deviations from the exponential law at large times, but these are not conclusively established. Cartwright (1983) has argued that the exponential law is an experimental fact and that this is an example of approximations to a general theory taking us closer to the truth.

6.9. The ensemble interpretation

Throughout this paper I have been concerned with an individual system, and have treated the state vector as if it was a property of an individual system. Eventually I arrived at a formulation of quantum mechanics in which the notion of the state of an individual system was foundational, and I claimed this as a merit. But there are problems in this formulation: isn't this an indication that the concept of state cannot apply to an individual system, but only to an ensemble of systems?

Another argument: the investigation was motivated by the shortcomings of the projection postulate, which gives absurd results in the context of continuous observation. Yet in the end I only replaced it by another projection postulate. Since it has always been a matter for debate whether the projection postulate is needed at all, surely it would be better to drop it completely?

Both these points have been put forward by Pearle (1980), who argues that the watched-pot paradox is a compelling reason for dropping the projection postulate, and that this can consistently be done in the ensemble interpretation of the state vector (Pearle 1967). The situation after a measurement, in this view, is that the normalized statevector of the system and apparatus is a superposition of the form $c_1 |\psi_1\rangle |\alpha_1| +$

$c_2 | \psi_2 \rangle | \alpha_2 \rangle$ ($|\psi_i\rangle$ being system eigenstates, $|\alpha_i\rangle$ the corresponding apparatus states), and it remains in this superposition, never being reduced. This has obvious kinship to Everett's relative-state interpretation (Everett 1957), but Pearle's interpretation of this state vector is that it refers to an ensemble of systems, and that in a proportion $|c_1|^2$ of them the apparatus shows the characteristics of $|\alpha_1\rangle$. There is then no difficulty in interpreting the state (6.28) of a time-dependent system and apparatus in the context of continuous observation: at time t the proportion of the ensemble in which the apparatus registers 'no decay' is $|a(t)|^2$, and that is all that needs to be said. There is never any state reduction, and so the watched pot theorem does not apply.†

This ensemble interpretation of the state vector is attractive, convincing, and paradox-free—and entirely equivalent to my interpretation of it as an objective property of an individual system. For what is an ensemble? How many systems does it comprise? If it is to be the subject of empirical statements it must contain only a finite number of systems; but in that case nobody believes the predictions of quantum mechanics about the ensemble, if these take the form 'in a fraction $|c_1|^2$ of the ensemble the result is such-and-such'. For a start this may not give a whole number of systems; more significantly, it is possible that the prediction will be wildly wrong as the result of a statistical fluctuation. Possible but unlikely, to be sure; but the force of that 'unlikely' can only be explained, in this approach, by an appeal to an ensemble of ensembles. A finite ensemble, in fact, constitutes an individual system, and quantum mechanics is not supposed to make predictions about individual systems.

So an ensemble must be an infinite affair. But in that case it has no physical reality; it is merely an imaginary picture used as an aid in thinking about the individual system. Moreover, the meaning of 'the proportion $|c_1|^2$' of this infinite set is going to have to be very carefully defined; and one will have to make sure that finite numbers of systems are selected from the ensemble just so as to (probably) reflect these proportions. Thus we end up with a statement of probability about individual systems: we might just as well have adopted 'probability' as an undefined property of the individual system in the first place, and the state vector with it. To say of an individual system that it belongs to an ensemble described by a state vector $c_1|\psi_1\rangle + c_2|\psi_2\rangle$, of which a

---

† Pearle insists that the state vector should never be reduced; other authors (e.g. Ballentine 1970; Belinfante 1975) are prepared to accept state reduction as a reflection of the observer's selecting certain members of the ensemble. On this view there is nothing mysterious about the sudden change of state vector, since the state vector describes an ensemble, and the subensemble in which a particular experimental result was found constitutes a different ensemble requiring a different state vector. Even with this domesticated projection postulate, however, I am not sure that the watched-pot paradox can be avoided.

proportion $|c_1|^2$ exhibit certain characteristics, is to make precisely the same statement as that the system itself is in that state, and the probability that it will exhibit those characteristics is $|c_1|^2$.

The attempt to avoid the projection postulate is equally nugatory. The object is to place the time development entirely under the rule of the Schrödinger equation and avoid the intervention of any other rule. This cannot succeed, for the Schrödinger equation is causal and cannot by itself cater for the unpredictability inherent in quantum mechanics. Some extra process such as is provided by the projection postulate is necessary to describe this unpredictability; interpretations like the many-worlds interpretation which try to do without it are attempts to run away from indeterminism. The ensemble interpretation, which incorporates indeterminism for individual systems, implicitly assumes that a process corresponding to projection happens to the ensemble in the course of a measurement, for its postulate that a proportion $|c_1|^2$ of the systems in the ensemble show certain characteristics requires that it is possible to distinguish some members of the ensemble from others: the ensemble has become inhomogeneous. Before the measurement, on the other hand, there was no difference between the members of the ensemble: this is the significance of describing the ensemble by a pure state. The description of the ensemble after the measurement ought to reflect the fact that it is composed of discernible subensembles; the ensemble should therefore be assigned a mixed state with the statistical operator (6.7). The operational rules of the ensemble interpretation force the usual projection postulate upon it.

In the context of continuous observation, the interpretation of a time-dependent state $c_1(t)|\psi_1\rangle + c_2(t)|\psi_2\rangle$ as meaning that at time t the proportion of the ensemble showing the characteristics of $|\psi_1\rangle$ is $|c_1(t)|^2$ implies that the ensemble is inhomogeneous and is described by the statistical operator

$$\rho(t) = \sum |c_i(t)|^2 |\psi_i\rangle \langle \psi_i| = \sum P_i |\psi_i(t)\rangle \langle \psi_i(t)| P_i, \qquad (6.34)$$

which, with $\rho_0 = |\psi_0\rangle\langle\psi_0|$, is precisely the same as the continuous projection postulate A.

This is part of a general programme of showing that all interpretations of quantum mechanics are saying the same thing in different ways.

References

Ballentine, L. (1970). The statistical interpretation of quantum mechanics. *Rev. Mod. Phys.* **42**, 358.

Belinfante, F. J. (1975). *Measurement and time reversal in objective quantum theory*. Pergamon, Oxford.

Burgos, M. E. (1984). An objective interpretation (and formulation) of orthodox quantum mechanics *Found. Phys.* **14,** 739 and 753.

Butt, D. K. and Wilson, A. R. (1972). A study of the radioactive decay law *J. Phys. A* **5,** 1248.

Cartwright, N. (1983). *How the laws of physics lie.* Clarendon Press, Oxford. Essay 6.

Chiu, C. B., Misra, B., and Sudarshan, E. C. G. (1977). Time evolution of unstable quantum states and a resolution of Zeno's paradox, *Phys. Rev.* D **16,** 520.

Daneri, A., Loinger, A., and Prosperi, G. M. (1962). Quantum theory of measurement and ergodicity conditions *Nucl. Phys.* **33,** 297.

Dirac, P. A. M. (1930). *The principles of quantum mechanics,* 1st edn. Oxford University Press, Oxford. p. 49.

Everett, H. (1957). Relative state formulation of quantum mechanics. *Rev. Mod. Phys.* **29,** 454.

Fonda, L., Ghirardi, G. C., Rimini, A., and Weber, T. (1973). On the quantum foundations of the exponential decay law *Nuovo Cim.* **15A,** 689 and **18A,** 805.

—— ——, and Rimini, A. (1978). Decay theory of unstable quantum systems. *Rep. Prog. Phys.* **41,** 587.

—— ——, Omero, C., Rimini, A., and Weber, T. (1979). Small-time behaviour of quantum non-decay probability and Zeno's paradox in quantum mechanics *Nuovo Cim.* **52A,** 421.

Kraus, K. (1981). Measuring processes in quantum mechanics I: continuous observation and the watchdog effect *Found. Phys.* **11,** 547.

Misra, B. and Sudarshan, E. C. G. (1976). The Zeno's paradox in quantum theory *J. Math. Phys.* **18,** 756.

Pearle, P. (1967). Alternative to the orthodox interpretation of quantum theory *Am. J. Phys.* **35,** 742.

—— (1980). *Continuous measurements and the rules of quantum theory.* Hamilton College, Clinton, N.Y.

Peres, A. (1980). Nonexponential decay law *Ann. Phys. N.Y.* **129,** 33.

Sudbery, A. (1984). The observation of decay *Ann. Phys. N.Y.* **157,** 512.

Von Neumann, J. (1932). *Mathematical foundations of quantum mechanics* Princeton University Press, 1955. Ch. VI.

7

Models for reduction

Philip Pearle

7.1. State vector reduction in quantum theory

When an experiment is described by quantum theory, the resulting state vector is

$$|\psi, t\rangle = \sum_{n=1}^{N} a_n(t) |\phi_n, t\rangle. \tag{7.1}$$

Here $|\phi_n, t\rangle$ describes one of the macroscopically distinguishable outcomes of the experiment. The squared amplitude $|a_n(t)|^2$ equals the probability of that outcome, and does not change with time after the measurement is completed.

What is meant mathematically by the reduction of the state vector is the replacement of (7.1) by

$$|\psi, t\rangle = 1 \cdot |\phi_n, t\rangle. \tag{7.2}$$

However, the physical reason for such a replacement is not so simply stated, and depends upon how you interpret quantum theory.

I will only describe the reduction of the state vector in the context of the interpretation espoused by Einstein (1949), because it is the only one I understand (Pearle 1967; Ballentine 1970; Belinfante 1975; Pearle 1984a). In this interpretation, the state vector describes an *ensemble* of identically prepared systems. Then, provided all relevant apparatus is included in the state vector description, *there is no need for state vector reduction*: the evolution of the state vector according to the Schrödinger equation is an adequate description of the ensemble's behaviour.

However, if a theoretical physicist is describing a sequence of experiments, he may know that the state vectors in the superposition (7.1) will not interfere in subsequent experiments. In that case he may, without error, just consider the further evolution of (7.2) if he is interested only in the future behaviour of a subset of the ensemble (the one with the nth outcome of the first experiment). In other words, the reduction of the state vector is just a 'convenience, not a necessity' (Pearle 1967). It is something done on a piece of paper by a theoretical physicist. *It does not describe a physical process*.

Then why should one construct dynamical equations for the reduction process, as is done in in this paper?

7.2. State vector reduction as a physical process

Suppose one wishes to interpret the state vector as corresponding to a single system in nature. The motivation for this comes, on the one hand, from Einstein (1949). He argued that single systems do exist in nature, that it is the job of physics to describe what exists in nature, and since quantum theory only describes ensembles and not single systems, quantum theory must be an incomplete description of nature.

Another motivation may be a desire to return to determinacy or at least cryptodeterminacy (determinacy but unpredictability due to complexity or lack of information). It is hard for some to believe in the quantum theory dictum that events (particle decays, scatterings, experimental outcomes) occur for no reason at all. It may be easier to believe that some mechanism in nature determines these events even though that mechanism may be outside an experimenter's control.

On the other hand, the great success of quantum theory suggests making a minimal alteration in the theory, only invoking those changes that the single system re-interpretation of the meaning of the state vector requires.

This re-interpretation immediately implies that quantum theory must be modified. This is because we do not believe that a single system (including observer) is described by a superposition of macroscopically distinct states (7.1) (although it is not really clear what this would entail: an observer's visual image like a multiple exposure photograph perhaps?); it is described by a macroscopically uniform state like (7.2). Since the Schrödinger equation only produces state vectors like (7.1), we must modify the Schrödinger equation so that it describes the evolution of state (7.1) into state (7.2).

The hypothesis of an instantaneous state vector reduction is such a modification. But, because the state vector is supposed to be in 1:1 correspondence with a physical system, and physical systems we know of do not evolve with time by making discontinuous jumps, we are motivated to look for a modified Schrödinger equation which describes the reduction as a continuous process.

7.3. Properties of a dynamical reduction theory

Before constructing a dynamical equation to describe state vector reduction, it is worth considering how the solutions of that equation ought to behave. A given initial state vector can end up as any one of a number of reduced state vectors, so we expect an ensemble of solutions.

To achieve this, the dynamical equation must contain additional variables which, together with the state vector amplitudes, determine the end-product of the reduction. In the first theory of this type, constructed by Bohm and Bub (1966), these are 'Wiener–Siegel hidden variables', random valued constants that guide the reduction process. In my initial work (Pearle 1976) it was the phases of the amplitudes, assumed to be initially randomly distributed, which were the additional variables. (Perhaps when the statistical behaviour of the solutions of differential equations is better understood this approach can be more vigorously pursued.) In subsequent work (Pearle 1979, 1982, 1984a,c; Gisin 1984) and here, the dynamical equation contains randomly fluctuating coefficients: a particular time dependence of these coefficients determines the reduction outcome. Thus

$$\text{one state vector} \rightarrow \text{infinite number of state vectors.}$$

The first property we shall require is imposed upon each individual state vector solution. Introducing the magnitudes x_n and phases θ_n as convenient variables

$$a_n(t) \equiv [x_n(t)]^{1/2} e^{i\theta_n(t)} \tag{7.3}$$

we write the *Bookkeeping property*:

$$\sum_{n=1}^{N} x_n(t) = 1 \tag{7.4}$$

Actually, eqn (7.4) entails no loss of generality, for if it were not true, we could work with the variables $x_n' \equiv x_n / \sum_{m=1}^{N} x_m$ which do satisfy the Bookkeeping property.

Unlike eqn (7.4), the next two properties are necessary. They define what is meant by a 'dynamical reduction theory'.

The second property, also imposed upon each individual solution, is the *Reduction property*:

$$\text{one } x_n(\tau) = 1, \quad \text{all other } x_{\neq n}(\tau) = 0 \tag{7.5}$$

where τ is the time at which the reduction is completed. This ensures that each state vector ends up as just one of the macroscopically distinguishable states, and not as a superposition.

The third property is imposed upon the ensemble as a whole, and ensures that the results after reduction is completed agree with the predictions of quantum theory. It is the *Fundamental property*:

$$\text{the fraction of solutions with } [x_n(\tau) = 1, x_{\neq n}(\tau) = 0] \text{ is } x_n(0). \tag{7.6}$$

We are assuming here that the state vector evolves according to the ordinary Schrödinger equation until time 0, and that if Schrödinger evolution continued the x_ns would not change, but that the evolution actually proceeds according to the reduction equation for time >0. Then $x_n(0)$ is the quantum prediction of the probability of the nth outcome, and the Fundamental property requires that it equals the frequency of nth outcome reduced states.

We could (and shall) impose additional conditions, but these are the minimum requirements for a dynamical reduction theory.

7.4. Equivalent statements of the Reduction and Fundamental properties

It is remarkable that the Reduction and Fundamental properties can be ensured by simple conditions involving the ensemble average of just the first and second moments of the x_ns:

Reduction property: $\langle x_n(\tau)x_m(\tau)\rangle = 0$ for $n \neq m$ (7.7)

Fundamental property: $\langle x_n(\tau)\rangle = x_n(0)$ for all n (7.8)

Although the Reduction property obtains for *individual* solutions, it can be expressed in terms of an *ensemble* average (7.7) because of the non-negative nature of the integrand

$$\langle x_n(\tau)x_m(\tau)\rangle = \int x_n x_m \rho(\mathbf{x}; \tau)\, dx_1 \ldots dx_n,\qquad (7.9)$$

where ρ is the joint probability distribution of $\mathbf{x}$ at time τ. The vanishing of eqn (7.9) and the Bookkeeping property (7.4) require ρ to be of the form

$$\rho(\mathbf{x}; \tau) = c_1\, \delta(1 - x_1)\, \delta(x_2) \ldots \delta(x_N) + \ldots + c_N\, \delta(x_1)\, \delta(x_2) \ldots \delta(1 - x_N)$$
(7.10)

(apart from non delta-function behaviour on a set of measure zero which can be ignored), where the c_ks are positive constants obeying $\sum_{k=1}^{N} c_k = 1$. Equation (7.10) is equivalent to eqn (7.5).

The proof that eqn (7.8) is equivalent to (eqn 7.6) proceeds as follows:

$$\langle x_n(\tau)\rangle = 1 \cdot \text{Prob}\{x_n(\tau) = 1\} + 0 \cdot \text{Prob}\{x_n(\tau) = 0\},\qquad (7.11)$$

since the Reduction property ensures that $x_n(\tau)$ takes on no other values but one or zero. Then eqns (7.8) and (7.11) imply

$$x_n(0) = \text{Prob}\{x_n(\tau) = 1\},\qquad (7.12)$$

which is the Fundamental property (7.6).

7.5. The Constant Mean hypothesis

We now observe that the Fundamental property (7.8) can be written in a more symmetrical form:

$$\langle x_n(\tau) \rangle = \langle x_n(0) \rangle \tag{7.13}$$

This is because all state vectors in the ensemble start out at $t = 0$ with identical amplitudes, and of course the ensemble average of a constant equals that constant.

Equation (7.13) suggests a simple way of ensuring that the Fundamental property is satisfied during the reduction process. It is to require

$$\frac{\mathrm{d}\langle x_n(t) \rangle}{\mathrm{d}t} = 0 \tag{7.14}$$

during the reduction process.

Equation (7.14), which we call the 'Constant Mean hypothesis' (Mathematicians call it the 'Martingale' property), goes beyond what is strictly necessary for a dynamical reduction theory. For example, it is not a property of the Bohm–Bub theory. Nonetheless, we shall assume it because, as we shall now show, it has important and reasonable physical consequences.

7.6. Experimental tests of the Constant Mean hypothesis

A dynamical reduction theory produces the predictions of quantum theory if the reduction is allowed to go to completion.

However, one can conceive of an 'interrupted reduction' experiment sequence, whereby a first experiment whose reduction starts at $t = 0$ is interrupted at time $T < \tau$ by a second experiment whose reduction is allowed to go to completion. (Of course, a longer sequence of 'interrupted reduction' experiments is also possible.) How do the predictions of quantum theory compare with the predictions of the dynamical reduction theory in this case?

We repeat the brief argument here (Pearle 1984b). According to quantum theory, the state vector at $t = 0$, T (just before the second experiment starts), and T^+ (when the second experiment ends) is

$$|\psi\rangle \rightarrow \sum_n a_n(0) |\phi_n\rangle \rightarrow \sum_n a_n(T) |\phi_n\rangle \rightarrow \sum_{n,m} a_n(T) b_{nm} |\phi_{nm}\rangle \tag{7.15}$$
$$t = \qquad\quad 0 \qquad\qquad\qquad T \qquad\qquad\qquad T^+$$

where the evolution $|\phi_n\rangle \rightarrow \sum b_{nm} |\phi_{nm}\rangle$ which takes place over the negligibly short (for simplicity) time interval $T^+ - T$ is due to the second experiment, and

$$|a_n(T)|^2 = |a_n(0)|^2 \tag{7.16}$$

First consider the overwhelmingly most common case where the different $|\phi_{nm}\rangle$ are orthogonal. This will occur if the first experimental result n is recorded or otherwise involves a macroscopic change in the apparatus. Then the quantum prediction of the probability of the n, m sequence is

$$P_{nm} = |a_n(T)b_{nm}|^2 = |a_n(0)|^2 \, |b_{nm}|^2, \qquad (7.17)$$

by eqn (7.16).

According to a dynamical reduction theory, the sequence (7.15) is still correct, but eqn (7.16) is not true because the squared amplitudes $|a_n(t)|^2$ change for $0 < t < T$ during the (incomplete) reduction following the first experiment. However, we can still calculate P_{nm} because, according to the Fundamental property, the fraction of solutions ending up as $1 \cdot |\phi_{nm}\rangle$ after the reduction following the second experiment is

$$P_{nm} = \int \rho(\mathbf{a}, T) \, \mathrm{d}\mathbf{a} \, |a_n(T)b_{nm}|^2 = \langle |a_n(T)|^2 \rangle \, |b_{mn}|^2, \qquad (7.18)$$

where $\rho(\mathbf{a}, T) \, \mathrm{d}\mathbf{a}$ is the fraction of solutions which at time T have amplitudes between $\mathbf{a}$ and $\mathbf{a} + \mathrm{d}\mathbf{a}$.

Now compare the prediction (7.18) of the dynamical reduction theory with the prediction (7.17) of quantum theory. We see that if and only if the Constant Mean hypothesis

$$\langle |a_n(T)|^2 \rangle = \langle |a_n(0)|^2 \rangle = |a_n(0)|^2 \qquad (7.19)$$

is obeyed, those predictions will be identical regardless of the time T chosen to perform the second experiment.

Although no experiment has been performed to rule out a dynamical reduction theory without the Constant Mean hypothesis (Papaliolios (1967) could have modified his experiment slightly to test this (Pearle 1984b)), it is clear that our result makes the Constant Mean hypothesis preferable.

Now consider the case where the $|\phi_{nm}\rangle$ are not orthogonal, so the second experiment measures interference among the states produced by the first experiment. Here we can find experimental tests which could distinguish quantum theory from a dynamical reduction theory with the Constant Mean hypothesis.

This is best illustrated by an example. Zeilinger et al. (1981, 1983) have obtained a double slit neutron interference pattern which agrees with the prediction of quantum theory to slightly better than 1 per cent accuracy. We may think of the neutron encountering the slits as the first experiment which initiates the reduction. The superposition of the two wavepackets exiting the slits is like the superposition of two 'pointer' states, as they are separated by a macroscopic distance (0.126 mm.), although the pointer has only the neutron's mass. It is not unreasonable to suppose that the rate of reduction depends upon separation and mass, and that a slow 'spontaneous' reduction may thus ensue. This slow reduction slightly enhances the amplitude of one packet at the expense of the other, until the second experiment, the detection of the neutron's position at the distant (about 5 m and 0.05 s away) screen, is performed. Thus a slight washout of the interference pattern is expected. The accuracy to which no washout was seen, and the long travel time of the neutron enabled Zeilinger *et al.* to put a lower limit $\tau > 8$ s. on the reduction time for this situation. Zeilinger (private communication) has indicated it may be possible to improve this result by better than a factor of 10 with ultra-slow neutrons.

In general, one expects that experiments which measure interference among the states of a macroscopic system (such as coherent flux tunnelling in a SQUID proposed by Leggett (1980 and in Chapter 3)) can be used to test dynamical reduction theories with the Constant Mean hypothesis against standard quantum theory.

7.7. Superluminal communication

The theory presented here is non-relativistic in that the reduction evolves in a universal time. It is not surprising that superluminal communication can take place in a non-relativistic theory. On the contrary, it is surprising that non-relativistic quantum theory does *not* allow superluminal communication to take place via correlated particles (e.g. EPR phenomena) when it can take place by other mechanisms (e.g. wavepacket travel or spread).

Notwithstanding these remarks, the fact that the Constant Mean hypothesis prevents superluminal communication in certain circumstances could be regarded as evidence in its favour.

Consider a state vector at $t = 0$

$$|\psi, 0\rangle = \sum_n a_n(0) |l_n\rangle |r_n\rangle \qquad (7.20)$$

describing widely separated correlated particles. The states $|l_n\rangle$ refer to particles off to the left, far distant from the particles off to the right described by $|r_n\rangle$.

Since, as explained in the previous section, it is only interrupted reduction experiments which might differ in outcome from quantum theory's predictions, we will only consider experiments of this type.

Suppose that the r-experimenter performs an experiment at $t = 0$, resulting in states $|r_n\rangle$ which are sufficiently macroscopically distinguishable to start the reduction process going. For example, the $|r_n\rangle$ could correspond to different spin-*z* wavepacket states of a massive particle just passed through a Stern–Gerlach apparatus, triggering a slow 'spontaneous' reduction. However, the *r*-experimenter chooses a time T before the reduction is completed, and then performs a second experiment measuring interference among states

$$|\bar{r}_m\rangle = \sum_n \langle r_n | \bar{r}_m\rangle |r_n\rangle. \tag{7.21}$$

(We are suppressing the time dependence of these state vectors, which have ordinary Schrödinger evolution.) For example, the above wavepackets could be recombined by a second Stern–Gerlach magnet and the *x*-spin measured using yet a third magnet. The reduction following the measurement of $|\bar{r}_m\rangle$ goes to completion.

While this is going on, the *l*-experimenter may or may not choose to perform an experiment making the states $|l_n\rangle$ sufficiently macroscopically distinguishable to affect the reduction rate.

We want to see whether either experimenter can be aware of the other's actions.

The density matrix at time T is

$$\rho = \sum_{ij} |l_i\rangle |r_i\rangle \langle a_i(T)a_j^*(T)\rangle \langle r_j| \langle l_j| \tag{7.22a}$$

$$= \sum_{ijnm} |l_i\rangle |\bar{r}_n\rangle \langle \bar{a}_{in}(T)\bar{a}_{jm}^*(T)\rangle \langle \bar{r}_m| \langle l_j|, \tag{7.22b}$$

where $\langle a_i(T)a_j^*(T)\rangle$ represents the ensemble average, and

$$\langle \bar{a}_{in}(T)\bar{a}_{jm}^*(T)\rangle \equiv \langle \bar{r}_n | r_i\rangle \langle a_i(T)a_j^*(T)\rangle \langle r_j | \bar{r}_m\rangle. \tag{7.23}$$

The form of eqn (7.22) is the same whether or not the *l*-experimenter has also been doing an experiment over the time interval 0 to T, although the magnitudes of the matrix elements can depend on this.

When the reduction has been completed following the second experiment, the off-diagonal density matrix elements vanish, while the diagonal elements are unchanged. This can be achieved by inserting $\delta_{ij}\delta_{nm}$ or δ_{nm} in the sum in (7.22b), depending upon whether or not the *l*-states are macroscopically distinct, i.e. whether $|\bar{r}_n\rangle |l_i\rangle$ or $|\bar{r}_n\rangle \sum_i \bar{a}_{in} |l_i\rangle/(\sum_j |\bar{a}_{jn}|^2)^{1/2}$ are the states that compete in the reduction 'game':

$$\rho = \sum_{ijn} |l_i\rangle |\bar{r}_n\rangle \langle \bar{a}_{in}(T)\bar{a}_{jn}^*(T)\rangle \langle \bar{r}_n| \langle l_j| \times (\delta_{ij} \text{ or } 1). \tag{7.24}$$

The *l*-experimenter's reduced density matrix is

$$\text{Tr}_r \rho = \sum_i |l_i\rangle \langle |a_i(T)|^2\rangle \langle l_i| \tag{7.25}$$

(using eqn (7.23)) whether or not there has been an *l*-experiment. If the

Constant Mean hypothesis is obeyed, this reduces to the quantum result

$$\text{Tr}_r \rho = \sum_i |l_i\rangle \, |a_i(0)|^2 \, \langle l_i|, \tag{7.26}$$

and no superluminal message could be sent from r to l.

However, if the hypothesis is not obeyed at time T, the l-experimenter will have a different reduced density matrix depending upon whether an r-experiment has been performed (eqn (7.25)) or not (eqn (7.26)). A superluminal message could be sent from r to l by this means.

The r-experimenter's reduced density matrix is

$$\text{Tr}_l \rho = \sum_{ni} |\bar{r}_n\rangle \langle \bar{r}_n | r_i \rangle \langle |a_i(T)|^2 \rangle \langle r_i | \bar{r}_n \rangle \langle \bar{r}_n|. \tag{7.27}$$

The *form* of (7.27) is independent of whether or not an l-experiment has been performed. However, if the Constant Mean hypothesis is not obeyed, one might expect the *magnitudes* $\langle |a_i(T)|^2 \rangle$ to depend upon the nature of the competing states, so superluminal communication could take place by this means.

If the Constant Mean hypothesis is obeyed (so $\langle |a_i(T)|^2 \rangle = |a_i(0)|^2$ in eqn (7.27)) no communication from l to r is possible, even though the result (7.27) generally differs from the prediction of quantum theory. Curiously, if the $|a_i(0)|^2$ are all equal, as in all spin-correlation experiments conducted to date, the result (7.27) reduces to that of quantum theory.

Although the Constant Mean hypothesis prevents superluminal communication in the cases described above, it is not sufficient to prevent superluminal communication in all circumstances. For example, if the state vector (7.20) is replaced by the more general

$$|\psi, 0\rangle = \sum_{nm} a_{nm}(0) |l_n\rangle \, |r_m\rangle, \tag{7.28}$$

then the r-result will depend upon the l-experimenter's actions in the above experiment. This conclusion also obtains with starting state vector (7.20) if the r-experimenter performs a sequence of *three* interrupted reduction experiments. In both of these cases there is no r to l communication, as the l-experimenter's reduced density matrix is identical to the quantum result.

Thus, the result of some interrupted reduction sequence experiments can be affected by a distant experiment. One might hypothesize a mechanism preventing interrupted reduction experiments, such as non-occurrence of slow spontaneous reductions due to a mass–distance threshold beyond which reduction proceeds too rapidly to be interrupted. Then superluminal communication could not occur. However, experiments of the type described in Section 7.6 could still produce disagreement with quantum theory if these rapid spontaneous reductions do take place for sufficiently simple systems.

7.8. The gambler's ruin game

We wish to find a mathematically and conceptually simple description of the reduction process. It is therefore of interest that there is a well-known non-quantum process that is precisely analogous to the reduction process in that the Bookkeeping, Reduction and Fundamental properties are all satisfied as well as the Constant Mean hypothesis. This is the gambler's ruin game (Feller 1950; Pearle 1982).

For simplicity, consider two gamblers G_1 and G_2 who start out with \$33 and \$67 respectively. A fair coin is tossed: heads G_1 wins \$1, tails G_2 wins \$1. They play until one gambler is 'wiped out'.

Let x_n equal the fraction of total money held by gambler G_n: i.e. before the game commences $x_1 = 0.33$, $x_2 = 0.67$. Then

Bookkeeping property: $x_1 + x_2 = 1$, because no money leaves the game.
Reduction property: the game ends with either $(x_1 = 1,\ x_2 = 0)$ or $(x_1 = 0,\ x_2 = 1)$.
Fundamental property: if repeated games are played with the same initial stakes, G_1 wins 33 per cent of them and G_2 wins 67 per cent of them.

The proof of this last remark is as follows. Let $P(x)$ equal the probability of G_1 winning if he has x of the money. Then

$$P(x) = \tfrac{1}{2}P(x - 0.01) + \tfrac{1}{2}P(x + 0.01) \tag{7.29}$$

i.e. on the next toss, with probability $\tfrac{1}{2}$, G_1 can lose but eventually win the game, or with probability $\tfrac{1}{2}$, G_1 can win and eventually win the game. The solution of the difference eqn (7.29) is $P(x) = Ax + B$, and with the boundary conditions $P(0) = 0$ (G_1 can't win if he has no money) and $P(1) = 1$ (G_1 has won if he has all the money). The solution is

$$P(x) = x. \tag{7.30}$$

To see that the Constant Mean hypothesis is satisfied, consider the diagram below charting the equally likely paths of G_1's fortune:

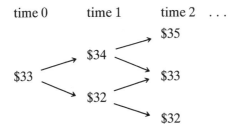

Clearly, after each toss, the average money possessed by G_1 in the ensemble of games is \$33. This is even true when the diagram shows games that have ended, provided we do 'proper' game counting. That is, we imagine that the coin is still being tossed and the number of different sequences is still being counted, even though the game has ended and money no longer changes hands.

Note that the Constant Mean hypothesis is equivalent to a 'fair game'.

We shall adopt a continuous-time, many-player version of the gambler's ruin game as our model for the state vector reduction process. The squared quantum amplitudes can be visualized as 'playing the game' to see which experimental outcome 'wins'. There are other processes which possess the Bookkeeping, Reduction, Fundamental, and Constant Mean properties, but I have found none so conceptually and mathematically simple.

7.9. The diffusion equation

Consider N gamblers playing against each other in pairs. Let the probability that any pair play for the wager Δx in an interval Δt be proportional to Δt, inversely proportional to $(\Delta x)^2$, and possibly depend upon the money possessed by the pair. Then it can be shown (Pearle 1982) in the limit $\Delta t, \Delta x \to 0$, that the probability density $\rho(x_1, \ldots, x_N; t)$ obeys the diffusion equation

$$\frac{\partial \rho}{\partial t}(\boldsymbol{x}; t) = \sum_{n,m=1}^{N} \left(\frac{\partial}{\partial x_m} - \frac{\partial}{\partial x_m}\right)^2 \alpha_{nm}(x_n, x_m)\rho(\boldsymbol{x}; t) \qquad (7.31)$$

where each positive quantity α_{nm} increases linearly with the n–m players' rate of play and increases as the square of their wager.

We will consider eqn (7.31) as the basic equation describing time evolution of the squared quantum amplitudes $x_n = |a_n|^2$ during the reduction process. In the rest of this chapter we will construct (stochastic) dynamical equations for the quantum amplitudes a_n, whose ensemble of solutions satisfies eqn (7.31). In this way, satisfactory reduction dynamics for the state vector will be achieved.

However, in this and the next section we will examine the behaviour of eqn (7.31).

From the way eqn (7.31) was constructed, it must possess the Bookkeeping, Reduction, and Fundamental (Constant Mean) properties. Still, it is nice to prove that directly. For definiteness, we shall select $\alpha_{nm} = \sigma_{nm}^2 x_n x_m$ (σ_{nm}^2 are constants) which is our preferred choice; one reason for this choice will appear in the next section.

Starting with

$$\frac{\partial \rho}{\partial t} = \sum_{n,m=1}^{N} \sigma_{nm}^2 \left(\frac{\partial}{\partial x_n} - \frac{\partial}{\partial x_m}\right)^2 x_n x_m \rho \qquad (7.32)$$

we multiply both sides by x_n, $x_n x_m$ respectively and integrate over all x. The right-hand side may be integrated by parts, and the boundary contributions vanish. (We shall not prove that here—see Pearle (1982).) Denoting $\int \rho(x, t) f(x) \, dx \equiv \langle f(x) \rangle$, we obtain

Fundamental (Constant Mean or 'Martingale') *property*:

$$\frac{d \langle x_n \rangle}{dt} = 0$$

Reduction property:

$$\frac{d}{dt} \langle x_n x_m \rangle = -2\sigma_{nm}^2 \langle x_n x_m \rangle \Rightarrow \langle x_n x_m \rangle = \langle x_n x_m \rangle_0 e^{-2\sigma_{nm}^2 t} \xrightarrow[t \to \infty]{} 0$$

Bookkeeping property:

$$\rho \sim \delta\left(1 - \sum_{n=1}^{N} x_n\right) \text{ is a solution}$$

The latter holds because $(\partial/\partial x_n - \partial/\partial x_m)$ commutes with the delta-function.

The two-state diffusion equation

$$\frac{\partial \rho}{\partial t}(x_1, x_2; t) = \sigma^2 \left(\frac{\partial}{\partial x_1} - \frac{\partial}{\partial x_2}\right)^2 x_1 x_2 \rho(x_1, x_2; t) \qquad (7.33)$$

is worth looking at. Writing $\rho(x_1, x_2; t) = \delta(1 - x_1 - x_2)\rho(x_1, t)$ and $x_1 = x$, we have

$$\frac{\partial \rho}{\partial t} = \sigma^2 \frac{\partial^2}{\partial x^2} x(1 - x)\rho. \qquad (7.34)$$

The solution to this equation that satisfies the initial condition $\rho(x, 0) = \delta(x - x_0)$ can be written using Legendre functions $P_n(x)$ as

$$\rho(x, t) = \theta(x)\theta(1 - x)f(x, t) + \delta(x) \int_0^t f(0, s) \, ds + \delta(1 - x) \int_0^t f(1, s) \, ds \qquad (7.35a)$$

$$f(x, t) = \frac{d}{dx} \int_{x_0}^1 dw \sum_{n=1}^{\infty} (2n + 1)e^{-n(n+1)\sigma^2 t} P_n(2x - 1)P_n(2w - 1). \qquad (7.35b)$$

The delta-function terms in (7.35a) actually belong in the solution of eqn (7.34). (The delta-functions inside the derivative are wiped out by multiplication by $x(1-x)$, but they reappear when the derivatives of the step-functions $\theta(x)\theta(1-x)$ are taken!) Incidentally, unlike the solution of the ordinary diffusion equation (eqn (7.34) with $x(1-x)$ replaced by 1) which allows a *choice* of boundary conditions (e.g. absorbing rather than reflecting), here the singular nature of the elliptic operator at $x = 0$, 1 (so-called exit boundaries (Wong 1971)) *forces* absorbing boundary conditions on this solution. For large t, eqn (7.35) becomes

$$\rho(x,t) = \theta(x)\theta(1-x)6x_0(1-x_0)e^{-2\sigma^2 t} + \delta(x)[1 - x_0 - 3x_0(1-x_0)e^{-2\sigma^2 t}]$$
$$+ \delta(1-x)[x_0 - 3x_0(1-x_0)e^{-2\sigma^2 t}] + O(e^{-6\sigma^2 t}) \quad (7.36)$$

which displays the asymptotic Reduction and Fundamental properties.

7.10. Reduction time

A more stringent requirement, which we have not yet imposed, is that the reduction process should take a finite time. A state vector remaining in a superposition of macroscopically different states for an infinite time is in this respect conceptually identical to standard quantum theory even if all squared amplitudes but one may get exponentially small. The aim of a dynamical reduction theory is to ensure that such superpositions only last for a 'brief' interval.

Consider the reduction time for a two-state quantum system. (The results are the same for the more general case (Pearle 1984d).) Starting with eqn (7.31), and going from two variables x_1, x_2 to one variable $x = x_1$ as in eqns (7.33) and (7.34), the diffusion equation is

$$\frac{\partial \rho}{\partial t} = \frac{\partial^2}{\partial x^2} \sigma^2(x)\rho(x). \quad (7.37)$$

From this equation one can derive an equation for the mean time $m(x_0)$ for a game which starts at $x = x_0$ to end (i.e. for x to reach 0 or 1). This equation, called Dynkin's equation (Schuss 1980), is

$$\sigma^2(x_0) \frac{d^2}{dx_0^2} m(x_0) = -1 \quad (7.38)$$

and must be supplemented by the boundary conditions $m(0) = m(1) = 0$ (since it takes no time to reach 0 or 1 if you are already there).

For our preferred choice eqn (7.34), $\sigma^2(x_0) = \sigma^2 x_0(1-x_0)$ and $m(x_0)$ is certainly finite, having the entropy-like expression

$$m(x_0) = -\sigma^{-2}[x_0 \ln x_0 + (1-x_0) \ln (1-x_0)]. \quad (7.39)$$

In general, if $\sigma^2(x)$ vanishes at the boundaries $x = 0$ or $x = 1$ as x^r or $(1-x)^r$ respectively, then for $r < 2$ there is a solution $m(x_0)$, so the mean reduction time is finite. For $r \geq 2$, m is infinite at the boundary, so the boundary conditions cannot be satisfied. It can be shown in the latter case (the so-called 'open boundary' case (Wong 1971)) that no sample diffusion actually reaches the boundary, so the reduction time for each sample is infinite. From the gambler's ruin point of view, no game ends because the rate of play near the end of the game proceeds too slowly (or alternatively, the wager gets too small).

For our preferred choice eqn (7.34), $r = 1$. Because the mean reduction time is finite, all sample diffusions (except a set of measure zero) reduce in a finite time. Indeed, this is one of the major reasons why we prefer it. (Another major reason is its close relation to the Schrödinger equation, which will be seen in Sections 13 and 14.)

The reduction time in the theory of Bohm and Bub (1966) is infinite. In the recent model of Gisin (1984), $r = 2$ so the reduction time is also infinite.

7.11. Stochastic differential equations

In the remainder of this paper we are going to write down differential equations of motion which contain randomly fluctuating coefficients. The solutions of such an equation of motion obey a diffusion equation which we will find. In this section and the next we briefly review how this is done. (See Arnold 1974; Schuss 1980; Wong 1971.)

The basic quantity to be used in modelling, because of its nice properties, is Brownian motion (random walk) $B(t)$. Over the ensemble of Brownian motions, $B(t)$ has a Gaussian distribution, with zero mean ($\langle B(t) \rangle = 0$), variance $\langle (B(t_2) - B(t_1))^2 \rangle = \sigma^2(t_2 - t_1)$, and $\Delta B \equiv B(t_2) - B(t_1)$ is statistically independent of $B(t_1)$. The central idea is that $\Delta B / \Delta t$ is used to model a function with large random fluctuations. However, in the limit,

$$\lim_{\Delta t \to 0} \Delta B / \Delta t \equiv \dot{B}(t)$$

(called white noise) has infinite values and so does not exist.

Although differential equations containing $\dot{B}$ do not exist, integrals containing ΔB do exist. This suggests writing differential equations as equations relating differentials, such as

$$dx = f(x)\, dB + g(x)\, dt, \tag{7.40}$$

rather than as equations relating derivatives.

What eqn (7.40) means is

$$x(t) - x(t_0) = \lim \sum_{k=0}^{k} \{f(x_k)[B(x_{k+1}) - B(x_k)] + g(x_k)[t_{k+1} - t_k]\} \tag{7.41}$$

(where $x_{k+1} = x(t_{k+1})$ and $t_{k+1} - t_k = (t - t_0)/K$). The sum $\sum f \Delta B$ in eqn

(7.41) is special in that its expectation vanishes because $f(x_k)$ depends on B at times earlier than t_k (f is a so-called non-anticipating function) so it is statistically independent of ΔB. This was recognized by Itô as being an especially useful definition of an integral, because then the first and second moments of dx can easily be found from eqn (7.40) to first order in dt:

$$\langle dx \rangle = g(x)\, dt, \qquad \langle (dx)^2 \rangle = \sigma^2 f^2\, dt. \qquad (7.42)$$

Equations (7.42) are all one needs to construct a diffusion equation describing the behaviour of the solutions of eqn (7.40). This is because Itô has shown that such solutions are a Markov process obeying the Fokker–Planck equation

$$\frac{\partial \rho}{\partial t} = -\frac{\partial}{\partial x}\frac{\langle dx \rangle}{dt}\rho + \frac{1}{2}\frac{\partial^2}{\partial x^2}\frac{\langle (dx)^2 \rangle}{dt}\rho. \qquad (7.43)$$

Thus, all one needs to do is to substitute the first and second moments (7.42) into the Fokker–Planck eqn (7.43) to obtain the diffusion equation associated with eqn (7.40):

$$\frac{\partial \rho}{\partial t} = -\frac{\partial}{\partial x}g(x)\rho + \frac{\sigma^2}{2}\frac{\partial^2}{\partial x^2}f^2\rho \qquad (7.44)$$

The two terms on the right-hand side of eqn (7.44) are called the drift and diffusion terms.

Another useful definition of an integral involving ΔB was given by Stratonovich. The stochastic differential equation

$$(S) \qquad dx = f(x)\, dB + g(x)\, dt \qquad (7.45)$$

(the S stands for Stratonovich) means

$$x(t) - x(t_0) = \lim_{\Delta t \to 0} \sum_{k=0}^{k} \left\{ f\left(\frac{x_{k+1}+x_k}{2}\right)[B(x_{k+1}) - B(x_k)] + g(x_k)(t_{k+1}-t_k) \right\} \qquad (7.46)$$

Here f is no longer a non-anticipating function. However, eqn (7.46) can be converted into an Itô equation by the following heuristic procedure. Expand to lowest order in Δt

$$f\left(\frac{x_{k+1}+x_k}{2}\right) = f(x_k + \tfrac{1}{2}\Delta x) = f(x_k) + \tfrac{1}{2}f'(x_k)\,\Delta x + \dots$$

$$= f(x_k) + \tfrac{1}{2}f'(x_k)f(x_k)\,\Delta B + O(\Delta t) \qquad (7.47)$$

using eqn (7.45) (heuristically, ΔB is of order $\Delta t^{1/2}$). Then substitute eqn (7.47) into eqn (7.46), replacing $(\Delta B)^2$ by $\langle (\Delta B)^2 \rangle = \sigma^2\,\Delta t$. Thus the

Stratonovich eqn (7.45) is equivalent to the Itô equation

$$dx = f(x)\,dB + \left[g(x) + \frac{\sigma^2}{2}f(x)f'(x)\right]dt, \qquad (7.48)$$

whose solutions obey the diffusion equation

$$\frac{\partial\rho}{\partial t} = -\frac{\partial}{\partial x}g(x)\rho + \frac{\sigma^2}{2}\frac{\partial}{\partial x}f(x)\frac{\partial}{\partial x}f(x)\rho. \qquad (7.49)$$

The drift term in eqn (7.49) differs from the Itô drift term in eqn (7.44) by the addition of

$$-\frac{\sigma^2}{2}\frac{\partial}{\partial x}f(x)f'(x)\rho, \qquad (7.50)$$

while the diffusion terms in eqns (7.44) and (7.49) are identical.

7.12. Stratonovich and Itô modelling

Suppose one encounters the equation

$$\frac{dx}{dt} = f(x)\alpha(t) + g(x), \qquad (7.51)$$

where $\alpha(t)$ is a randomly fluctuating function with a correlation time that is short compared to the characteristic time over which x changes. Then one might suppose that $\alpha(t)$ can be modelled by white noise $\dot{B}$ which has zero correlation time. However, $\dot{B}$ does not exist, so the equation

$$\frac{\partial x}{\partial t} = f(x)\dot{B}(t) + g(x) \qquad (7.52)$$

must be interpreted as a stochastic differential equation—but, which one?

The answer is, the Stratonovich eqn (7.45). Wong and Zakai (1965) have shown that if $\alpha(t)$ is replaced by a sequence of functions which converge to white noise, the solution of eqn (7.51) in the limit is the solution of eqn (7.48). Van Kampen (1976) has shown that the solutions of eqn (7.51) obey the diffusion equation (7.49) (omitting terms which vanish with the correlation time), with σ^2 replaced by $\int_0^\infty d\tau\langle\alpha(t+\tau)\alpha(t)\rangle$.

A second modelling situation occurs in the context of the finite difference equation

$$\Delta x_k = f(x_k)\alpha(t_k)\,\Delta t + g(x_k)\,\Delta t. \qquad (7.53)$$

$(\Delta x_k \equiv x_{k+1} - x_k,\ \Delta t \equiv t_{k+1} - t_k)$ where $\alpha(t_k)$ is a random variable whose values at successive times are completely uncorrelated. Then one might expect to be able to model eqn (7.53) by Itô's stochastic differential eqn (7.40). It has been shown by Khasminskiy (1969) that if the value of x between the discrete times t_k is interpolated by straight lines connecting the x_ks, in the limit $\Delta t \to 0$, the solution of eqn (7.53) (with $\alpha \, \Delta t$ replaced by ΔB) converges to the solution of Itô's equation (7.40).

In the next two sections we will present two models for the reduction process, a Stratonovich model and an Itô model.

7.13. Nonlinear Schrödinger equation

Consider the nonlinear Schrödinger equation

$$i\frac{da_n}{dt} = \sum_m H_{nm} a_m^* \frac{a_n}{a_n^*}, \qquad (7.54a)$$

where the Hermitian matrix components H_{nm} fluctuate randomly with a short correlation time (compared to the characteristic time for change of the amplitudes a_n) and are statistically independent.

According to the previous section, this equation can be modelled by

$$i\frac{da_n}{dt} = \sum_m \dot{B}_{nm} a_m^* \frac{a_n}{a_n^*}. \qquad (7.54b)$$

(where $\dot{B}_{nm}$ is a Hermitian matrix of complex white noise, i.e., $\dot{B}_{nm} = \dot{R}_{nm} + i\dot{I}_{nm}$, where $\dot{R}_{nm} = \dot{R}_{mn}$ and $\dot{I}_{nm} = -\dot{I}_{nm}$ are independent white noise 'functions', $\langle \dot{B}_{nm}(t')\dot{B}_{rs}(t)\rangle = 2\sigma_{nm}^2 \delta_{ns} \delta_{mr} \delta(t'-t)$) or rather, by the associated Stratonovich stochastic differential equation. From this equation the diffusion equation can be constructed. After a change of variables from a_n and a_n^* to squared magnitude x_n and phase θ_n (eqn 7.3), this diffusion equation becomes

$$\frac{\partial \rho}{\partial t} = \sum_{n,m} \sigma_{nm}^2 \left[\left(\frac{\partial}{\partial x_n} - \frac{\partial}{\partial x_m} \right)^2 x_n x_m \rho + \frac{1}{2} \frac{x_m}{x_n} \frac{\partial^2 \rho}{\partial \theta_n^2} + \frac{1}{2} \frac{\partial^2 \rho}{\partial x_n \partial x_m} \right] \qquad (7.55)$$

(Pearle 1979). After integration over θ_n, eqn (7.55) becomes identical to the gambler's ruin diffusion eqn (7.32). Thus eqn (7.54b) satisfactorily describes the reduction process, and this is our first model.

It is interesting to compare this result with that obtained by considering the usual Schrödinger equation with a random Hamiltonian

$$i\frac{da_n}{dt} = \sum_m H_{nm} a_m, \qquad (7.56a)$$

modelled by

$$i\frac{da_n}{dt} = \sum_m \dot{B}_{nm} a_m. \qquad (7.56b)$$

The diffusion equation obtained by the same procedure can be written as

$$\frac{\partial \rho}{\partial t} = \sum_{n,m} \sigma_{nm}^2 \left[\left(\frac{\partial}{\partial x_n} - \frac{\partial}{\partial x_m} \right) x_n x_m \left(\frac{\partial}{\partial x_n} - \frac{\partial}{\partial x_m} \right) \rho + \frac{1}{2} \frac{x_m}{x_n} \frac{\partial^2 \rho}{\partial \theta_n^2} + \frac{1}{2} \frac{\partial^2 \rho}{\partial x_n \partial x_m} \right].$$

(7.57)

This differs from the right-hand side of eqn (7.55) by a non-zero drift term

$$-\sum_{n,m} \frac{\partial}{\partial x_n} \sigma_{nm}^2 (x_m - x_n) \rho.$$

(7.58)

The lack of a drift term in eqn (7.55) is responsible for the Constant Mean behaviour during reduction. On the other hand, the drift term (7.58) in eqn (7.57) is responsible for the asymptotic ergodic behaviour of the solution of eqn (7.57): as $t \to \infty$, the probability density ρ approaches a constant value over the simplex-shaped x space ($x_n \geq 0$, $\sum x_n = 1$) (Pearle 1976). This is dramatically different from the reduction behaviour of eqn (7.55), where all the probability ends up at the 'corners' ($x_n = 1$, $x_{\neq n} = 0$) of the simplex. Speaking loosely, *the Schrödinger equation without the drift reduces the state vector*.

The close similarity between the reduction equation (7.54a) and the Schrödinger equation (7.56a) is even more apparent when one expresses both differential equations in terms of x_n and θ_n:

$$\frac{dx_n}{dt} = \pm \frac{1}{i} \sum_m [H_{nm} \rho_{mn} - \rho_{nm} H_{mn}],$$

(7.59a, b)

$$\frac{d\theta_n}{dt} = -\frac{1}{2x_n} \sum_m [H_{nm} \rho_{mn} + \rho_{nm} H_{mn}]$$

$$\rho_{nm} \equiv \sqrt{x_n x_m}\, e^{i(\theta_n - \theta_m)}.$$

(7.59c)

The upper sign in eqn (7.59a) corresponds to the Schrödinger equation (7.56a). The lower sign corresponds to the reduction equation (7.54a). Speaking loosely, *only a sign separates Schrödinger behaviour from reduction behaviour* in our Stratonovich modelling. In the next section, we will see that *even the sign difference can be made to disappear* in Itô modelling.

7.14. The Schrödinger difference equation

Suppose you wanted to solve the Schrödinger equation (7.56a) by computer. You might write a difference equation for the amplitude, or rather, for the real and imaginary parts of the amplitude $\sqrt{x_n} \cos \theta_n$ and $\sqrt{x_n} \sin \theta_n$, and solve the equations by iteration. However, that would be a mistake, because the important Bookkeeping property $\sum x_n = 1$, the only conservation law of the Schrödinger equation variables, would be violated to order $(\Delta t)^2$ in each iteration. Thus one might naturally turn to difference equations for the squared amplitude and phase which preserve the Bookkeeping property at each iteration:

$$\Delta x_n = -2 \Delta t \sum_m |H_{nm}| \sqrt{x_n x_m} \sin(\theta_n - \theta_m + \phi_{nm}) \qquad (7.60a)$$

$$\Delta \theta_n = -\Delta t \sum_m |H_{nm}| \sqrt{x_n^{-1} x_m} \cos(\theta_n - \theta_m + \phi_{nm}), \qquad (7.60b)$$

(where $H_{nm} \equiv |H_{nm}| \exp(i\phi_{nm})$, $\Delta t \equiv t_{k+1} - t_k$, $\Delta x_n \equiv x_n(t_{k+1}) - x_n(t_k)$ and all variables on the right-hand side are evaluated at t_k).

If H_{nm} varies slowly compared to the time increment Δt, eqn (7.60) is of course equivalent to the ordinary Schrödinger equation. But if H_{nm} varies randomly and rapidly compared to Δt, eqn (7.68) can be modelled by the Itô equation

$$dx_n = i \sum_m \sqrt{x_n x_m} \, [dB_{nm} e^{i(\theta_n - \theta_m)} - dB_{mn} e^{-i(\theta_n - \theta_m)}] \qquad (7.61a)$$

$$d\theta_n = -\frac{1}{2} \sum_m \sqrt{x_n^{-1} x_m} \, [dB_{nm} e^{i(\theta_n - \theta_m)} + dB_{mn} e^{-i(\theta_n - \theta_m)}]. \qquad (7.61b)$$

The diffusion equation corresponding to eqn (7.61) is eqn (7.55), so eqn (7.61) reduces the state vector![†]

This remarkable result, that the Schrödinger equation itself, expressed as a difference equation in appropriate variables, produces reduction behaviour with a sufficiently rapidly fluctuating Hamiltonian, is given here for the first time. It encourages one to pursue this model further. A sketch of one possible approach occupies the remainder of this chapter.

7.15. Brownian motion analogy

Difference dynamical equations have not so far appeared in physics as fundamental equations. Equation (7.60) could be fundamental if there were a graininess in time. However, one's first inclination is to regard a difference dynamical equation as an approximation to a more detailed time-continuous dynamics, and to search for that dynamics.

A suggestive analogy is the random walk of a Brownian particle in a molecular medium. A random walk can be modelled by the difference

---

[†] If the sign on the right-hand side of eqn (7.61a) is changed, then eqn (7.61) is the Itô equation associated with the Stratonovich nonlinear Schrödinger eqn (7.54b). That is, either sign leads to the diffusion eqn (7.55).

equation $\Delta x = \alpha \, \Delta t$ which in turn is modelled by the Itô equation $dx = dB$, with solution $x = B$. The associated diffusion equation is

$$\frac{\partial \rho}{\partial t} = \frac{\sigma^2}{2} \frac{\partial^2 \rho}{\partial x^2}. \tag{7.62}$$

However, physical considerations lead one to consider the molecular battering of the Brownian particle as causing a viscous force upon which is superimposed a random force. Thus a better model for Brownian motion is the second-order equation of motion, the Langevin equation

$$\ddot{x} = -\beta \dot{x} + \beta \alpha(t) \tag{7.63}$$

(where $\dot{x} \equiv dx/dt$, and β is a constant that is an increasing function of the temperature).

Since $\alpha(t)$ is a rapidly randomly varying function, it can be modelled by white noise $\dot{B}(t)$, and eqn (7.63) modelled by the Stratonovich equations

$$dx = v \, dt, \qquad dv = -\beta v + \beta \, dB \tag{7.64}$$

(which incidentally are the same as the Itô equations). The associated diffusion equation is

$$\frac{\partial \rho}{\partial t} = -\frac{\partial}{\partial x} v\rho + \beta \frac{\partial v\rho}{\partial v} + \tfrac{1}{2}\beta^2 \sigma^2 \frac{\partial^2}{\partial v^2} \rho. \tag{7.65}$$

Equations (7.63–65) describe a more detailed dynamics: the position and velocity relax exponentially with time constant $\sim\beta^{-1}$ to a random walk and the Maxwell–Boltzmann distribution respectively. The question arises, how can one readily see that this second-order dynamics is nonetheless equivalent, in the limit of large damping β, to the first-order random walk dynamics?

The answer is provided by the Smoluchowski–Kramers approximation (Chandrasekhar 1943; Schuss 1980): starting with the solution $\rho(x, v; t)$ of eqn (7.65), it can be shown that $\rho(x; t) \equiv \int dv \rho(x, v; t)$ satisfies eqn (7.62) with neglect of terms of order β^{-1}.

A heuristic way of showing how the second-order behaviour damps out, leaving behind the first-order behaviour, is to integrate eqn (7.63) directly, obtaining

$$v = v_0 e^{-\beta t} + \beta \int_0^t e^{-\beta(t-\tau)} \alpha(\tau) \, d\tau. \tag{7.66}$$

In the limit of large β, using $\beta e^{-\beta(t-\tau)}\theta(t-\tau) \to \delta(t-\tau)$, we obtain the first-order equation $v = \alpha(t)$.

The success of this more detailed modelling suggests one further modelling effort. One would like to write down the (second-order) Newtonian equations of motion of the Brownian particle interacting with a gas of molecules in thermal equilibrium, and to show that the solution satisfies the Langevin eqn (7.63) to a good approximation (Mazo 1978). Then finally the random walk of the Brownian particle emerges as the result of a calculation based upon fundamental physical equations.

7.16. Generalized Smoluchowski–Kramers approximation

The argument of the previous section can be generalized as follows.
  Consider the generalized Langevin equation

$$\ddot{x}_n = -\beta \sum_m b_{nm}(x)\dot{x}_m + \beta K_n(x) + \sum_{r,s} \dot{x}_r H^n_{rs}(x)\dot{x}_s + \beta \sum_m g_{nm}(x)\dot{B}_m, \quad (7.67)$$

with Stratonovich (and Itô) stochastic differential equations

$$dx_n = v_n\,dt$$
$$dv_n = -\beta \sum_m b_{nm}v_m\,dt + \beta K_n\,dt + \sum_{r,s} v_r H^n_{rs}v_s\,dt + \beta \sum_m g_{nm}\,dB_m. \quad (7.68)$$

The associated diffusion equation is

$$\frac{\partial}{\partial t}\rho(x,v;t) = -\sum_n \frac{\partial}{\partial x_n}v_n\rho + \beta \sum_{n,m}\frac{\partial}{\partial v_n}b_{nm}v_m\rho - \beta\sum_n\frac{\partial}{\partial v_n}K_n\rho$$

$$-\sum_{nrs}\frac{\partial}{\partial v_n}v_r H^n_{rs}v_s\rho + \frac{\beta^2}{2}\sum_{n,m}\frac{\partial^2}{\partial v_n\,\partial v_m}G_{nm}\rho \quad (7.69)$$

(where $G_{nm} \equiv \sum_s g_{ns}g_{ms}$).
  We will state the generalized Smoluchowski–Kramers approximation without proof. (We hope to publish the proof in the future.) Define

$$\rho(x;t) \equiv \int dv\rho(x,v;t). \quad (7.70)$$

Then $\rho(x,t)$ satisfies the diffusion equation

$$\frac{\partial\rho}{\partial t} = \sum_{n,m}\left[-\frac{\partial}{\partial x_n}b^{-1}_{nm}K_m - \frac{1}{2}\frac{\partial}{\partial x_n}b^{-1}_{nm}\mathrm{Tr}(H^m M)\right.$$

$$\left.+\frac{1}{2}\sum_s\frac{\partial}{\partial x_n}b^{-1}_{ns}\frac{\partial}{\partial x_m}M_{sm}\right]\rho + O\left(\frac{1}{\beta}\right) \quad (7.71)$$

(where $2G = bM + M\bar{b}$ defines M) in the limit of large damping.

7.17. Damped Schrödinger equation

Suppose we treat the Schrödinger difference eqn (7.60 or 61) as arising from a damped second-order equation of motion, in analogy with Brownian motion. We may regard the quantum amplitudes as physical variables representing something real in nature (although what they represent is not clear) rather than just as mathematical tools for making predictions about ensembles—their usefulness in the latter capacity might be considered to be something of an accident, arising from their reduction dynamics. It is natural to expect physical variables to satisfy second-order dynamical equations. Furthermore, such an approach gives the promise that the random fluctuations will be understandable in terms of the interaction of the quantum amplitudes with other dynamical variables.

This motivates the dynamical equations

$$\ddot{x}_n = -\beta \dot{x}_n - \beta i \sum_m [H_{nm}\rho_{mn} - \rho_{nm}H_{mn}] \tag{7.72a}$$

$$\ddot{\theta}_n = -\beta \dot{\theta}_n - \beta(2x_n)^{-1} \sum_m [H_{nm}\rho_{mn} + \rho_{nm}H_{mn}] \tag{7.72b}$$

$$\rho_{mn}(x, \theta) \equiv \sqrt{x_n x_m} \; e^{i(\theta_n - \theta_m)}. \tag{7.72c}$$

When the matrix elements H_{nm} fluctuate so rapidly that they can be modelled by Brownian motion B_{nm}, then eqns (7.72) are of the form of the generalized Langevin eqns (7.67), and we can follow the procedure outlined there: Stratonovich stochastic differential equations $\rightarrow$ associated diffusion equation $\rightarrow$ diffusion equation in the limit of large damping. The result is that the latter diffusion equation is identical to the reduction diffusion eqn (7.55).

Thus the dynamical eqns (7.72) lead to reduction behaviour when the Hamiltonian fluctuates rapidly, and they also lead to ordinary Schrödinger behaviour, when the Hamiltonian varies slowly (by the argument associated with eqn (7.66)).

Dynamical equations for the amplitudes (and the complex conjugate amplitudes) with these same properties are readily obtained. It follows from eqn (7.72) that

$$\ddot{a}_n = -\beta a_n - \dot{a}_n \frac{\dot{a}_n^*}{a_n^*} + \frac{\beta}{i} \sum_m H_{nm}a_m \tag{7.73}$$

plus the extra terms on the right-hand side:

$$\frac{1}{2}\left[\frac{\dot{a}_n^2}{a_n^2} - \frac{\dot{a}_n^{*2}}{a_n^{*2}}\right]a_n, \tag{7.74}$$

which we eliminate because their contribution vanishes in the limit of large damping. (Their $\mathrm{Tr}(H^m M)$ contribution in eqn (7.71) vanishes.) The procedure of Section 7.16 applied to eqn (7.73) produces the reduction diffusion eqn (7.55). Moreover, eqn (7.73) can be written as the integral equation

$$\dot{a}_n(t) = e^{-\beta t}\dot{a}_n(0)\frac{a_n^*(0)}{a_n^*(t)} + \frac{\beta}{a_n^*(t)}\int_0^t d\tau\, e^{-\beta(t-\tau)}a_n^*(\tau)\frac{1}{i}\sum_m H_{nm}(\tau)a_m(\tau)$$

(7.75)

which, in the limit of large β, reduces to the ordinary Schrödinger equation because $\beta e^{-\beta(t-\tau)} \to \delta(t-\tau)$ as in eqn (7.66).

7.18. Concluding remarks

The final step in this modelling program is to construct plausible equations of motion describing the interaction of the quantum amplitudes with other degrees of freedom, and to show how this accounts for the damping and for the randomly fluctuating Hamiltonian. We will not attempt that here. It requires physical insights that have so far not appeared.

What is the preferred basis into which the state vector can reduce? Some discussion has been given (Kubler and Zeh 1973; Bedford and Wang 1975, 1977; Pearle 1979; Deutsch 1980; Zurek 1981; Page 1984), but I believe this is very much an open question. It may be that the preferred basis is really determined by the dynamical equations. This brings us to a central question: what physical interaction is responsible for reduction behaviour?

Penrose (1981, and in Chapter 9) has argued that a consistent quantum theory of gravitation should have spontaneous state vector reduction built into it. Károlyházy et al. (1982, and in Chapter 8) have presented a picture of a fluctuating metric responsible for reduction to localized wavepacket states. Gravity may be intimately related to state vector reduction, but more physical ideas are needed in this area.

What is especially needed is an understanding of the mechanism that triggers the reduction process. We have assumed that the interaction between macroscopic states governed by the diffusion coefficients σ_{nm}^2 becomes larger as the states become more macroscopically distinguishable, but we have presented no theory behind this behaviour. In Brownian motion (eqn (7.62)), the diffusion coefficient σ^2 is proportional to the square root of the temperature. If the modelling program described in the beginning of this section preserves the analogy with Brownian motion, the question might become: what heats up the degrees of freedom with which the quantum amplitudes interact, thereby bringing about the reduction process?

I close by remarking that I find it delightful that Brownian motion, which was first elucidated by Einstein, might play a key role in fulfilling the program outlined above which was also inspired by Einstein (1949), who wrote:

'Assuming the success of efforts to accomplish a complete description, the statistical quantum theory would, within the framework of future physics, take an approximately analogous position to the statistical mechanics within the framework of classical mechanics. I am rather firmly convinced that the development of theoretical physics will be of this type, but the path will be lengthy and difficult.'

References

Arnold, L. (1974). *Stochastic differential equations: theory and applications.* Wiley, New York.

Ballentine, L. E. (1970). *Rev. Mod. Phys.* **42,** 358.

Bedford, D. and Wang, D. (1975). *Nuovo Cimento* **B26,** 313.

—— —— (1977). *Nuovo Cimento* **B37,** 55.

Belinfante, F. (1975). *Measurements and time reversal in objective quantum theory.* Pergamon, Oxford.

Bohm, D. and Bub, J. (1966). *Rev. Mod. Phys.* **38,** 453.

Chandrasekhar, S. (1943). *Rev. Mod. Phys.* **15,** 1.

Deutsch, D. (1980). *Quantum theory as a universal physical theory.* University of Texas Center for Theoretical Physics report.

Einstein, A. (1949). In *Albert Einstein, philosopher scientist* (ed. P. A. Schilpp). Harper and Row, New York.

Feller, W. (1950). *An introduction to probability theory and its applications.* Wiley, New York.

Gisin, N. (1984). *Phys. Rev. Lett.* **52,** 1657.

Károlyházy, F., Frenkel, A., and Lukacs, B. (1982). In *Physics as natural philosophy* (ed. A. Shimony and H. Feshbach). M.I.T. Press, Massachusetts.

Khasminskiy, R. Z. (1969). *Stability of systems of differential equations with random parameters.* Nauk, Moscow.

Kubler, O. and Zeh, H. (1973). *Ann. Phys. N.Y.* **76,** 405.

Leggett, A. (1980). *Suppl. Prog. Theor. Phys. N.Y.* **76,** 405.

Leggett, A. (1980). *Suppl. Prog. Theor. Phys.* **69,** 80.

Mazo, T. M. (1978). *Stochastic processes in nonequilibrium systems* (ed. L. Garrido, P. Geglar and P. J. Shepherd) Springer-Verlag, Berlin. p. 54.

Page, D. (1984). *Information basis of states for quantum measurements.* To be published in proceedings of SUNY-Albany conference on Fundamental Questions in Quantum Mechanics (ed. A. Inomata, J. Kimball, and L. Roth).

Papaliolios, C. (1967). *Phys. Rev. Lett.* **18,** 622.

Pearle, P. (1967). *Am. J. Phys.* **35,** 742.

—— (1976). *Phys. Rev.* D **13,** 857.

—— (1979). *Internatl J. Theor. Phys.* **18,** 489.

—— (1982). *Found. Phys.* **12,** 249.

—— (1984a). In *The wave particle dualism* (ed. S. Diner, D. Fargre, G. Lochak, and F. Sellers). Reidel, Dordrecht.

—— (1984b). *Phys. Rev.* D **29,** 235.

—— (1984c). *State vector reduction as a dynamical process.* To be published in proceedings of SUNY-Albany conference on Fundamental Questions in Quantum Mechanics (ed. A. Inomata, J. Kimball, and L. Roth).

—— (1984*d*). On the time it takes the state vector to reduce. To be published in *J. Stat. Phys.*

Penrose, R. (1981). In *Quantum gravity* 2, *A second Oxford symposium*. (ed. C. J. Isham, R. Penrose, and D. W. Sciama). Oxford University Press, Oxford.

Schuss, Z. (1980). *Theory and applications of stochastic differential equations*. Wiley, New York.

Van Kampen, N. G. (1976). *Phys. Rep.* **24**, 171.

Wong, E. (1971). *Stochastic processes in information and dynamical systems*. McGraw-Hill, New York.

—— and Zakai, M. (1965). *Ann. Math. Stat.* **36**, 1560.

Zeilinger, A., Gaehler, R., Shull, C. G., and Treimer, W. (1982). *AIP Conf. Proc., Neutron Scattering*. (ed. J. Faber). American Institute of Physics, New York. p. 93.

——, Horne, M. A., and Shull, C. G. (1983). In *Proc. Int. Symp. on Foundations of Quantum Mechanics in the Light of New Technology* (eds. S. Kamefuchi, H. Ezawa, Y. Murayama, M. Namiki, S. Nomura, Y. Ohnuki, and T. Yajima). Physical Society of Japan, Tokyo. p. 289.

Zurek, W. H. (1981). *Phys. Rev.* D **24**, 1516.

8

On the possible role of gravity in the reduction of the wave function

F. Károlyházy, A. Frenkel, and B. Lukács

8.1. The problem of the superposition principle

In the past decades beautiful experiments have been performed which confirm the validity of the superposition principle in quantum mechanics. The wave function of a microscopic system consisting of one or two elementary particles was separated into two parts which moved away from each other to macroscopic distances. In the one-particle experiments the two parts were brought together and interference patterns were looked for. In the two-particle experiments the persistence of spin–spin correlations at macroscopic distances were checked. In all these experiments—sometimes against the expectations—the predictions of orthodox quantum theory were corroborated. No breakdown of the superposition principle could be observed; the microsystems in question did not show any sign of classical behaviour.

On the other hand, it occurred seriously to no-one, during all these years, to investigate the limits and to look for signs of a breakdown of the superposition principle in the behaviour of macro-objects. The reason for this seems to be quite understandable. We all know that we never observe an actual situation in nature that stands in a one-to-one correspondence with a state vector describing a superposition of macroscopically different states. The pure state $2^{-1/2}|\text{cat alive}\rangle + 2^{-1/2}|\text{cat dead}\rangle$ of Schrödinger's paradox is for all practical purposes, either by itself or because we sooner or later have to look at it, equivalent to a mixture of its components, something that is definitely not true for a superposition like $2^{-1/2}|\text{spin up}\rangle + 2^{-1/2}|\text{spin down}\rangle$ of a single spin. To put it differently, at a high enough level of complexity of the total system involved a stochastic element enters the time evolution of the (single) system, a stochastic element that does not seem to be present when one thinks, for example, of the regular precession of a spin in a homogeneous magnetic field. Yet it may appear pointless to ask the following question: at what stage (and what kind) of complexity of a process that, according to the Schrödinger equation, tends to develop into a superposition of macroscopically different states, does the reduction of the pure state to a mixture—or, in other words, the entering of the stochastic element in the time evolution of the single system—occur? The answer, according to a widespread view, is that the confirmation that a superposition genuinely describes a pure state and is 'not yet' equivalent to a mixture, can only be accomplished through interference effects, and the detection of such interference effects between macroscopically different components would be practically impossible even if they existed in principle. First, the interactions of a macro-object with its surroundings destroy the phase correlations that could give rise to the interference. Second, the fine details of the initial state of a macro-object are not controllable so that the initial state cannot be reproduced with sufficient accuracy. To stress this pointlessness: everybody would smile at the idea of an attempt to perform a two slit type interference experiment not with electrons but with macroscopic balls. To begin with, one should isolate the ball from the rest of the universe for long enough that the wave function of its centre of mass could spread out like the wave function of an electron in a vacuum tube. But interaction with the environment, if only through the emission of heat radiation, would continuously take place. A separate wave function for the centre of mass of the ball would not even exist, as was pointed out very clearly, for example, by H. D. Zeh (1970). Thus, if one agrees with the tacit assumption that the only consequence of the breakdown of the superposition principle would be the deletion of interference effects, then it is indeed natural not to pay attention to experiments other than those dealing with only a few microparticles (i.e. controllable phase relations), probing into the 'macroscopic world' merely by the macroscopic distances involved.

However, we would like to show that an opposite view may be taken: *it is quite conceivable that precisely in the case of macro-objects the breakdown of the superposition principle, if it does occur, leads to some unexpected consequences that can be observed under suitable but realistic conditions.*

In order to have a clear basis for our subsequent speculations, let us tentatively adopt the following position with reference to the role of the state vector in quantum mechanics. The state vector (or wave function) of any single system—including, if necessary, all relevant parts of its surroundings—should always remain in a close correspondence with the actual state of the system. Similarly to the causal time evolution described by the Schrödinger equation, the stochastic element encountered in the behaviour of *single systems* should also be recognized as an integral part of the law of propagation of the *wave function*, inherent in nature, not as an exception, connected somewhat mysteriously with the rather accidental act of observation. This would not mean the drastic abandonment of the causal aspect in the wave propagation, i.e. of the Schrödinger equation—not even in the motion of everyday macroscopic objects. (The sight of a flying stone should not tell us simply that 'classical mechanics' is valid; it should tell us that Schrödinger's equation is valid for not too long time intervals.) Instead, the causal and stochastic aspects in the wave propagation should permanently combine, the former always tending to create superpositions with macroscopically different components (e.g. by continuously increasing the uncertainty of the position of the centre of mass of the flying stone), the latter always tending to 'reduce' them (e.g. by keeping the actual spread in the position of the centre of mass of the stone small). Such a program— together with the elegant mathematical tools to handle it—has been very clearly formulated, on a general basis, by P. Pearle (1984a,b).

The question now arises: what is there in nature—if not the consciousness of the observer—with which to associate the stochastic aspect (reduction) in the wave propagation?

8.2. The possible role of gravity

We have already mentioned the *practical* impossibility of the complete isolation (in the microscopic sense) of a highly composite system from its surroundings. It is often stated that the *inevitable* microscopic interactions of a macroscopic system with its surroundings are basically responsible for the breakdown of the superposition principle, the system being 'continuously measured' by the 'rest of the universe'. In spite of the fact that the strength of the interaction of an object with its environment is greatly variable (for example, with the progress of technology we can go to very low temperatures, etc.), that idea cannot be excluded. But one has to admit that there is nothing in the present formalism of statistical physics or of quantum mechanics that would forbid us in principle from visualizing an ideal isolation for any complex system, or that would give the slightest indication as to the degree (and kind) of complexity at which the superposition principle should break down.

There seems to be only one other theory that may play a role in our problem. That is the theory of general relativity, because it says something about the structure of space–time—the stage on which the quantum mechanical wave function propagates. In what follows we shall investigate the idea that the stochastic reduction of the wave function is connected with gravity.

The very thought of such a connection should make it clear that what we have in mind is not some kind of 'regular' quantum gravity. Quantization of the gravitational field, in the 'conventional' sense would mean an extension of the realm of the Schrödinger equation, not a step toward its restriction. (Such an extension might, in itself, be completely justifiable; gravitons may turn out as real as photons, but our present concern shall be the behaviour of much larger masses.)

The idea that gravity may play a role in the breakdown of the superposition principle is not new. However, the Planck length Λ constructed from the universal constants $\hbar$, G, and c, is desperately small

$$\Lambda \equiv \sqrt{\frac{\hbar G}{c^3}} \approx 10^{-33} \text{ cm}, \qquad (8.1)$$

whereas the universal mass m_Λ that corresponds to it is as large as

$$m_\Lambda = \frac{\hbar}{\Lambda c} \approx 10^{-5} \text{ g} \qquad (8.2)$$

i.e., according to our intuition, it is by far the mass of a macro-object. Thus it is not at all obvious how Λ can affect the balance between the causal and the stochastic sides of the wave propagation. We shall see that Λ is not the only length parameter that enters the formulae determining that balance.

Now, we know that it is the large masses that contribute perceptibly to the curvature of space–time and we also know from experience that it is the large bodies that 'move classically', i.e. that refuse to develop large quantum uncertainties in position and velocity. Therefore we start with the general assumption that nature, in some not yet fully understood way, tries to reconcile the *classical* Einstein equations with quantum mechanics as much as possible.

Space–time then has a fairly definite metric structure, its curvature being determined mainly by the large masses with fairly definite positions. But the metric must not be completely sharp; it has to render a slight amount of 'built in' haziness, some kind of imprecision, in order to avoid contradiction with (in other words, to accommodate, to allow for) the basic quantum mechanical aspect of the motion of the masses. According to our general assumption, even the 'classical' motion of the macroscopic bodies cannot be *perfectly* classical: the uncertainties Δx and Δv in the position and velocity of the centre of mass of the body have to satisfy the relation

$$\Delta x \Delta v \gtrsim \frac{\hbar}{2m}, \qquad (8.3)$$

otherwise there could be no talk about an approximate validity of the Schrödinger equation for not too long time intervals.

On the other hand, the haziness in the space–time metric, while it gives 'elbow-room' to the bodies to obey eqn (8.3), also yields a rather forceful clue for the accommodation of the stochastic element in the wave propagation. We shall show that whenever the wave packet of the centre of mass of a body, initially sufficiently narrow, spreads out, following the Schrödinger equation, into a space domain the linear dimension of which is larger than a critical value (the latter depending on the body), the coherence between the distant parts of this wave function is apt to get destroyed, because of the imprecision of the space–time structure. We shall *interpret* this decay of coherence as a signal for a stochastic reduction of the extended wave function to one of its smaller (coherent) parts, executed by nature itself, i.e. without reference to any observer. Thus, by successive reductions the actual quantum mechanical spread in the centre of mass of a large body would be kept small, thereby preventing the body from adding more uncertainty to the metric than was put into it at the beginning.

Let us repeat: when we seek to associate the striking difference between microscopic and macroscopic behaviour (unlimited validity of the superposition principle against nearly classical motion) with gravity, we avoid the problem of the dynamics of microscopic perturbations (gravitons) in the space–time structure. Instead, having larger masses in mind, we try to utilize the assumption that by a slight *mutual* restriction of validity, nature tries to keep both the classical Einstein equations and the Schrödinger equation valid.

What we have to do then is:

(1) to put the 'proper amount' of haziness into the space–time structure, and

(2) to investigate the wave propagation on the smeared space–time.

We do not hesitate to confess that in neither regard are we able to present anything near a complete theory. With reference to (1) we can only give heuristic estimates; with reference to (2) we have created an admittedly very crude mathematical model. What, under such circumstances, can be the justification for our speculations? First, they lead to fairly quantitative and sensible results concerning the transition between microscopic and macroscopic behaviour. Second, the bonus, perhaps, to the naivety of our model is that it deviates as little as possible from well-established concepts.

8.3. The imprecision in space–time structure

Suppose we wish to talk about a world-line segment of length $s = cT$ in some flat space–time domain (see Fig. 8.1). With what precision can we implement ('realize') such a world-line segment by some physical construct?

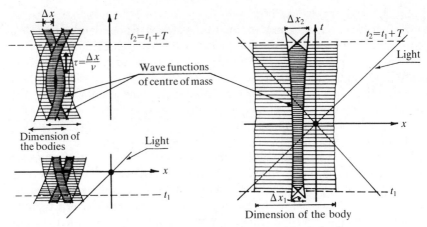

Fig. 8.1. World-lines drawn by the centres of mass of large bodies

There could be a dispute as to what kind of realization we should accept. We may insist on having two bodies ('pointer' and 'dial'), oscillating against each other and thereby producing distinguishable configurations. Let the mass of, say, the 'pointer' be M, the spread of its position at any moment (averaged over the period T) Δx, and its average velocity v. Then obviously in whatever way we wish to identify a definite configuration, it can not be done with greater accuracy than within the time interval $\tau = \Delta x/v$ (Fig. 8.1(a)). Therefore $\Delta s = c\tau$ is the inaccuracy in s. But Δs also gives the largest reasonable spatial extension l of the 'pointer'. Indeed, whatever kind of interaction we imagine to establish the configuration, for example the coincidence of the centre of mass of 'pointer' and 'dial', it has to take effect within the time interval τ. It would be unreasonable to regard distant masses that can in no way be reacted within this time belong to the pointer.

The uncertainty in the energy of the pointer is $v\,\Delta p = v\hbar/\Delta x = c\hbar/\Delta s$, and the corresponding uncertainty in mass is $\Delta M = \hbar/c\,\Delta s$. We may assume that the 'dial' plays a symmetric role. Therefore in the following estimate we may regard M and ΔM simply as the mass and mass uncertainty of our 'clock'.

Now, the mass of the clock changes the structure of space–time slightly in the vicinity of the clock. If the length of our world-line segment is $s = cT$ when projected, as usual, via fictitious light on a fictitious distant world-line (Fig. 8.2), its length s' in the vicinity of the clock is

$$s' \approx \left(1 - \frac{r_s}{2r}\right)cT, \tag{8.4}$$

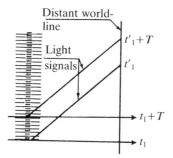

Fig. 8.2. Light signals to distant world-lines

where $r_s = 2GM/c^2$ and $r \approx l \approx \Delta s$. The uncertainty of M then gives rise to an uncertainty $\Delta s'$ in s':

$$\Delta s' \approx \frac{G}{c^2} \Delta M \frac{1}{l} cT \approx \frac{G\hbar}{c^3 \Delta s^2} s = \frac{\Delta^2}{\Delta s^2} s. \qquad (8.5)$$

The greatest accuracy is reached when $\Delta s' \approx \Delta s$. Then we have

$$\Delta s^3 \approx \Lambda^2 s \qquad (8.6)$$

or

$$(\Delta T)^3 \approx \frac{\Lambda^2}{c^2} T \qquad (8.6')$$

for the smallest conceivable uncertainty in s or T.

In our heuristic derivation we have used the concept of the centre of mass. In connection with relativistic many-particle wave equations this concept is not quite clear. Therefore (and also for the sake of further calculations) we shall restrict ourselves to situations in which all material bodies move slowly relative to each other.

Our contention is that when we use the concept of space–time with a sharp (e.g. exactly Minkowskian) metric in order to deal with 'ordinary' quantum mechanics, forgetting all about gravity, we push an idealization too far. Physical space–time should incorporate an imprecision expressed by eqn (8.6) in its very concept.

It is perhaps worthwhile to mention that we may assume, seeking an obligatory imprecision in $s = cT$, a more liberal attitude toward the question of what does constitute a physical 'realization' of a world-line segment of length s. We may regard as such a much less tangible construction, one that relies more directly on the mathematical in-gredients of the theory: simply the corresponding segment of the centre of mass wave function of a single body (Fig. 8.1(b)). The uncertainty of the starting point and end point of the world-line segment along its direction can be optimally identified with the spatial spread at times t_1 and $t_2 = t_1 + T$. (No 'knife' can cut across the world tube of the wave function of the centre of mass quicker than light.) Starting with a minimum wave packet and using eqn (8.3), we see that Δx_2 cannot be smaller than either Δx_1 or $\Delta v T \approx \hbar T / \Delta x_1 M$, where M is the mass of the single body. For a given T and M we get the minimal Δs if we choose

$$\Delta x_1 \cong \frac{\hbar}{\Delta x_1 M} T, \tag{8.7}$$

from which we obtain

$$\Delta s^2 \approx \frac{\hbar}{Mc} s. \tag{8.8}$$

This relation still contains M. We might wish to increase M, in order to diminish Δs. But we must not forget that in this type of approach that the 'realization' of the world-line segment is rather mathematical. We have not bothered with coincidences to stake out the interval. By the same token, because we wish to implement mathematically the world-line segment in a flat space–time domain, we should feel obliged not to spoil the flatness of the domain by too large a mass M, except to a degree compatible with the imprecision expressed by eqn (8.8). That is, we can again write down the equation corresponding to eqn (8.4):

$$s' \approx \left(1 - \frac{r_s}{2R}\right) cT, \tag{8.9}$$

with $r_s = 2GM/c^2$ and R being the radius of the body, but now we have to demand that $|s' - s|$ should not exceed Δs in eqn (8.8). Also, it is easy to convince ourselves that the largest acceptable extension of the body is $R \approx cT$ if we want to attach any reasonable significance to the centre-of-mass wave function. Making use of $|s' - s| \approx \Delta s$ and $R \approx cT$ we again arrive at the rather remarkable result (8.6).

We make the following remarks about eqn (8.6):

(a) The uncertainty expressed by eqn (8.6) is absolute (as compared to $\Delta x \Delta p \approx \hbar/2$).

(b) ΔT and T are related to each other by Λ/c, i.e. only by universal constants of nature.

(c) The uncertainty is very small: e.g., for $T \approx 1$ s, $\Delta T \approx 10^{-29}$ s.

(d) The laws of high-energy particle physics were not taken into account in the derivation of eqn (8.6), so it should not be applied when $cT < 10^{-13}$ cm, that is when $T < 10^{-23}$ s.

In order to be able to deal with the wave propagation, we have devised the following simple mathematical model for a physical space–time domain of nearly Minkowskian metric with a smear in its structure corresponding to eqn (8.6).

We introduce a family $\{g_{\mu\nu}(x^\lambda)_\beta\}$ of matter-free metrics each very close to the Minkowski metric

$$g_{00} = -g_{11} = -g_{22} = -g_{33} = 1; \qquad g_{\mu\nu} = 0 \quad \text{if} \quad \mu \neq \nu.$$

We can regard the index β that labels the members of the family as a random variable. The proper length $s = cT$ between two world points x_1^λ and x_2^λ will be defined as the mean value of the lengths s_β calculated with their corresponding metrics $(g_{\mu\nu})_\beta$. Thus

$$s = \bar{s}_\beta, \tag{8.10}$$

the bar denoting the average over β. At the same time we define Δs as

$$\Delta s = [\overline{(s - s_\beta)^2}]^{1/2}. \tag{8.11}$$

We will show presently that one can choose the family $\{(g_{\mu\nu})_\beta\}$ such that eqns (8.10) and (8.11) will satisfy eqn (8.6) for any pair of points x_1^λ and x_2^λ lying on a world-line that corresponds to a motion with velocity $v \ll c$. We have already accepted this restriction which means, among other things, that our speculations cannot be applied to massive bodies the relative velocities of which are relativistic. But, clearly, neither is the condition that the $(g_{\mu\nu})_\beta$ should be close to the Minkowskian metric Lorentz invariant. It does not seem preposterous to assume that the imprecision of our space–time domain is somehow connected with the matter distribution of the universe. Therefore we assume that we have found a co-ordinate system in which macroscopic bodies move slowly.

Let us now construct the family $\{(g_{\mu\nu})_\beta\}$ leading to eqn (8.6). Because of the restriction $v \ll c$, only $(g_{00})_\beta$ will be needed. We assign the value $\beta = 0$ of the set $\{\beta\}$ to the Minkowskian metric: that is, we write

$$(g_{00})_0 = 1. \tag{8.12}$$

The small deviation of $(g_{00})_\beta$ from $(g_{00})_0$ will be denoted by $\gamma_\beta(x, t)$:

$$(g_{00})_\beta = 1 + \gamma_\beta(x, t) \quad \text{for} \quad \beta \neq 0. \tag{8.13}$$

We construct the γs through their Fourier series

$$\gamma_\beta(x, t) = \frac{1}{\sqrt{l^3}} \sum_k [c_\beta(k)e^{i(kx - \omega t)} + \text{c.c.}]. \tag{8.14}$$

In (eqn 8.14), l is the length of the edge of an arbitrarily chosen large box, and

$$k = \frac{2\pi}{l} n \quad \text{with} \quad n_x, n_y, n_z \text{ integers.} \tag{8.15}$$

Also, we have

$$\omega = c\,|\mathbf{k}| \equiv ck \tag{8.16}$$

from

$$\Box\,\gamma_\beta = 0, \tag{8.17}$$

the latter equation reflecting the assumption that—apart from test bodies—our space–time is matter-free.

Let us now choose an integer

$$N_k > 2 \tag{8.18}$$

for each $\mathbf{k}$ and introduce the random variable $\mathscr{C}(\mathbf{k})$ which may take N_k values

$$\mathscr{C}(\mathbf{k}) = \frac{2\pi}{N_k}[0, 1, \ldots, N_k - 1]. \tag{8.19}$$

A particular choice of the values of the $\mathscr{C}(\mathbf{k})$ gives a particular set $c_\beta(\mathbf{k})$ through the definition

$$c_\beta(\mathbf{k}) = f(k)\mathrm{e}^{i\,\mathscr{C}(\mathbf{k})}, \tag{8.20}$$

with the function $f(k)$ yet to be determined. In other words, each set of the $\mathscr{C}(\mathbf{k})$ corresponds to some non-zero value of the set $\{\beta\}$. We give the same weight to each value of β. Then it is easy to see that

$$\overline{c_\beta(\mathbf{k})} = 0, \qquad \overline{c_\beta^2(\mathbf{k})} = 0 \tag{8.21}$$

and

$$\overline{|c_\beta(\mathbf{k})|^2} = (f(k))^2. \tag{8.22}$$

Consider now two world points A, B lying on a world-line $x^\lambda(t)$. The proper length s_β between them is given by

$$s_\beta = \int_{t_A}^{t_B} \mathrm{d}t\left[(g_{\mu\nu})_\beta\frac{\mathrm{d}x^\mu}{\mathrm{d}t}\frac{\mathrm{d}x^\nu}{\mathrm{d}t}\right]^{1/2}; \qquad x^0 \equiv ct. \tag{8.23}$$

For world-lines along which $\mathbf{v} \equiv \mathrm{d}\mathbf{x}/\mathrm{d}t = 0$ we find

$$s_\beta = c\int_{t_A}^{t_B} \mathrm{d}t[1 + \gamma_\beta(\mathbf{x}, t)]^{1/2} \approx c\int_{t_A}^{t_B} \mathrm{d}t[1 + \tfrac{1}{2}\gamma_\beta(\mathbf{x}, t)]. \tag{8.24}$$

Equation (8.21) shows that $\gamma_\beta(\mathbf{x}, t) = 0$, that is,

$$s \equiv \bar{s}_\beta \approx c(t_B - t_A). \tag{8.25}$$

Furthermore, eqns (8.21) and (8.22) give

$$\Delta s^2 \equiv \overline{(s - s_\beta)^2} = \frac{1}{(2\pi)^3}\int \mathrm{d}k\,\frac{(f(k))^2}{\omega^2}(1 - \cos\omega(t_A - t_B)), \tag{8.26}$$

and we see that eqn (8.6) is reproduced if

$$f(k) = \Lambda^{2/3} k^{-5/6}. \tag{8.27}$$

For world-lines along which $v \ll c$, eqn (8.6) remains valid.

Only the averages (8.21) and (8.22) are important for this derivation. It turns out that only these enter the calculations of the propagation of the quantum mechanical wave functions, too.

We wish to stress that the only role of the family $\{(g_{\mu\nu})_\beta\}$ is to provide us with a mathematical model of a hazy space–time. No physical significance should be attached to the individual members of the family. Furthermore, eqn (8.6) obviously loses its meaning when s becomes very small. However, as the calculation shows, our results are reasonably insensitive to the details of the behaviour of the $c_\beta(k)$ in the $k^{-1} \to 0$ limit. For instance, the results are practically unchanged if instead of eqn (8.27) we apply a cut-off $f(k) = 0$ from, let us say, $k^{-1} < 10^{-13}$ cm (see our remark (d) on p. 116).

8.4. Wave propagation

In 'ordinary' quantum mechanics we almost always work with the Minkowskian metric to which the value $\beta = 0$ of the set β has been assigned. Let the corresponding non-relativistic Schrödinger wave function propagating in flat space be denoted by ψ_0. We shall first assume that ψ_0 describes a possibly complicated, yet isolated system (e.g. a macroscopic body consisting of many atomic particles). The Schrödinger equation for ψ_0 is then the usual one:

$$i\hbar \frac{\partial}{\partial t} \psi_0(t) = H\psi_0(t), \tag{8.28}$$

with H being

$$H = \sum_{i=1}^{N} \left(-\frac{\hbar^2}{2m_i} \Delta_i \right) + \sum_{i>k}^{N} V_{ik}, \tag{8.29}$$

if the system consists of N particles. (The spin is neglected in eqn (8.29), but it could be included). The particles may be the nucleons and the electrons of a body, or the molecules of that body, depending on the nature of the system and on the approximation used. The V_{ik} are the interaction energies between the particles.

The natural way to generalize the quantum description for the hazy space–time is to introduce a family $\{\psi_\beta\}$ of wave functions, ψ_β corresponding to the metric $(g_{\mu\nu})_\beta$. For a single (e.g. scalar) elementary particle it is easy to derive, via the relativistic equation

$$\frac{1}{\sqrt{-g}}(\sqrt{-g}\, g^{\mu\nu}\varphi_{,\nu})_{,\mu} - \left(\frac{mc}{\hbar}\right)^2 \varphi = 0,$$

the non-relativistic approximation

$$i\hbar \frac{\partial}{\partial t}\psi_\beta(\boldsymbol{x}, t) = \left(-\frac{\hbar^2}{2m}\Delta + V_\beta\right)\psi_\beta(\boldsymbol{x}, t), \tag{8.30}$$

where the small 'perturbation' V_β is given by

$$2V_\beta(\boldsymbol{x}, t) = mc^2[(g_{00}(\boldsymbol{x}, t))_\beta - 1] = mc^2\gamma_\beta(\boldsymbol{x}, t). \tag{8.31}$$

This can immediately be generalized into a non-relativistic many-particle equation paralleling eqns (8.28) and (8.29) for each $(g_{\mu\nu})_\beta$. We shall have, denoting by X the point $(\boldsymbol{x}_1, \boldsymbol{x}_2, \ldots, \boldsymbol{x}_N)$ of the configuration space (where N may go up to 10^{23} or more),

$$i\hbar \frac{\partial}{\partial t}\Psi_\beta(X, t) = [H + U_\beta(X, t)]\Psi_\beta(X, t) \tag{8.32}$$

with

$$2U_\beta(X, t) = c^2 \sum_{i=1}^{N} m_i \gamma_\beta(\boldsymbol{x}_i, t). \tag{8.33}$$

Although it is often difficult or impossible to find the solutions of eqn (8.32) even when U_β is set equal to zero, it is possible to study how the U_β perturb these poorly known solutions and to arrive at quantitative conclusions in a number of interesting cases. The general method can be outlined as follows:

Let the system be represented at some 'initial' moment on all the metrics $(g_{\mu\nu})_\beta$ by the same wave function

$$\Psi_\beta(X, 0) \approx \Psi_0(X, 0). \tag{8.34}$$

(It would be more appropriate to start with slightly different Ψ_β on the different $(g_{\mu\nu})_\beta$. The result would be unchanged (Károlyházy 1974)). At some later time $t > 0$ the Ψ_β will no longer be equal to each other. Instead, owing to the structure and the smallness of the U_β we find, to a good approximation, that

$$\Psi_\beta(X, t) \approx \Psi_0(X, t)e^{i\phi_\beta(X, t)}, \tag{8.35}$$

where

$$\phi_\beta(X, t) = -\frac{1}{\hbar}\int_0^t dt'\, U_\beta(X, t'). \tag{8.36}$$

The spread

$$\{\overline{[\phi_\beta(X^{(1)}, t) - \phi_\beta(X^{(2)}, t)]^2}\}^{1/2} \tag{8.37}$$

in the relative phase of the wave functions at two points $X^{(1)}$, $X^{(2)}$ of the configuration space can be calculated as a function of these points and of time.

It turns out that the spread in the relative phase depends—except for negligibly small values of t—only on the separation in configuration space. With increasing separation the spread in the relative phase may reach the value π. As we have already hinted, these features suggest a natural interpretation of the physical state represented by the set $\{\Psi_\beta\}$ and enable us to grasp the difference and especially the transition region between microsystems and macrosystems.

8.5. Microscopic and macroscopic behaviour

Let us first consider, for the sake of comparison, the propagation of a stable elementary particle of mass M. The configuration space is identical with the ordinary space of the single co-ordinate x. (We neglect the spin.) The perturbing potential (8.31) is known from eqns (8.14,) (8.20) and (8.27), and the spread $\Delta(a)$ in the relative phase (with $a \equiv |x_1 - x_2|$) can be easily calculated. For small values of a we find $\Delta(a) \ll \pi$, for a sufficiently large critical value a_c

$$\Delta(a_c) \approx \pi \tag{8.38}$$

will be reached. The relative phase between two points separated by a distance equal to or larger than a_c will have a spread of order π.

Let us call a domain of linear dimension a_c in space a *coherence cell* and let the space be divided into such cells. (The arbitrariness that goes with this division does not affect our conclusions. For a discussion, see (Károlyházy 1974.) Suppose that at $t = 0$ the particle is confined to a single cell (i.e. $\psi_\beta(x, 0) = 0$ outside the cell for all β). If the particle remained in this single cell for all times, the coherence of states represented by the set β would be preserved for all times. The system would then behave as if it had a single coherent wave function, the ψ_β belonging, to a good approximation, to a single ray in the Hilbert space.

However, it is well known that the Schrödinger equation for a single free particle (and also for the centre of mass co-ordinate of an isolated system) leads to an expansion of the wave function. If the original width of a (minimum) wave packet is a_c, then after a time

$$\tau_c \approx \frac{ma_c^2}{\hbar} \tag{8.39}$$

the wave functions $\psi_\beta(x, 0) = \psi_0(x, 0)$ will extend over a region of linear dimension $2a_c$. Then for points belonging to different cells the set $\{\psi_\beta\}$

will no longer behave as a single coherent wave function; the relative phases will change violently with β. This means that the system has developed a state that is no longer adequately represented by a single ray. Whenever this stage is reached, that is, whenever the original coherent set has developed incoherent parts of comparable weights (i.e. right after the time τ_c given by eqn (8.39) in our present example), we interpret the situation as a signal for a stochastic jump to take effect; in other words, we interpret it as a signal for a reduction of $\{\psi_\beta\}$ to one of its coherent parts with the corresponding weight to occur, so that $\{\psi_\beta\}$ will again be confined to a single cell after the reduction.

Thus the wave propagation in our model can be described in terms of expansion–reduction cycles. In the cycle, both the causal (Schrödinger equation) and the stochastic (reduction) parts of the propagation are included. The stochastic part of the wave propagation is now correlated with a physical cause—the spread in the space–time metric—and therefore is conceived as the description of a physical process that takes place independently of any observer. The wave function $\{\psi_\beta\}$ always remains in an (approximate) one-to-one correspondence with the actual state of the object.

Our heuristic estimates can only indicate how strongly the stochastic aspect influences the causal aspect of the propagation. For a smooth incorporation of the former into the latter the mathematical equations are lacking. In other words, in our model we are forced to make the reductions 'by hand', just as in the orthodox theory. However, we now know when (how often) the reductions have to take place, and we can draw important conclusions.

For an elementary particle of mass m a straightforward calculation leads to

$$a_c \approx \frac{\hbar^2}{G}\frac{1}{m^3} = \left(\frac{L}{\Lambda}\right)^2 L, \qquad L \equiv \frac{\hbar}{mc}. \tag{8.40}$$

For an electron, eqns (8.40) and (8.39) give

$$a_c \approx 10^{35}\,\text{cm}, \qquad \tau_c \approx 10^{70}\,\text{s}. \tag{8.41}$$

These values exceed astronomical scales. Thus there is no chance for an isolated electron to expand into a region corresponding to two cells and then undergo a reduction, and there is no chance for us to observe that phenomenon. We come to the result that for an isolated electron only the causal behaviour is important, and accordingly its adequate mathematical representative is a set $\{\psi_\beta\}$ in which all the ψ_β belong to the same ray represented, for example, by $\psi_0(x, t)$. The situation is the same for any known elementary particle: for example, for a proton $a_c \approx 10^{25}$ cm, $\tau_c \approx 10^{53}$ s. For obvious reasons we may call such systems microsystems.

We see now that according to our theory it is indeed absolutely hopeless to observe the breakdown of the superposition principle by increasing the distance between the split beams of microparticles in an interference experiment.

Let us now turn to complex systems. In particular, we want to investigate the motion of macroscopic bodies.

When describing the motion of a massive body as a whole, one usually represents it by a single degree of freedom—its centre of mass co-ordinate—and one associates the whole mass with this degree of freedom. This cannot be done without circumspection when we wish to calculate the effects of space–time fluctuations on our ball. In particular, eqn (8.40) cannot be applied to massive systems just by saying that a_c refers to the centre of mass co-ordinate and L is the Compton wavelength corresponding to the total mass. The reason is that the 'gravitational perturbation' U_β given by eqn (8.33) is picked up from an extended region in space. Yet it can be shown to affect only the phase of the wave function of the centre of mass. Therefore, the concepts of the coherence cell and coherence length a_c can in a straightforward manner be applied to the centre of mass co-ordinate. However, unlike the case for an elementary particle, the expression obtained for a_c will now depend not only on the mass M of the object, but also on its size R (or on its density ρ) because, as already mentioned, the imprecision of the metric is felt over the whole volume of the body. For normal terrestrial densities a simple calculation leads to

$$a_c \approx \left(\frac{\hbar^2}{G}\right)^{1/2} \frac{R^{2/3}}{M} = \left(\frac{R}{\Lambda}\right)^{2/3} L, \quad \text{where again} \quad L = \frac{\hbar}{Mc}. \qquad (8.42)$$

For a ball of $R \approx 1$ cm this gives

$$a_c \approx 10^{-16}\,\text{cm}, \qquad \tau_c \approx 10^{-4}\,\text{s}. \qquad (8.43)$$

The imprecision of the metric does not allow the centre of mass coordinate of the ball to develop an uncertainty larger than $\sim 10^{-16}$ cm. For all practical purposes, the centre of mass seems to be a classical point as expected. However, at a closer look, its evolution shows a strongly curtailed quantum behaviour, which is different not only from the classical one, but also from the behaviour predicted by the orthodox quantum theory. According to the latter, the waver function of the centre of mass co-ordinate of the isolated ball would expand smoothly as required by the Schrödinger equation until an observer condescends to look at it. In our model the wave function undergoes 10^4 expansion–reduction cycles per second, and at the end of each cycle its mean momentum performs a jump Δp_c of the order

$$\Delta p_c \approx \frac{\hbar}{a_c}, \tag{8.44}$$

which corresponds to a small change Δv_c in its mean velocity:

$$\Delta v_c \approx \frac{\hbar}{a_c M} \approx \frac{a_c}{\tau_c} \approx 10^{-12} \text{ cm s}^{-1}. \tag{8.5}$$

Owing to these small but repeated kicks, the world tube of the wave function shows a cactus-like shape in the (x, t) plane, and the centre of mass of the isolated ball performs an anomalous Brownian motion in the smeared space–time.†

We shall see that under suitable conditions the elongation due to this anomalous motion may surpass the perturbing effects of the environment and may become observable. Complete isolation is neither possible nor necessary to observe the remnants of the quantum behaviour of our ball.

Before discussing the proposed experiments, let us indicate where the transition region between micro- and macrobehaviour lies.

For our ball of $R \approx 1$ cm, we obviously have $a_c \ll R$. For the opposite case $a_c \gg R$ the derivation leads to the same result as for the microparticles, that is to eqn (8.40). Noticing that when $a_c = R$ eqn (8.42) coincides with eqn (8.40), we realize that we have the following situation:

$$a_c \gg R \ \left(\text{i.e.} \ \frac{\hbar^2}{G} \gg M^3 R \right) \quad \begin{matrix} \text{microbehaviour} \\ \text{dominates} \end{matrix}$$

$$a_c \approx R \ \left(\text{i.e.} \ \frac{\hbar^2}{G} \approx M^3 R \right) \quad \text{transition region} \tag{8.46}$$

$$a_c \ll R \ \left(\text{i.e.} \ \frac{\hbar^2}{G} \ll M^3 R \right) \quad \begin{matrix} \text{macrobehaviour} \\ \text{dominates} \end{matrix}$$

Taking into account that $\rho \approx 1 \text{ g cm}^{-3}$ for terrestrial bodies, we find

$$a_0^{tr} \approx R^{tr} \approx 10^{-5} \text{ cm}, \qquad \tau^{tr} \approx 10^3 \text{ s} \tag{8.47}$$

and

$$M^{tr} \approx 10^{-14} \text{ g}, \tag{8.48}$$

instead of the straightforward but excessively large $10^{-5} \text{ g} \approx \hbar/\Lambda c$. Our

---

† The tiny discrepancy in the energy conservation connected with the anomalous Brownian motion ($\sim \hbar^2/M a_c^2$ per cycle, which amounts to $\sim 10^{-3}$ erg during the whole life of the universe in the case of our ball) remains—except for refined situations—utterly undetectable amidst the usual thermal agitation. The eventual reaction of the motion anomaly of the ball on the smeared metric is not taken into account in our model. This would need to go beyond our phenomenological treatment of space–time.

M^{tr} is the mass of a colloidal grain, containing $\sim 10^9$ molecules. If such a grain could be isolated with sufficient accuracy for a few hours, then, according to eqn (8.47), its wave function would undergo several reductions without any measurement being accomplished.

Let us remark that the transition region is characterized by the maximal value of the uncertainty $\Delta v_c \approx a_c/\tau_c$ of the mean velocity. Indeed, we have

$$\frac{\Delta v_c}{c} \approx \begin{cases} 10^{-45} & \text{for an electron} \\ 10^{-19} & \text{for a colloidal grain} \\ 10^{-22} & \text{for a ball of } R \approx 1 \text{ cm} \end{cases}$$

Δv_c is a measure of the departure of the object from causal behaviour. The departure is the largest in the transition region. Going from there towards macro-objects, we approach classical determinism (Newton's equation takes over), while going towards micro-objects, we approach quantum determinism (the Schrödinger equation takes over). In both cases Δv_c decreases.

8.6. Possible experiments

(i) A ball suspended in a dilute gas
(ii) A dumb-bell in a space laboratory

(i) A ball suspended on a thin thread in a gas acts as a pendulum which receives stochastic kicks from the molecules of the gas, inducing a tiny normal Brownian motion (nBm) of the ball. If our speculations are sound, the ball also should perform the anomalous Brownian motion (aBm) due to the velocity jumps at the end of each expansion–reduction cycle; of course, the dilute gas provides us with a weak damping force, too.

The mean square elongations of the two Brownian motions are given by the equations

$$\langle x_n^2 \rangle \approx \frac{1}{\omega^2} \frac{k_B T}{\rho R^3}, \tag{8.49}$$

$$\langle x_a^2 \rangle \approx \frac{1}{\omega^2} (\hbar G^4 R^7)^{1/3} \rho^2 \eta^{-1}, \tag{8.50}$$

where $T \approx 300$ K is the temperature of the gas, $\omega \approx \sqrt{g/l}$ is the frequency of the pendulum ($g \approx 10$ m s^{-2}), $\eta \approx 10^{-4}$ g s^{-1} cm^{-1} is the viscosity of the gas. (We apologize for denoting by T both the time interval and the temperature.)

In order to observe the aBm, we need $\langle x_a^2 \rangle \gg \langle x_n^2 \rangle$, and $\langle x_a^2 \rangle$ should be large enough. It turns out that with the above values of T and η we have $\langle x_a^2 \rangle \approx \langle x_n^2 \rangle$ if $R \approx 1$ cm, $\rho \approx 1$ g cm^{-3}. Since the elongation of the aBm increases rapidly with R and ρ, while that of the nBm decreases, it is sufficient to take e.g. $R \approx 5$ cm to arrive at

$$\langle x_a^2 \rangle \approx 1000 \langle x_n^2 \rangle. \tag{8.51}$$

The value of $\langle x_a^2 \rangle$ still depends on ω. With $\omega = 0.3$ s^{-1} we find

$$\sqrt{\langle x_a^2 \rangle} \approx 10^{-6} \text{ cm}. \tag{8.52}$$

The time the pendulum needs to develop this average elongation is given by the relaxation time t_r equal to the inverse of the damping coefficient γ:

$$t_r = \gamma^{-1} = \frac{M}{6\pi R \eta} \approx 5 \times 10^4 \text{ s}. \tag{8.53}$$

The main difficulty in this experiment comes from possible 'microfrictions' in the thread, which may prevent the ball from developing a discernible elongation. The best way to get rid of the problem of the support seems to be to put the ball together with its gas container on an orbiting spacecraft.

(ii) The trouble then is that together with the suspension we get rid of the harmonic force providing a convenient 'position zero'. In order to recover it, we propose to work with a 'dumb-bell', that is with two heavy balls connected by a light rigid rod. Then the tidal torque will try to force the axis of the dumb-bell to point towards the focus of the orbit, while the Brownian motions will make it oscillate around this direction. The calculation shows (Károlyházy *et al.* 1982) that for a dumb-bell 1 metre long with balls with $R \approx 3$ cm and total mass ≈ 1000 g, the anomalous angular elongation will surpass the normal one about a hundred times and its value will be around 10″, which should be an observable effect.

It may well be that other experiments can be thought of to detect the anomalous Brownian motion if it exists. The above estimates seem to indicate that the hunt for it is not hopeless.

8.7. Conclusion

Our considerations† show that if one takes into account the small but unavoidable mutual limitations of quantum theory and general relativity, one may arrive at a picture in which all objects obey the same, generally

---

† For a more detailed exposition, see Károlyházy *et al.* (1982).

valid law of propagation on a space-time with a slightly smeared metric. In this picture there is no need of idealized, purely classical objects in order to interpret quantum mechanics. Indeed, both the causal (Schrödinger equation) and the stochastic (reduction) aspects of the propagation of the wave function can be incorporated in the proposed law. In our model the wave function of any single system, initially confined to a single 'coherence cell', starts out obeying the Schrödinger equation and tends to spread out in configuration space; but because of the haziness of the space–time structure, the coherence between sufficiently distant components of the wave function (between those belonging to different coherence cells) gets destroyed. Whenever this happens, we interpret it as a signal that in nature a stochastic reduction of the wave function to one of its coherent components takes place, independently of any observer. In other words, the time evolution goes through expansion–reduction cycles. The rate of these cycles does depend on the mass, size and structure of the system considered. Without introducing new parameters, we arrive at sensible quantitative estimates. Microparticles (electrons, protons, etc.) effectively do not feel the haziness of the metric and propagate as if it were sharp. As a result, the wave function of an isolated microparticle (or of a few microparticles) would remain coherent during astronomical times and over astronomical distances. In contrast, massive objects feel the imprecision of the space–time structure strongly and quickly develop incoherent parts in their wave functions. The resulting rapid reductions prevent the centre of mass of the body from spreading out noticeably. However, the trajectory is not purely classical; the deviation from it appears as an anomalous Brownian motion, which under suitable circumstances should be observable.

The measurement process now ceases to hold a special status. When a microsystem interacts with a macrosystem, the interaction may induce a sufficiently large change in the mass distribution of the latter, making its 'interaction took place this way' state and 'interaction took place that way' state incoherent. The state of the total system will then be reduced, and thereby the state of the microparticle also undergoes the appropriate reduction. Such interactions and reductions occur around us continually without being observed by anybody. Man-made measuring apparatuses behave in the same way.

The mathematical description of the successive reductions in our model is admittedly awkward. However, we do not think that the final solution will necessarily consist of a return to some purely causal mechanism. The experimental evidence against local hidden variable theories (Clauser and Shimony 1978) strengthens our view that the stochastic aspect of the wave propagation is a basic feature of nature.

References

Clauser, J. F. and Shimony, A. (1978). *Rep. Prog. Phys.* **41**, 1881.

Károlyházy, F. (1974). Gravitation and quantum mechanics of macroscopic bodies. *Magyar Fizikai Folyóirat* **12**, 24. (Thesis, in Hungarian).

——, Frenkel, A. and Lukács, B. 1982. On the possibility of observing the eventual breakdown of the superposition principle. In *Physics as natural philosophy, essays in honor of László Tisza on his seventy-fifth birthday* (ed. A. Shimony and H. Feshbach). MIT Press, Cambridge, USA.

Pearle, P. M. (1984a). *Phys. Rev.* D**29**, 235.

—— (1984b). Dynamics of the reduction of the state vector. In *The wave–particle dualism—a tribute to Louis de Broglie on his 90th birthday* (ed. S. Diner *et al.*) Reidel, Dordrecht.

Zeh, H. D. (1970). *Found. Phys.* **1**, 67.

9

Gravity and state vector reduction

Roger Penrose

At the Nuffield-supported Oxford conference, held in March 1984, on which the chapters in this book are based, many stimulating ideas and illuminating results were put forward. It would surely be foolish for me to attempt to summarize them. Nevertheless, I am certain that they have had a definite influence on my own particular viewpoint. I hope that this influence will not seem to be too subtle to be noticeable as actual changes in my expressed attitudes! These I shall be elaborating shortly.

But first, in order to put people into an appropriate frame of mind, I am providing a picture (Fig. 9.1) showing a concrete realization of the dynamic collapse of a wave function ψ. Provided that the state being described by ψ is sufficiently tiny or uncluttered, then ψ can float freely without any need for support. But a big hefty state needs all sorts of crutches to keep it up. Yet all is to no avail—it collapses anyway! (I was stimulated to show this by noticing that in Abhay Ashtekar's talk, a ψ was missing in one of his transparencies. It occurred to me that ψ must have collapsed!)

Despite this introduction, I should make clear that the rest of what I have to say actually *is* in deadly earnest.

9.1. Quantum theory and objective reality

I should begin by expressing my general attitude to present-day quantum theory, by which I mean standard non-relativistic quantum mechanics. The theory has, indeed, two powerful bodies of fact in its favour, and only one thing against it. First, in its favour are all the marvellous agreements that the theory has had with every experimental result to date. Second, and to me almost as important, it is a theory of astonishing and profound mathematical beauty. The one thing that can be said against it is that it makes absolutely no sense!

Its nonsensical aspect is, of course, the strange picture of 'reality' that it presents, where systems are described as evolving according to one or other of two apparently incompatible procedures: deterministic unitary

Fig. 9.1. Dynamical collapse of a wave function.

Schrödinger evolution (or, completely equivalently, Heisenberg evolution, but I shall, for convenience only, phrase my discussion entirely in the Schrödinger picture) and probabilistic state vector reduction. There is no clear-cut rule for deciding the stages at which unitary evolution must be suspended and replaced by the reduction procedure. This procedure has to have been invoked at least by the time an 'observation' is deemed to have been made by any conscious observer.[†] But state vector reduction may well be normally considered to be applicable a good while before this, at a stage where some 'irreversible' or 'macroscopic' change has been induced. Though there is no clear-cut rule telling us exactly when the reduction procedure is to be invoked, it is one of the powerful facts about the formalism of quantum mechanics that it makes *no difference* what stage the procedure is actually applied (cf. von Neumann (1932) and Chapter 6), provided that this is after any stage at which interference phenomena can plausibly be measured, and provided that it

---

[†] It is necessary to mention the many-worlds interpretation, however, (cf. Everett (1957) and Chapters 13 and 14), according to which reduction is supposed not to be mirroring anything which actually takes place physically at all, but which only appears to do so by some quirk of one's conscious perception. I have a number of serious reservations about this interpretation, but would not be appropriate for me to try to express these here (see Chapter 12 for a critique of the many-worlds view). Another viewpoint, in which, also, reduction is regarded as not taking place, assumes that the probabilistic behaviour normally described by the reduction procedure can be attributed to uncertainty in knowledge about the external environment, it being presumed that unitary Schrödinger evolution actually holds at all times. I do not regard this viewpoint as greatly plausible and I shall not attempt to give a discussion of it in this chapter.

is not beyond the stage of conscious perception. This powerful fact is indeed vital for the general applicability of the theory, and it obviates the need for any detailed description of a physical reduction procedure.

However, there is the unfortunate consequence, in my opinion, that people are easily led to take too subjective a view of the quantum mechanical picture of things. Since it seems to make so little difference when reduction is deemed to have taken place it is a commonly adopted view that in some way reduction is illusory and represents merely a shift in our knowledge about a physical system. (This leaves aside other possible attitudes, such as the many-worlds view according to which this illusion resides in the way a conscious observer would—allegedly— actually perceive just one world from among the linearly superposed infinitude of alternative possibilities.) The state vector itself is, accord- ingly, viewed as providing merely a 'state of knowledge' and not an objective description of (a part of) physical reality. The 'ensemble' viewpoint, according to which the state vector describes properties of ensembles of systems and not of a single physical system is similar, since again the state vector is taken as not describing an actual reality of the physical world. Such views illustrate a strong reluctance to take the state vector seriously—or sometimes, even, to take *reality* seriously!

However, it seems to me that so long as one accepts the mathematical procedures of quantum mechanics as valid for making predictions, one must also accept a certain physical objectivity for the state vector, at least as it applies to isolated systems. Let us envisage that the content of some isolating container C is initially prepared, at time $t = 0$, in the (normal- ized) quantum state $|\psi_0\rangle$, and then is allowed to evolve, by unitary (Schrödinger) evolution to the state, at time t,

$$|\psi_t\rangle = U(t)\,|\psi_0\rangle \tag{9.1}$$

($U(t)$ being a unitary operator). The question I wish to raise is: 'does the state vector $|\psi_t\rangle$ provide an objective description of an actual *physically real* state within C?''. It seems to me that the answer to this question, *provided* that one accepts the normal procedures of quantum mechanics, must be an unequivocal 'yes'; for at any time $t > 0$ one has the option of performing an experiment which measures the (bounded) Hermitian operator

$$E_t = |\psi_t\rangle\langle\psi_t| . \tag{9.2}$$

The state vector $|\psi_t\rangle$ is (up to a phase) the *only* state with the property that with certainty it yields the result unity when the experiment is performed. The contents of C have to be 'ready' for the possibility that at any moment t, the experiment E_t *might* actually be performed—even when it is not going to be so measured (at least in any standard version of the theory). So, it would seem to be clear that $|\psi_t\rangle$ (at least up to a phase multiplier) represents—at *any* time t (until some other measurement is made)—an *objective* property of the contents of C.

There is actually nothing conceptually problematical about this—as far as it goes. Why, then, is there such reluctance to accept the state vector as describing an actual physical reality? I am aware of at least three arguments. The first is the problem of 'Schrödinger's cat' (or some essential equivalent such as 'Wigner's friend', etc.) It seems hard to believe in any actual 'reality' at the level of a cat which requires that such states of death and life can co-exist. The situation is, of course, even more puzzling for Wigner's friend. Thus one may prefer to adopt the view that reality is no longer being described by the state vector, and the state vector now provides merely a calculational aid for computing probabilities of results to experiments. But what of our possible measurement E_t which *requires* that the cat be in a state neither dead nor alive? The standard argument would presumably be that, in this case E_t is a totally impossible measurement to perform in practice. But the formalism of quantum mechanics does not *in itself* have anything to say about what is practical and what is not. According to *strict quantum mechanics*, it seems to me, ψ_t—representing some complex linear combination of a live cat and a dead one—should still have to provide an objective description of a 'reality' contained within C.

My own view is that this is a clear indication that physical reality and the Schrödinger-evolved state vector have simply diverged from one another and, at the level of a cat, that state vector has indeed ceased to describe reality. Thus, on this view, the Schrödinger equation has become in fact grossly inaccurate in its description of reality. (There are, of course, many other views!) But the form of deviation from Schrödinger's equation that would be needed has a remarkable property, namely that the Schrödinger-evolved state vector remains, until the next 'observation' is made, an accurate tool for computing the probabilities of the possible outcomes of that 'observation'. For this reason, the 'mere computational tool' view of the state vector has its undoubted value.

A second argument against the objectivity of the state vector—or, more specifically, against the reality of state vector reduction—is that it seems not to be relativistically invariant. The difficulty is made manifest in experiments of the Einstein–Podolsky–Rosen type, and has been presented to us most forcefully as a fact of nature by the beautiful experiment of Aspect and his collaborators (cf. Chapter 1). Here one has a state consisting of a photon pair, the spins of which are observed at two events A, B with a *space-like* separation of some twelve metres. It would seem that the observation at each one of A and B should effect a state vector collapse, but the collapse which takes place *earlier*—say that at

A—provides the state which is to be observed in the *other* observation— say at B. However, the *temporal order* in which these collapses take place—i.e. whether A or B is considered to be the 'earlier'—depends upon the overall references frame and, therefore, according to relativity, is not an objective property. (In this I am making the assumption that each collapse takes place simultaneously in this 'overall reference frame'. If the collapses were to take place along, say, the future light cones of A and B then there would be an inconsistency with the correlations that Aspect and his co-workers actually observed, since such retarded collapse would be equivalent to using a *local* hidden variable theory.)

It does, however, make *no difference* to the correlations whether the collapse due to A or that due to B is deemed to have occurred 'first'. There is no inconsistency of the actual observations with relativity whichever choice is taken. There is, nevertheless, an inconsistency with the *spirit* of relativity *whichever* choice is taken if we wish to regard these collapsing state vectors as providing a picture of physical reality. One could, of course, simply ignore this fact and take the view that 'physically real' collapse takes place in some 'absolute' frame, determined perhaps cosmologically. This would be consistent with relativity, but it is a viewpoint that I should, myself, be most reluctant to adopt. It seems to me that the reduction process—which I believe to be, in some appropriate sense, *real*—must involve some very subtle and non-local space–time (or perhaps 'beyond' space–time†) ideas (cf. also Chapter 12).

Similar remarks may apply in relation to the third argument against state vector objectivity that I have in mind. This concerns the question of time-reversibility. The Schrödinger-evolution of the state vector which is normally envisaged (and which was argued for in the above discussion supporting state vector objectivity) requires that the evolution is such that the state is an eigenstate of the observation performed at the *beginning* of this evolution, not of the one at the end. This description is not time-symmetric; but precisely the same relative probabilities would be obtained if one took the 'perverse' view that it is the backwards-Schrödinger-evolved state vector that is relevant, this being the appropriate eigenstate of the operator describing the observation at the *future* end

† This kind of non-locality had, quite early on, supplied some of the motivations for the point of view of twistor theory (compare Penrose (1986)). It should be pointed out, however, that the violations of Bell's inequality that these experiments confirm, do not *in themselves* have anything to do with space–time or non-locality; it is just these particular *manifestations* of the violations of the inequality which do. But it is such manifestations that are most puzzling to our intuitions.

On this general topic, I might raise a further issue which occurred to me. In relation to Lee Smolin's talk, in which non-local hidden variables referring to *pairs* of particles were described, one might ask whether Bell-type inequalities for particle *triples* (or quadruples etc.) might exist for such theories, which could be violated in some quantum-mechanical experiment.

of the evolution, rather than the past (see Penrose (1979), p. 584). If we believe that time-symmetry ought to be a feature of an objective state vector reduction, then this might present us with a problem, over and above that presented in the preceding paragraph. However, my own view, as I shall present in the next section, will be seen to involve an essential time-asymmetry—and my preference will be that, to a good degree of approximation, physical reality is described by state vectors evolving in the normal future direction from some 'eigenstate' until some (spontaneous) reduction takes place.

9.2. Singularities and the Second Law

An essential feature of the physics involved in the reduction process must be the Second Law of Thermodynamics. If it is impossible *in principle* to measure, for example, the unitarily evolved state which represents Schrödinger's half dead and half alive cat, then it is because somehow the essential phase relations have become 'irretrievably lost'. One envisages that these phase relations have become spread over a large number of degrees of freedom; so, in some appropriate sense the entropy has 'gone up'. This irreversibility, however, has to be of some absolute and objective character if the phase relations are to be actually 'lost' in principle and not simply retrievable only with extraordinary difficulty. Thus, it would seem that we require some entropy concept with a greater physical objectivity than those that one normally encounters in physics.

In my view, an essential clue to this is to be found in the Bekenstein–Hawking *black hole entropy*, which is given by the remarkable formula

$$S_{bh} = \frac{kAc^3}{4\hbar G},\tag{9.3}$$

where A is the surface area of the black hole's even horizon, and where c, G, k, and $\hbar$ are, respectively, the velocity of light, the gravitational constant, Boltzmann's constant, and Planck's constant divided by 2π. Note that here an entropy measure is set equal to a precise and seemingly *objective* quantitative feature of space–time geometry. I shall try to argue that the black hole entropy, as given by eqn (9.3), represents merely an extreme case of a *gravitational entropy*; and that it is this concept of gravitational entropy that has the needed objectivity and which supplies the physical ingredient involved in the triggering of state vector reduction. In order to motivate this claim, I shall need to outline a somewhat roundabout argument that I have given on earlier occasions concerning the origin of the Second Law, and its relation to the structure of space–time singularities. For further details I refer the reader to earlier publications (Penrose 1979, 1981).

The basic puzzle that the Second Law presents us with is, it seems to me, that at some time in the distant past, some very strong constraint must have been placed on the structure of the universe. This constraint resulted in a universe with a very low entropy, and the universe has been evolving to the more 'probable' situations of higher entropy every since—hence the Second Law. The time at which this constraint was imposed was presumably that of the Big Bang. We must ask what the nature of this constraint was, and try to understand it in relation to other aspects of physics.

One point that should first be made is that if one assumes deterministic evolution, then, in a certain sense, the state of the universe remains just as 'peculiar' at late times as it was at early times. But at late times this peculiarity takes the form of very precise *correlations* between particle motions. These correlated motions, when evolved backwards in time, produce the most manifest type of peculiarity, at early times, that is described as 'low entropy'. Thus, the form of peculiarity that our universe exhibits is, according to the Second Law: low entropy at early times, and precise correlations at late times. My point of view will always be to regard the presence of such correlations as secondary to (i.e. a consequence of) the low entropy, and not the other way about. Had it been the case in our universe that correlations of the opposite type which *reduce* entropy were *frequently* present (e.g. shattered glasses of water miraculously assembling themselves, etc., being commonplace occurrences) then I believe we should naturally have adopted a teleological type of terminology which asserts that the future low-entropy states are the *causes* of the earlier correlations (e.g. that the future existence of an assembled water glass provides the *reason* for the earlier precise particle motions in the pieces of glass and split water). Of course, *our* world is not like that, which is why this kind of teleological terminology seems unnatural. However it is a helpful form of words for the discussion of the Second Law, since it is not biased with regard to one direction of time or the other.

It is frequently argued that the Second Law arises *because* of the absence of precise entropy-reducing correlations in the initial state—and that what needs to be explained is the absence of such correlations. But my own view, as is implicit in the above, is that the presence or absence of such correlations is *secondary* to the entropy question, and is not the place to look for an explanation of the Second Law. In any case, it is the *low-entropy* (or manifest peculiarity) of the initial state which needs explanation. Given that, the absence of correlations which might reduce the entropy still further is no surprise. The vast majority of states compatible with the initial low entropy must indeed be uncorrelated in the sense that higher entropy states then evolve from them. To evolve to even lower entropy, the initial state would have had to have been even more special than the requirement of initial low entropy already constrains it to be. That would have needed, in my way of looking at it, some further macroscopic constraint to have been imposed at some *later* time.

One apparent paradox is the fact that the Big Bang, though the universe-state of *lowest* entropy (by the Second Law), is treated as a state in which the matter is in thermal equilibrium, i.e. at *maximum* entropy! It should be made clear that the resolution of this paradox does *not* lie simply in the fact that the universe is expanding. (That is a commonly believed fallacy!) If one imagines the universe to be collapsing, one can still envisage that the Second Law holds. Galaxies fall in towards one another; they unite; the black holes at their cores unite; their stars fall together into more and more massive black holes, which all unite together into one final all-embracing singularity—the *Big Crunch*. Here the final singularity represents maximum entropy for the universe. Reversing time we obtain an expanding universe with a Big Bang whose entropy is a *maximum* and for which the opposite of the Second Law holds. Thus, the expansion of the universe can in no way explain the rise in entropy and is thus not at all the resolution† of the aforementioned paradox.

The actual resolution of the paradox lies in the fact that *gravitational degrees of freedom* (i.e. Weyl conformal curvature) are not taken into account in the 'thermal equilibrium' that seems to have been closely approximated in the very early stages of our actual universe. In the final stages of a generic 'Big Crunch', the Weyl curvature would diverge to infinity (as it does at the singularity even of the symmetrical Schwarzschild black hole), whereas the closely (Friedmann–) Robertson–Walker form that is assumed for the Big Bang has a Weyl curvature which is set equal to zero. Thus, I propose that the constraint on the Big Bang which sets the Second Law on its course is that initially the Weyl curvature must be set to zero.

This hypothesis has the implication that not only has the early universe its initial low‡ entropy, as is required for the Second Law, but the

---

† There seems to be a current misconception that the inflationary models somehow provide an explanation for the evident specialness of the initial state. But this is just another version of the same fallacy. These models all rely on time-symmetric physics, and the above argument still holds. There are vastly more modes of disappearance for a collapsing universe into a final singularity then those modes which resemble the time-reverse of an 'inflationary scenario'. Inflation does not explain why those vastly more probable modes of *reappearance* do not occur.

‡ Again there appears to be a common fallacy. The entropy per baryon figure of $10^8 k$–$10^9 k$ that is observed in the microwave background is often regarded as a 'high' value, whereas for a universe model with 10^{80} baryons, say, the entropy per baryon figure rises to $10^{43} k$ in the Big Crunch. Of course, an explanation for the observed figure is still needed.

uniformity and isotropy that is observed on a large scale is also a rough consequence. With gravitation it is the uniform states which have low entropy and the clumpy ones, high. Initially the gravitational contribution to the entropy is set to zero, but the entropy of the matter is at its maximum. Owing to the extreme weakness of the gravitational coupling, it is a long while before this entropy imbalance can take effect. But gradually the matter begins to clump as the gravitational entropy contribution rises. Galactic masses are formed, and matter collects into stars. The hot spot in our sky which is our Sun, and upon which all life on this planet depends for a needed powerful entropy sink, owes its existence to the clumping effect of gravity. The clumping together of masses rather greater than our Sun leads to black holes, these achieving the maximum entropy possible for a given mass (in a limited volume). For mass m, the maximum surface area for a black hole's event horizon is the Schwarzschild value

$$A = 16\pi G^2 c^{-4} m^2, \tag{9.4}$$

which, with the formula (9.3), yields the entropy

$$4\pi \hbar^{-1} kGc^{-1} m^2. \tag{9.5}$$

Note that it is entropically favourable for collections of small black holes to collect together into larger ones. This is consistent with the Second Law and our picture of the final collapse to a Big Crunch.

This is perfectly satisfactory as far as it goes. But it seems to me to be unsatisfactory simply to postulate for the Big Bang that its Weyl curvature should vanish, as a once-only condition unrelated to the other constraints given to us as physical laws. Space–time singularities are places where 'the known laws of physics break down'. But I do not believe that physics itself breaks down at a space–time singularity. It is just that the laws that govern their structure are presently unknown to us. The very uniformity of the structure of the Big Bang singularity is in itself an indication of the existence of a powerfully effective law. However this law cannot apply to the singularities of collapse, since that would lead to a violation of the Second Law and to the existence of peculiar teleological effects preventing the formation of local black holes. It is perhaps significant that (assuming Cosmic Censorship holds true; cf. Penrose 1978, 1979) space–time singularities can be separated into two disjoint classes: *past* singularities (singular TIFs) from which matter and signals can escape—e.g. the Big Bang or the singularities of white holes—and *future* singularities (singular TIPs) into which matter and signals can fall—e.g. the Big Crunch or the singularities of black holes. Thus, rather than applying the hypothesis of vanishing Weyl curvature specifically at the occasion of the Big Bang and nowhere else, we can hypothesize that it applies in a more uniform universal way:

HYPOTHESIS: *The Weyl curvature vanishes at all past singularities, as the singularity is approached from future directions.*

This has the advantage that white holes, with their unpleasant anti-thermodynamic behaviour, are excluded. (The qualification about approaching from future directions is to attempt to cover possible violations of Cosmic Censorship that could occur at the final explosion in the Hawking evaporation of a black hole; cf. Penrose (1981).)

This hypothesis is time-asymmetric, as indeed could have been anticipated, since it yields the time-asymmetric Second Law. I shall indicate, moreover, that this time-asymmetry cannot be regarded as separate from all the other procedures of physics, so one must also anticipate some role for time-asymmetry elsewhere.

Envisage the following *gedanken* situation (introduced by Hawking 1976). Consider a large isolating container—with perfectly reflecting walls—the total energy content of the container being sufficient that the maximum entropy state involves a black hole (but not so great that the container cannot contain the hole). The phase space $\mathscr{P}$ of the contents is schematically illustrated in Fig. 9.2. The region of $\mathscr{P}$ representing configurations for which there is a black hole in the container is denoted

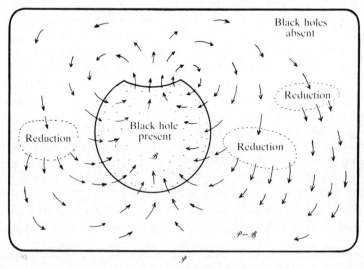

Fig. 9.2. Phase space $\mathscr{P}$ of a large isolated container with perfectly reflecting walls.

by $\mathscr{B}$. White holes (the time-reverses of black holes) are excluded since their singularities violate the Weyl curvature hypothesis. This time-asymmetric constraint on what can happen inside the container leads to some interesting consequences. However, I should make clear that not everyone would be in agreement about this picture. In particular, Hawking (1976) takes the view that black holes and white holes are to be regarded as physically *identical,* and so he would reject the hypothesis. In my own opinion this identification carries with it some severe difficulties, mainly because the classical space–time geometries of black and white holes are not at all the same, especially when one considers their modes of formation and disappearance; cf. Penrose (1979, 1982a). Hawking considers that this problem can be resolved if one takes the view that the geometries cannot be adequately considered as classical and must be taken to be subject to some decidedly large quantum uncertainties.)

I am being deliberately vague as to whether the phase space $\mathscr{P}$ depicted in Fig. 9.2 is to be regarded as a classical or quantum phase space (i.e. as a symplectic manifold or a Hilbert space). It is hard to see how the black hole geometries can (as of now) be treated other than classically; yet one is also concerned with the Hawking process, which is quantum mechanical, as providing the only way that black holes can disappear once they are formed. I prefer to think of the phase space as being, in reality, some sort of combination of a quantum and a classical one. But for the purposes of the present discussion I shall just remain somewhat vague about the matter. We note, however, that since the container has finite volume and a bounded energy (mass) content, $\mathscr{P}$ must, if classical, be a *compact* symplectic manifold and, if quantum mechanical, a *finite-dimensional* Hilbert space.

We recall that whereas black holes can form by means of classical collapse, they can disappear only quantum mechanically, the mass being reduced through the process of Hawking evaporation. Since according to the view to which I adhere, black holes and white holes are physically distinct objects, the classical collapse modes are not time-reverses of anything which can be achieved by means of Hawking evaporation. The implication seems to be that there are more ways for a black hole to form than there are for it to disappear. In terms of Fig. 9.2, this means that there are more routes (in terms of phase space volume or quantum mechanical state count) entering the region $\mathscr{B}$ than there are leaving it. As far as the region $\mathscr{B}$ itself is concerned the violation of Liouville's theorem or of quantum mechanical unitarity that this implies is not paradoxical because a space–time singularity is present for the states within $\mathscr{B}$. Information can be destroyed at this singularity in violation of the normal assumptions underlying Liouville's theorem and unitarity, and this accounts for the loss of phase-space volume or of quantum states within $\mathcal{B}$. This situation occurs because the singularities are future-singularities (i.e. information absorbing singularities), all past-singularities (i.e. information creating ones) that otherwise might occur within the container having been excluded by the hypothesis on p. 138.

However, with regard to the remaining region $\mathcal{P} - \mathcal{B}$ there is a definite problem: some effective violation of Liouville's theorem or unitarity seems to be implied, yet no singularities occur for that region. More routes *leave* the region $\mathcal{P} - \mathcal{B}$ (which has finite total volume) than enter it. Somehow, phase space volume is *created* within $\mathcal{P} - \mathcal{B}$, yet the region is concerned with situations of 'ordinary' physics, where no exotic phenomena associated with the presence of black holes or space–time singularities occur. My contention is that this effective violation of Liouville's theorem or unitarity is an objective *state vector reduction*. This is the other side of the coin from our hypothesis and seems to be needed for a consistent picture. My suggestion has been that the origin of the Second Law is to be regarded as a (quantum) gravitational phenomenon (i.e. determined from space–time singularity structure) and now we appear to need the compensating idea that state vector reduction process is also a (quantum) gravitational phenomenon. For a good many years, Károlyházi has argued on quite different grounds that an objective physical reduction process should occur by virtue of quantum gravitational considerations (see Károlyházy 1966, 1974, and Chapter 8; also Komar 1969). Here we appear to have some significant independent support of this idea.

Indeed, without some spontaneous reduction taking place within the container it is difficult to make any sense of the geometry within it whatsoever. Note that, according to my picture, this reduction must (at least sometimes) be a time-asymmetric process. As a single quantum state evolves via this reduction procedure, it becomes a collection of possible alternative states, only one of which is realized in the actual world (though it could be a different one on the next 'Poincaré cycle'). We get an effective spreading of phase space volume only because if we reverse the procedure fewer alternatives are available. To understand what is involved here, imagine a simple quantum experiment. A photon is aimed at a half-silvered mirror and photocells detect whether it has been reflected or transmitted (Fig. 9.3). One state goes in; then after reduction, one of two alternative states then takes place. The initial state vector is referred to a new basis and the two constituents of this initial state with respect to this new basis are both allowed. But reverse the process in time and we find that only one of the two resulting states is reasonable and the other is 'absurd'—the reasonable one corresponding to the photon emitted from the laboratory source and the 'absurd' one, to that where the photon is ejected from the laboratory wall. It is 'absurd'

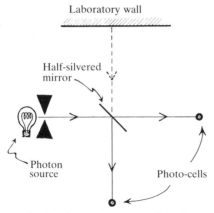

Fig. 9.3. State vector reduction in a simple case: is it time-reversible?

only for thermodynamic reasons: in ordinary time, the ejection process violates the normal progression of increasing entropy. From my own point of view, this 'absurdity' would (almost always) entail that if we evolve it backwards in time to the beginning of the universe, we get a contradiction with our hypothesis, since that is the origin of the Second Law. The reduction process and the Second Law would seem to have to be intimately connected with one another, which is a strong reason for believing that gravity must have something to do with reduction, since (on almost any point of view) it has a special role to play in connection with the Second Law.

9.3. Gravitational entropy and reduction

I shall argue that a key ingredient in the process of spontaneous state vector reduction is the concept of *gravitational entropy*. The idea is that reduction will occur when the decrease in entropy involved in the reduction effect is at least compensated by a corresponding increase in the gravitational entropy. Imagine the wave function of some particle spreading out through space, say from the decay of a radioactive atom. The geometry this produces will not have any significant Weyl curvature. But if the wave function becomes coupled to some larger system so that the different particle positions produce sufficiently clumped configurations of mass, then each individual clumped configuration may have sufficient Weyl curvature to represent a significant rise in the gravitational entropy. Under such circumstances, I would contend, reduction to one of these individual clumped geometric configurations could take place.

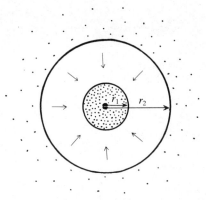

Fig. 9.4. Formation of a droplet—for the rough estimation of gravitational entropy increase.

Of course, such a suggestion requires a change in the standard rules of ordinary linear quantum evolution. But the 'rules' are themselves somewhat unclear in the context of curved space–time theory. One requires some analogue of the space- and time-translations to provide the 'natural' co-ordinate variables in terms of which quantization is to be carried out. In a general curved space–time, when such symmetries are absent, there are essential ambiguities inherent in the quantization procedure (Komar 1971). If one envisages differing space–time geometries being linearly superposed, then one must imagine that the differing 'quantization procedures' that are being adopted, for the various space–times must also be in some sense superposed. It is, to say the least, somewhat obscure what the 'standard rules' of quantum theory actually *are* when space–times of differing intrinsic geometry are superposed. So 'a change in the standard rules' for such circumstances is not such a shocking innovation as it might otherwise appear to be!

I envisage that nature feels uncomfortable about 'linearly superposing' space–time geometries which differ significantly from one another and, instead, prefers to settle for (i.e. to 'reduce' to) essentially just one of the geometries involved. We may think of this reduction as perhaps taking place via some nonlinear instability. But I would not expect that the geometries are ever quite precisely defined as classical structures. Each classical geometry would itself be some sort of average of contributing quantum states. Then, roughly speaking, the logarithm of the number of quantum states involved would provide a measure of the *gravitational entropy* that is to be associated with the classical geometry. The idea is that 'reduction' can take place when the apparent lowering of entropy that such a process leads to is actually more than compensated by a rise in this gravitational entropy.

Though I have no clear-cut mathematical expression for this gravitational entropy for a general classical geometry, some guides to the rough form of such an expression can be suggested. In the first place, we demand agreement with the Bekenstein–Hawking formula (9.3) for a stationary black hole. But in the general case, we imagine that the entropy measure could be 'time-dependent', i.e. it would depend upon the choice of some space-like hypersurface† $\mathcal{H}$ within the space–time $\mathcal{M}$. I envisage (perhaps) some integral expression involving the Weyl curvature of $\mathcal{M}$ at $\mathcal{H}$, where the 'density' to be integrated over $\mathcal{H}$ has the rough form: integral of Weyl curvature times second integral of Weyl curvature (i.e. something at the level of 'Christoffel symbols times metric'), but where something of the nature of a positive/negative frequency split may first have to be performed. The motivation for this is that such an expression exists in the source-free weak-field (linearized) limit of the gravitational field (cf. Fierz 1940) where one is here using an estimate of 'graviton number' in a classical field as a rough guide to its entropy. The aforementioned expression provides the expectation value of the number operator.

I wish to attempt a crude estimate of this gravitational entropy in a situation of relevance to the detection of a particle in a cloud chamber or bubble chamber. When a droplet condenses in a cloud chamber, associated with its more clumped state will be a space–time geometry with a slightly higher gravitational entropy than that of the uniform distribution of uncondensed gas. It is only the gravitational contribution to the entropy that concerns us here, and one might, at first, anticipate that this contribution would be insignificantly tiny. But for an idealized spherically symmetrical uniform droplet forming (see Fig. 9.4) in which a region of radius r_2 condenses to a smaller one of radius r_1, we can imagine that a non-zero Weyl curvature occurs only between the radii r_1 and r_2, the (linearized) Schwarzschild solution holding there (assuming, for crude calculational purposes only, that a vacuum is left between r_1 and r_2). The entropy 'density' to be integrated thus occurs between the radii and has

---

† The view has been expressed by Hawking and his collaborators that only when there is an absolute event horizon should there be a non-zero gravitational entropy. However, because of this dependence upon data at a (space-like) hypersurface it is not a clear-cut matter whether such a horizon exists or not. The black hole metrics are not, at $\mathcal{H}$, clearly distinguished from those that are not black holes. (If an apparent horizon is present on $\mathcal{H}$ then, with the weak energy condition holding, a horizon *will* exist. But this is not the correct *criterion* for a horizon. In any case, the 'turning on' of an entropy only when some inequality is satisfied seems to me to be quite inappropriate.)

the rough form of a constant times

$$\frac{m}{r^2} \times \frac{m}{r}, \tag{9.6}$$

where m is the mass contained within the (smaller) radius, so we anticipate a total gravitational entropy of something like

$$(\text{numerical constant}) \times \left(\frac{kG}{\hbar c}\right) m^2 \log\left(\frac{r_2}{r_1}\right). \tag{9.7}$$

This expression diverges for $r_1 \to 0$, but we expect a cut-off at the Schwarzschild radius $2mGc^{-2}$. The formula (9.7) is presumably only approximate and should go over to the Bekenstein–Hawking value (9.3) at roughly this cut-off radius.

Despite these various uncertainties and the obvious crudeness of eqn (9.7), we can use eqn (9.7) to give us a very rough estimate of the gravitational entropy in a cloud chamber droplet. Similarly we can use it for a bubble in a bubble chamber, where in this case we regard it as resulting from a condensation of negative mass over and above the positive-mass background. Because m occurs squared in eqn (9.7), the entropy contribution is still positive. (The formation of a bubble is still a *positive* clumping.) We find that if we need, say, an entropy of $100k$ to reduce the wave function of a particle which expands spherically symmetrically from a radioactive source in the laboratory to one localized in a track in the cloud chamber or bubble chamber, then either one rather large droplet or bubble is needed, of perhaps a millimetre in diameter, or, more appropriately, a string of somewhat smaller ones. This all seems rather on the large side, and one might have anticipated that a physically real state vector reduction ought to have already taken place before this. But the results are *not absurdly* unreasonable—as they could very easily have turned out to be! Had the logarithmic expression (9.7) been, instead, something like the $Gm^2 (r_1^{-1} - r_2^{-1})$ that one obtains for the gravitational *self-energy* contribution, then there would have been no chance whatever of being able to produce a significant value on the scale of a laboratory experiment.

I regard these results as providing some general encouragement for the point of view that I am promoting. But clearly we need a very considerable improvement in the gravitational entropy expression. The 'correct' expression (assuming that there is one) must clearly be of a non-local character. This is already true of gravitational *energy*, but it does not prevent the possibility of a precise formula at the quasi-local level, (cf. Penrose 1982b; Penrose and Rindler 1985). For entropy the non-locality may be one stage more 'nebulous' since there is one further degree of integration involved in the weak-field limit expression. The hope is that this non-locality can be matched with the kind of non-locality that quantum phenomena so frequently exhibit (cf. Chapter 4). This is yet a matter of speculation, but it opens up a number of intriguing possibilities.

Acknowledgements

I am grateful to many friends and colleagues for enlightening discussions, most particularly, to I. Bialynicki-Birula, P. Pearle, I. C. Percival, and D. Deutsch.

Postscript

In subsequent discussions, R. L. Wald has more-or-less persuaded me that my tentative expression (9.7) should not be interpreted as an entropy in the ordinary sense. I think that it is more correctly just a measure of the increase in *graviton number* and that the whole discussion should perhaps be rephrased in such terms. Recently, A. Ashtekar has informed me of a calculation he has performed, indicating that the numerical constant in expression (9.7) should be something like

$$768\pi^3 \doteq 2.4 \times 10^4.$$

This is gratifyingly large: it means that the droplet or bubble size, for reduction to take place, would be considerably smaller than I had previously estimated.

I am now suggesting that reduction takes place whenever *one* new ('longitudinal') graviton is produced.

References

Everett, H. (1957). *Rev. Mod. Phys.* **29,** 454.

Fierz, M. (1940). *Helv. Phys. Acta* **13,** 45.

Hawking, S. W. (1976). *Phys. Rev.* D**13,** 191; **14,** 2460.

Károlyházy, F. (1966). *Nuovo Cimento* A**42,** 390.

——, F. (1974). Gravitation and quantum mechanics of macroscopic bodies. *Magyar Fizikai Polyoirat* **12,** 24 (Thesis, in Hungarian).

Komar, A. B. (1969). *Internatl. J. Theor. Phys.* **2,** 257.

—— (1971). Semantic foundation of the quantization program. In *Studies in the foundations, methodology and philosophy of science,* Vol. 4, *Problems in the foundations of physics* (ed. M. Bunge). Springer, Berlin.

Penrose, R. (1978). Singularities of space–time. In *Theoretical principles and astrophysics and relativity* (ed. M. R. Lebovitz, W. H. Reid, and P. O. Vandervoot). Chicago University Press, Chicago.

—— (1979). Singularities and time-asymmetry. In *General relativity—An Einstein centenary survey.* (ed. S. W. Hawking and W. Israel). Cambridge University Press, Cambridge.

—— (1981). Time asymmetry and quantum gravity. In *Quantum gravity* 2. Oxford University Press, Oxford.

—— (1982a). Some remarks on gravity and quantum mechanics. In *Quantum structure of space and time* (ed. M. J. Duff and C. J. Isham). Cambridge University Press, Cambridge.

—— (1982b). *Proc. R. Soc. Lond.* A**381,** 53.

—— (1986). On the origins of twistor theory. In *Gravitation and geometry,* I. Robinson 60th Birthday volume, (ed. W. Rindler and A. Trautman) Bibliopolis, Naples.

—— and Rindler, W. (1985). *Spinors and space–time, Vol. 2; Spinor and twistor methods in space–time geometry.* Cambridge University Press, Cambridge.

Von Neumann, J. (1932). *Mathematical foundations of quantum mechanics* (Princeton University Press, 1955) Ch. VI.

10

Stochastic mechanics, hidden variables, and gravity

Lee Smolin

10.1. Introduction

In this chapter I should like to report on some recent work which was motivated by the expectation that the ultimate solution to the problem of quantum gravity will require some modification in the fundamental ideas we hold about quantum mechanics. While it appears that the basic principles of quantum mechanics can be applied meaningfully to certain special situations in which gravitational interactions are relevant, such as in the description of Hawking radiation or the scattering of a finite number of quanta on a flat background, all of these successful applications depend on recourse to a preferred time coordinate which is either available in an asymptotic region or is picked out by a symmetry of the background space–time. In more general circumstances, regions of strong fields without asymptotically flat regions or special symmetries, where there are no preferred time co-ordinates, the standard quantization procedures become ambiguous (Ashtekar and Geroch 1974; Ashtekar and Magnon-Ashtekar 1975; Kay 1978; Fulling 1973; Unruh 1976; DeWitt 1975; Kuchar 1982). This is a serious problem which goes to the foundations of quantum field theory and which rests, ultimately, on the conflict between the very different roles time plays in quantum mechanics and general relativity (Kuchař 1982). Thus, the point of view which I will pursue here is that, while we may be able to learn something about quantum gravity by trying to construct perturbatively sensible quantum field theories to describe the interactions of gravitons with matter on flat backgrounds, ultimately, in order to construct a viable and completely general theory bringing together quantum and gravitational phenomena, we may have to be prepared to meddle with the foundations of quantum mechanics.

There are several different results which to me suggest that the standard principles of quantum mechanics are breaking down in the presence of gravitational effects. They are: (1) The apparent loss of quantum coherence in black hole evaporation (Hawking 1976, 1982, 1984); (2) the mixing of the effects of quantum and thermal fluctuations in the presence of gravitational fields and under general co-ordinate transformations or, to put it differently, the lack of a *local* co-ordinate invariant distinction between real statistical fluctuations and virtual quantum fluctuations (Sciama 1979; Sciama *et al.* 1981); (3) the dependence of the Fock space construction of Hilbert space on the choice of a preferred time, which leads to the absence of an unambiguously and invariantly defined quantum theory in the absence of either a time-like killing field or an asymptotically flat region (Ashtekar and Geroch 1974; and others) and (4) the apparent indistinguishability, by physically realizable measurements, of pure and mixed states of the linearized gravitational field (Smolin 1984*a,b*).†

I have discussed these examples at length elsewhere (Smolin 1982), and will not go further into them here. I feel that they are serious enough to merit a search for an alternative approach to the quantum theory which agrees with the standard quantum mechanics in the absence of gravitation, but which might extend more naturally to processes in arbitrary background gravitational fields in which a preferred notion of time is not available. What I would like to do here is to mention briefly several considerations which I believe indicate a direction in which we might search for such a theory, and then to describe several of the results which have so far been found.

The principle conclusions which I draw from the examples just mentioned are the following: (1) the distinction between what is a real statistical fluctuation, by which we mean a random fluctuation in the pure state of the system, and what is a virtual quantum fluctuation, by which we mean a fluctuation in the measured value of some operator, due to the fact that the state is not one of its eigenstates, becomes frame-dependent whenever one goes beyond a description in terms of globally inertial observers in the absence of gravitational fields; (2) in some situations virtual quantum fluctuations in one region of space–time, for example near the event horizon of a black hole, can lead to real thermal fluctuations in some other region of space–time, in this case far from the black hole at late times. What is meant by this is that the randomness in the thermal distribution at late times is not due to the randomization of

---

† A possible counter-example to a conjecture made in these papers, that it is impossible to build a box which contains gravitational radiation, was found by D. Garfinkle and R. Wald (1985). This involves a charged shell with $e \approx M$ close to its gravitational radius. However, Dell (1985) has shown that this example, if large enough to be stable against spontaneous electron–positron creation, is unstable in the presence of external electromagnetic or gravitational radiation with wavelength less than 100 kilometres. It is thus not a candidate for an efficient detector for gravitational radiation.

the distribution of energy in the field, as it is distributed among many degrees of freedom, and certainly not due to ignorance about the state of the field. It is instead due to the fluctuations which would be observed due to the uncertainty principle, if we measured some field operator near the horizon. (In this sense a black hole is like a microscope which magnifies and makes real the virtual quantum fluctuations in the field near the horizon.)

Now, whenever we have a situation in which the distinction between two kinds of physical phenomena becomes frame-dependent, such as electric and magnetic fields or the effects of gravity and inertia, Einstein taught us that the distinction must not reside in the phenomena, but only in the combination of the phenomena and a particular observer. In these cases we should search for a single physical concept to encompass both phenomena, along with a co-ordinate invariant description of it, such that the distinction can only be introduced with respect to an explicitly indicated frame. What I would like to suggest is that we should try to do the same for the distinction between the effects of quantum and statistical fluctuations: accept the fact that the absolute distinction which holds conventionally between quantum and thermal effects is valid only with reference to a set of globally preferred frames in the absence of gravity, and look for some more general notion of a fluctuation which will allow us a co-ordinate invariant description of quantum processes, while allowing us always the possibility of drawing the distinction with respect to some choice of frame, should we find it useful to introduce it.

Now, how are we to do this? To begin with we need to find a concept which can encompass both the concept of a virtual quantum fluctuation and that of a real statistical fluctuation. It seems to me that the most straightforward way to do this is to adopt an interpretation of quantum mechanics in which virtual quantum fluctuations *are* ordinary statistical fluctuations. Such an interpretation, called the statistical interpretation of the wave function, has been around for a long time; it was in fact the view advocated by Einstein (1949). (For an excellent presentation of the statistical interpretation, see Ballentine (1970)). Its fundamental tenet is that the wave function is not a description of an individual system, rather it is a description of an ensemble of similarly prepared systems, where similarly prepared means prepared by identical macroscopic devices. In this view the distinction between a virtual and a quantum fluctuation is not an ontological distinction, but only a distinction as to the cause of the fluctuation. One must assume the existence of some universal source of fluctuations, and this source must impart some unusual properties to the correlations of these quantum fluctuations—chief among them being the existence of non-local correlations. The advantage of this view is that the question as to why these correlations, which are at the heart of most that is strange about quantum mechanics, exist becomes something which might be explained by new physical hypotheses, rather than something which must merely be accepted as a consequence of the linear structure of Hilbert space.

In order to see that this is the most straightforward direction to pursue consider the following argument. In order for it to be possible for the distinction between two phenomena to be seen as observer dependent, they must be phenomena of the same type. Given the conventional Copenhagen interpretation of quantum mechanics, in which the quantum state is postulated to be a complete description of an individual system there is a difficulty, because thermal fluctuations refer to fluctuations within an ensemble of distinct physical systems, while quantum fluctuations refer to uncertainty in the outcome of a measurement which might be performed on a single individual system. These are very different kinds of entities, and the distinction between them must remain absolute unless either the interpretation of the quantum state or the interpretation of the thermodynamic ensemble is altered. However, the possible alterations we might consider are severely restricted by the requirement that the interpretation of both the thermal ensemble and the quantum state must remain consistent with the conventional usage in the absence of gravity, where the distinction may be chosen unambiguously. Clearly we cannot alter the interpretation of the thermodynamic ensemble so that it refers only to an individual system. The only sensible alternative I am aware of is to modify the interpretation of the quantum state so that it refers to an ensemble of physical systems.

Once we adopt this interpretation there is no conceptual barrier to the distinction between thermal and quantum fluctuations being observer-dependent, or to quantum fluctuations being the cause of thermal fluctuations at some later time. However, this is only the first step. In order to proceed we need a mathematical formulation of the quantum theory in which the quantum state is explicitly formulated as a description of an ensemble of systems and where the quantum fluctuations are explicitly described in terms of a more general class of fluctuations. At least one such formulation does exist (the stochastic quantum mechanics of Fényes (1952) and Nelson (1966)), and it is for this reason the basis of the work I will describe in the following. In particular, Nelson's derivation of the Schrödinger equation from the more general theory of Brownian processes gives several conditions which distinguish quantum behaviour from other types of stochastic processes, and it is to an explanation of these conditions that we may look for a deeper understanding of what distinguishes quantum from thermal fluctuations.

There is another reason which might be mentioned for basing an extension of quantum mechanics to circumstances in which gravitation is present on the stochastic mechanics. The difficulties which have been encountered in the extension of quantum field theory to curved space–time may be seen to arise partly as a result of the rather indirect connection between the fundamental mathematical quantities on which quantum mechanics is based, which are an algebra of operators on Hilbert space, and what is measured, which are probability distributions. This connection involves several steps, one or more of which becomes problematical when gravitational fields are present. For example, one result of this is the lack of an invariant definition of a conserved inner product for quantum field theory on an arbitrary curved background (Ashtekar and Geroch 1974; and others). On the other hand, in stochastic mechanics the fundamental mathematical quantities on which the theory is based are the observed probability distributions. The usual order of things is reversed in that conservation of probability is built in from the beginning, and it is the linear Hilbert space structure which is derived in a series of steps. Thus, what we expect will happen if the steps connecting the Hilbert space to the observed probability distribution become problematic in the presence of gravitation is that in stochastic mechanics conservation of probability will be maintained, but it may not be possible to interpret what is measured in terms of a linear Hilbert space structure. Given a choice, this is clearly better than maintaining the linear Hilbert space structure, but at the cost of losing an unambiguous connection between it and physical measurements.

We may express this situation more simply as follows. In ordinary quantum mechanics we have both the conservation of probability and the superposition principle. However, it may not be possible to maintain both when we generalize to an arbitrary curved space–time, because the connection between them makes use of special properties of Minkowski space–time such as the existence of global inertial frames. If this is the case then it is better to give up the superposition principle to maintain conservation of probability than the reverse.

Indeed, results which I will discuss later show that just this kind of thing does happen when stochastic mechanics is generalized to describe the motion of a free particle in a background gravitational field. Conservation of probability is maintained, and the probability distribution obeys well defined laws of evolution. But, in general, these laws can only be expressed as linear evolution equations in the weak field limit.

Before turning to a discussion of these results I would like to comment on the basic difficulties that I believe underlie the various problems that we have been discussing. Perhaps the most basic of these, which was touched on above, is the conflict between the need for a globally preferred time co-ordinate in quantum theory and the purely local character of time in general relativity. This problem has many aspects, some of which were touched on in presentations at the Oxford meeting by Ashtekar, Barbour, and Sparling. However, at least in part, this problem may be seen as closely related to another basic problem. This is the conflict between the purely local description of physics in general relativity and the purely global character of the pure quantum state. Indeed, it must be emphasized that even in ordinary quantum field theory, the distinction between a pure and a mixed state can only be made with respect to the entire global state at a given time. For example, many states can be constructed which appear to describe a thermal ensemble in some region, but which are actually pure states, because measurements made in that region are correlated through non-local EPR type correlations to measurements made in other regions (cf. Chapter 22 for a related discussion).

Indeed, these considerations suggests that it might be true that there is no way to distinguish a pure state of a quantum field from a mixed state by local measurements. A consequence of this is then that there is locally no operational way to distinguish the effects of a quantum fluctuation from those of a thermal fluctuation. If this is the case then it is not surprising if the distinction between them depends locally on a choice of frame, and becomes ambiguous in the absence of preferred global frames.

Several different topics are covered in the following sections. In the next section a review of the formalism which is used to describe Brownian motion is presented. This is not complete, but should allow the reader to understand the details of the following material. However, the reader will not need to be familiar with the details of this material to understand the essential points of what follows. In the third section a review of Nelson's formulation of quantum mechanics is given, and in the one following an experimental test of one of Nelson's assumptions is discussed. Section 10.5 contains the generalization of Nelson's formulation to a free particle propagating in an arbitrary background gravitational field, and the last section contains a brief presentation of a class of hidden variable theories.

10.2. Review of the theory of Brownian motion

Let us begin with a brief review of Nelson's formulation of stochastic mechanics. While there is not enough space here for a complete presentation of the theory of stochastic processes, I will define the basic quantities and write down the basic equations that they satisfy. For those interested in more detail there are some excellent reviews (Nelson 1966, 1967, 1979; for an introduction to the subject of stochastic processes, see the papers collected in Wax (1954); a good review of Nelson's theory is in Seiler (1984).

In stochastic mechanics, quantum processes are seen to be special cases of Brownian processes, so we begin with an outline of the theory of Brownian motion. The basic assumption of the theory is that the motion of a particle, $x_i(t)$ is subject to a random stochastic motion $\Delta x_i(t)$. These random motions will be specified probabilistically. Thus, the basic quantities of the theory will be an ensemble of paths in an n-dimensional space, which will represent possible motions for the particle. The notation $x_i(t)$ then represents an ensemble of motions, and for each member of the ensemble the change in position in a time dt (>0) is given by

$$Dx_i = b_i(x(t), t)\, dt + \Delta x_i, \tag{10.1}$$

where $b_i(x(t), t)$ is a given function but Δx_i is a stochastic variable which varies over the ensemble. Δx_i is specified by its average over the ensemble; in particular the motion is called Brownian if

$$\langle \Delta x_i \rangle = 0 \tag{10.2}$$

$$\langle \Delta x_i\, \Delta x_j \rangle = 2v\delta_{ij}\, dt, \tag{10.3}$$

where v is called the diffusion constant. In particular, eqn (10.3) implies that the random change in $x_i(t)$ in a time dt is given by

$$|\Delta x_i| = \sqrt{\langle \Delta x_i^2 \rangle} \approx \sqrt{dt}\ .$$

That the differentials of the spatial variables are proportional to the square root of the time differential is characteristic of the theory of Brownian motion. As a result the motions $x_i(t)$ are represented by continuous but non-differentiable functions. Because of this it is important also to consider the change in the position of the particle in a small negative increment of time. This is given, for $dt < 0$, by

$$D^* x_i(t) = -b_i^*(x(t), t)\, dt + \Delta^* x_i(t), \tag{10.4}$$

where $\Delta^* x_i(t)$ satisfy

$$\langle \Delta^* x_i(t) \rangle = 0 \tag{10.5}$$

and

$$\langle \Delta^* x_i(t)\, \Delta^* x_j(t) \rangle = -2v\delta_{ij}\, dt \tag{10.6}$$

for $dt < 0$. In general, for Brownian motion $b_i \neq b_i^*$ and $\Delta x_i \neq \Delta^* x_i$.

At any time t we can define the probability density $\rho(x_i, t)$, which tells us the probability of finding the particle in a unit volume centred on the point x_i. Thus, the expectation value, in the ensemble of paths of some function $F(x_i, t)$, is given by

$$\langle F(x_i(t), t) \rangle = \int d^n x \sqrt{q(x)}\, \rho(x, t) F(x, t), \tag{10.7}$$

where q is the determinant of the metric on n-space. In particular, $\rho(x_i(t), t)$ satisfies

$$\int d^n x \sqrt{q(x)} \, \rho(x, t) = 1. \tag{10.8}$$

In addition, we may define the propagation kernel $P(x, t_2 | y, t_1)$ which is the probability of finding the particle in a unit volume centred on the position x_i at time t_2 if it was in a unit volume around y_i at time t_1. By definition it must satisfy

$$\sqrt{q(x)} \, \rho(x, t_2) = \int d^n y \sqrt{q(y)} \, P(x, t_2 | y, t_1) \rho(y, t_1) \tag{10.9}$$

$$\int d^n x \sqrt{q(x)} \, P(x, t_2 | y, t_1) = 1. \tag{10.10}$$

In addition we shall assume that $P(x, t_2 | y, t_1)$ satisfies the Smoluchowski equation

$$P(x, t_3 | y, t_1) = \int d^n z \sqrt{q(z)} \, P(x, t_3 | z, t_2) P(z, t_2 | y, t_1) \tag{10.11}$$

where $t_3 > t_2 > t_1$. In the rest of this section we shall assume for simplicity $q = 1$. However, we shall need these more general formulae in Section 10.5.

From these equations, one may derive the Fokker–Planck equations

$$\frac{\partial \rho}{\partial t} = -\frac{\partial}{\partial x_i} (\rho b_i) + v \nabla^2 \rho \tag{10.12}$$

$$\frac{\partial \rho}{\partial t} = -\frac{\partial}{\partial x_i} (\rho b_i^*) - v \nabla^2 \rho \tag{10.13}$$

It is very useful to define the current velocity, $v_i = \frac{1}{2}(b_i + b_i^*)$ and the osmotic velocity $u_i = \frac{1}{2}(b_i - b_i^*)$. These satisfy the equations

$$\frac{\partial \rho}{\partial t} = -\frac{\partial}{\partial x_i} (\rho v_i) \tag{10.14}$$

$$u_i = v \nabla_i LN \rho. \tag{10.15}$$

Finally, we need to define the forward and backwards stochastic time derivatives of a general function $F(x_i(t), t)$. These are given by

$$DF(x(t), t) \equiv \lim_{dt \to 0} \left\langle \frac{F(x(t + dt), t + dt) - F(x(t), t)}{dt} \right\rangle_{x(t)} \tag{10.16}$$

$$D^* F(x(t), t) \equiv \lim_{dt \to 0} \left\langle \frac{F(x(t), t) - F(x(t - dt), t - dt)}{dt} \right\rangle_{x(t)} \tag{10.17}$$

where the averages are over all trajectories that go through the point $x(t)$ at the time t.

10.3. Nelson's derivation of quantum mechanics as a Brownian motion process

In order to understand the significance of Nelson's derivation, it is useful to begin by decomposing the Schrödinger equation into a conservation equation and a dynamical equation. If we write

$$\psi = \sqrt{\rho}\, e^{iS/\hbar} \tag{10.18}$$

then Schrödinger's equation

$$i\hbar \frac{\partial \psi}{\partial t} = -\frac{\hbar^2}{2m}\nabla^2\psi + U\psi \tag{10.19}$$

decomposes into the conservation eqn (10.14), with the current velocity defined as

$$v_i = \frac{1}{m}\nabla_i S, \tag{10.20}$$

and the dynamical equation

$$\frac{\partial S}{\partial t} = \frac{-1}{2m}(\partial_i S)^2 + V + \frac{\hbar^2}{2m}\frac{\nabla^2\sqrt{\rho}}{\sqrt{\rho}}. \tag{10.21}$$

This equation has the form of a Hamiltonian–Jacobi equation for the motion of a particle in a potential V, plus an additional term

$$V_{\text{quantum}} = (\hbar^2/2m)\nabla^2\sqrt{\rho}/\sqrt{\rho}. \tag{10.22}$$

This term is rather strange from the point of view of probability theory, as it says that a quantum particle moves as if it were subject, in addition to external potentials, to a potential which is a function of its own probability distribution. However, as this term is now the only place where $\hbar$ appears, everything that distinguishes quantum mechanics from a probabilistic description of classical particle motion must be a consequence of the presence and form of this term. Thus, any attempt to explain quantum mechanics as arising from a probabilistic description of some more fundamental level of dynamics must come down to an explanation of this term.

In Bohm's (1952) hidden variable theory, the assumption is made that the particle and the wave function are separate and real entities. The wave function is assumed to obey Schrödinger's equation, and the particle is assumed to obey classical mechanics, with the additional assumption that it couples to the wave function through a potential which is exactly of the form of V_{quantum} with ρ assumed to be equal to $\bar{\psi}\psi$. This works, in the sense that one can then show that under some plausible assumptions, ρ will actually be the probability distribution of the particle. However, it is rather unsatisfying, as one just postulates that the wave function and the particle are coupled by V_{quantum}, and this is, to say the least, a very unusual form for a coupling between a particle and a field to have.

In stochastic mechanics, the term V_{quantum} is derived, and to some extent, explained, from the general theory of Brownian motion outlined above, by specifying that the Brownian motion processes satisfy three additional conditions. These conditions are as follows:

(1) The current velocity is irrotational. Thus, there exists a function $S(x, t)$ such that

$$mv_i = \nabla_i S. \tag{10.23}$$

(2) In spite of the fact that the particle is subject to random alterations in its motion there exists a conserved energy, defined in terms of its probability distribution, as

$$E = \int d^3x \rho(x, t)[\tfrac{1}{2}mv^2 + \tfrac{1}{2}mu^2 + V(x)]. \tag{10.24}$$

(3) The diffusion constant is inversely proportional to the inertial mass of the particle, with the constant of proportionality being a universal constant $\hbar$:

$$v = \frac{\hbar}{m} \tag{10.25}$$

The derivation of Schrödinger's equation then proceeds as follows. One can show that condition (2) is equivalent to a stochastic Newton's law,

$$\frac{m}{2}(D^*D + DD^*)x_i = -\frac{\partial V}{\partial x_i}. \tag{10.26}$$

Using the identities in Section 10.2, one may derive the relation

$$\tfrac{1}{2}(D^*D + DD^*)x_i = \frac{\partial v_i}{\partial t} + (v_j\nabla_j)v_i - (u_j\nabla_j)u_i - v\nabla^2 u_j. \tag{10.27}$$

Using eqns (10.23) and (10.25), one finds an equation which is exactly the gradient of eqn (10.21). An argument about boundary conditions allows one to remove the gradient and this completes the derivation.

As this may seem a bit of black magic, several comments are in order. The most important is that this is not a constructive derivation. It says that, *given* an ensemble of stochastic processes which satisfy conditions (1)–(3), the evolution of the probability distribution will be governed by the Schrödinger equation. To put it, perhaps loosely, into words, an ensemble of Brownian processes which are so delicately correlated that an exactly conserved energy of the form (10.24) may be defined in terms of their probability distribution (and which also obey conditions (1) and (2)) will behave as if each member of the ensemble is coupled to the probability distribution of the whole ensemble by eqn (10.21).

Thus, the quantum potential is explained, but at the cost of a very special set of assumptions which the ensemble of Brownian processes must obey. We might put it like this: consider the whole set of ensembles which satisfy the general conditions which define Brownian motion. Now, consider the very special subset of ensembles which also satisfy conditions (1)–(3). These ensembles may be labelled by the value of the conserved quantity E, and perhaps also by the values of other conserved quantities (although except for special cases these do not uniquely specify the ensemble). The paths which comprise each of these ensembles are very delicately correlated, so that (1) the evolution of the probability distribution is governed by a *linear* equation, in terms of the complex function ψ; and (2) the evolution of each path in the ensemble is governed by the probability distribution of the whole ensemble.

Clearly, such ensembles, if they exist, are very special. One thing that Nelson's derivation does not do is to tell us how to construct an enemble which satisfies these properties, nor does it tell us if any such ensembles actually exist. Indeed, it is clear that any theory which aims to explain quantum mechanics as the statistical mechanics of some underlying dynamics must do exactly this. It must explain why ensembles with these special properties exist, and it must tell us how to construct them from some underlying dynamics. Clearly Bohm's theory does this, although by putting the result in by hand by hypothesizing that the particles are coupled to the quantum potential. In the last section I will describe another way to do this, at least to a certain degree of approximation.

10.4. Testability of Nelson's assumptions

It is interesting to ask whether we can loosen any of Nelson's assumptions to construct a class of theories which deviate from quantum mechanics in a controlled way (Smolin 1982). By subjecting these deviations to experimental constraints we can make a statement about how well Nelson's assumptions are satisfied in nature. Also, we shall find an intriguing analogy to the equivalence principle.

Nelson's second assumption is just the conservation of energy, and I will not try to weaken that. The first assumption is difficult to weaken in a

controlled way as, as soon as v_i is no longer a gradient, the four
equations which determine the evolution of the probability density and
the current density cannot be reduced to a single complex equation. (In
the next section we will see an example where this happens in the
presence of the gravitational field.) The third condition, however, is easy
to relax. Indeed, it is quite natural to do this, as *a priori* there is no
reason to believe that the two quantities—the inertial mass that comes
into the dynamical condition (10.24) (or (10.26)) and the diffusion
constant, which deermines the correlation length over which the quantum
fluctuations of the particle's motion are strongly correlated—are at all
related. If we assume instead of eqn (10.25) that $v = \hbar/(m + b)$, where b
is a new constant with the dimensions of mass, then we find, instead of
the Schrödinger equation, a nonlinear equation,

$$i\hbar \frac{\partial \psi}{\partial t} = -\frac{\hbar^2}{2m} \nabla^2 \psi + V\psi - \frac{b\hbar^2}{2m(m+b)} \frac{\nabla^2 \sqrt{\bar{\psi}\psi}}{\sqrt{\bar{\psi}\psi}} \psi. \qquad (10.28)$$

If $b/m \ll 1$, then the nonlinear term will produce a shift in the energy of a
stationary quantum system, which can be estimated by expanding eqn
(10.28) in powers of b/m. We find that the first-order shift in energy is

$$\Delta E = \frac{\hbar^2 b}{8m^2} \int d^3x \frac{(\nabla_i \rho)^2}{\rho}. \qquad (10.29)$$

This term will make a contribution to the Lamb shift, from which we can
get an experimental bound on b/m. Using eqn (10.28) the splitting of the
2s from the 2p orbital is found to be

$$\Delta E_{2s} - \Delta E_{2p} = -\frac{11}{12} \frac{\hbar^2 b}{m^2} \frac{1}{a_0^2} + O\left(\frac{b^2}{m^2}\right) \qquad (10.30)$$

where a_0 is the Bohr radius. If we demand that this effect is less than the
present experimental uncertainty Δ_{exp} for the Lamb shift in hydrogen, we
find that,

$$\left|\frac{b}{m}\right| \lesssim \left(\frac{12}{11\alpha^2}\right)\left(\frac{\Delta_{exp}}{mc^2}\right). \qquad (10.31)$$

From the present experimental uncertainty of $\pm 0.06\,\text{MHz}$ (Triebwasser
et al. 1953; Robiscoe and Shyn 1970) we find that

$$\left|\frac{b}{m}\right| \lesssim 4 \times 10^{-13}. \qquad (10.32)$$

Thus, to an extremely high degree of accuracy, we know that one
constant, the inertial mass of a particle, actually determines two of its

other properties—its coupling to gravitation, and the correlation length associated with its quantum fluctuations. It is intriguing to contemplate whether this could happen were all these phenomena not closely connected in some fundamental sense.

10.5. Stochastic mechanics in a background gravitational field

In this section I would like to indicate briefly what happens when we try to extend the stochastic formulation of quantum mechanics to the case of a free particle propagating in an arbitrary background gravitational field. What we would like to do can be described in the following: we seek a set of equations which determine the evolution of a probability density for a particle moving freely in an arbitrary background space–time which satisfies three conditions: (1) probability conservation is maintained; (2) when the space–time becomes flat and velocities become non-relativistic the equations become equivalent to Schrödinger's equation for an appropriate choice of diffusion constant; (3) when the diffusion constant is set to zero the equations describe an ensemble of non-interacting particles moving geodesically.

We find that this can be done, but at a cost. This is that we must explicitly specify a preferred frame with respect to which the motion is Brownian. This is necessary because, for a Brownian process, $|dx| \approx \sqrt{dt}$; for this to be true we have to have an invariant distinction between dx and dt. Thus the dependence of the quantization procedure on a choice of time is present in the stochastic formulation as well.

This means that we do not have a complete theory unless we can introduce some equations which allow the preferred frames to be determined in terms of the other dynamical variables of the theory. We will see that there is not an obvious way to do this for the single particle theory in an arbitrary background gravitational field. For this reason, and because what we are generalizing is a single particle equation, it is best to regard the following as a warm-up for an extension of a stochastic formulation of field theory to the case of a background gravitational field.

We begin by extending the formalism presented in Section 10.2 to describe Brownian motion in a background space–time (M, g). In order to do this we shall give a preferred global $3 + 1$ slicing, which will be specified by a unit time-like hypersurface-orthogonal vector field W^μ. We will define $q_{\mu\nu} = -g_{\mu\nu} + W_\mu W_\nu$ to be the three-metric on the slices orthogonal to W^μ. Coordinates in the surfaces will be denoted by x^i and $d\tau$ will refer to an interval of proper time in the rest frame of W^μ. We define a probability density $\rho(x)$ relative to the slicing such that $\sqrt{q(x)}\, \rho(x, t)$ (where $q = \det q_{ij}$) is the probability of finding the particle in a unit spatial volume around the point x^i on the slice labeled by t. We will require that ρ satisfy eqn (10.8) on each of these slices. Similarly, we can carry over the definition of the propagation kernel $P(x, t_2 \,|\, y, t_1)$ to this context, and it will satisfy, by definition eqns (10.9) and (10.10). We will also demand that the Smoluchowski eqn (10.11) is satisfied for any three slices orthogonal to W^μ. The ensemble of Brownian processes is defined by giving the forwards and backwards drift equations:

$$\mathrm{d}x^\mu = b^\mu(x, t)\frac{\mathrm{d}\tau}{b \cdot W} + \Delta x^\mu, \qquad \mathrm{d}\tau > 0 \tag{10.33}$$

$$\mathrm{d}^*x^\mu = -b^{*\mu}(x, t)\frac{\mathrm{d}\tau}{b^* \cdot W} + \Delta^*x^\mu, \qquad \mathrm{d}\tau < 0, \tag{10.34}$$

where $b \cdot W = b_\mu W^\mu$ and b^μ and $b^{*\mu}$, the forward and backwards drift velocities, have been defined so as to transform as four-vectors. The fluctuating terms now satisfy

$$\langle \Delta x^\mu \rangle = \langle \Delta^* x^\mu \rangle = 0 \tag{10.35}$$

$$\langle \Delta x^\mu \Delta x^\nu \rangle = -\langle \Delta^* x^\mu \Delta^* x^\nu \rangle = 2vq^{\mu\nu}\,\mathrm{d}\tau. \tag{10.36}$$

We see that we have Brownian motion only for the components of the trajectories which are orthogonal to W^μ. In addition, since the difference between b^μ and $b^{*\mu}$ is due only to the action of these Brownian fluctuations we shall assume that $W^\mu(b_\mu - b_\mu^*) = 0$. Using these relations we can derive a pair of relativistic Fokker–Planck equations,

$$\frac{\mathrm{d}\sqrt{g}\,\rho}{\mathrm{d}\tau} = -\sqrt{q}\,D_i\!\left(\frac{\rho b^i}{b \cdot W}\right) + v\sqrt{q}\,q^{ij}D_iD_j\,\frac{\rho}{b \cdot W} \tag{10.37}$$

$$\frac{\mathrm{d}\sqrt{q}\,\rho}{\mathrm{d}\tau} = -\sqrt{q}\,D_i\!\left(\frac{\rho b^{i*}}{b^* \cdot W}\right) - v\sqrt{q}\,q^{ij}D_iD_j\,\frac{\rho}{b^* \cdot W}, \tag{10.38}$$

where D_i is the covariant associated with the three-metric q_{ij}.

Taking the sum and difference of eqns (10.37) and (10.38), we find

$$\frac{\mathrm{d}\sqrt{q}\,\rho}{\mathrm{d}\tau} = -\sqrt{q}\,D_i\!\left(\rho\frac{v^i}{v \cdot W}\right) = -\sqrt{q}\,q^{\mu\lambda}\nabla_\lambda\!\left(\rho\frac{q_{\mu\nu}v^\nu}{v \cdot W}\right) \tag{10.39}$$

$$u^\mu = vq^{\mu\nu}\nabla_\nu LN\!\left(\frac{\rho}{v \cdot W}\right), \tag{10.40}$$

where we have defined $v^\mu = b^\mu + b^{*\mu}$ and $u^\mu = b^\mu - b^{*\mu}$.

These equations allow us to define a covariantly conserved probability current. Recalling that ρ was defined to be the probability density seen by an observer moving with W^μ, we write

$$\rho = \hat{\rho} v \cdot W, \tag{10.41}$$

where $\hat{\rho}$ is independent of W^μ. The reader may then verify that eqn

(10.39) can be rewritten as

$$\nabla_\mu(\hat\rho v^\mu) = 0. \tag{10.42}$$

We now want to extend Nelson's three conditions to this case. To begin with we shall define the diffusion constant as we did before by eqn (10.25). However, we should note that for a free particle this equation does not have the significance it had before, as m does not appear in the equation of motion.

We now want to go on to generalize the dynamical condition, expressed by eqns (10.24) and (10.26). In order to do this we need to define appropriate covariant generalizations of the stochastic time derivatives (10.16) and (10.17). It turns out that the correct definition, which gives the correct reduction to the Schrödinger equation in the non-relativistic limit, is to choose to take the forward stochastic derivative in terms of the proper time seen by b^μ and the backwards time derivative in terms of $b^{*\mu}$. This definition gives us $Dx^\mu = b^\mu$ and $D^*x^\mu = b^{*\mu}$, and in general gives

$$DF(x) = b^\mu\nabla_\mu F + vq^{\mu\nu}\nabla_\mu q^\alpha_\nu\nabla_\alpha F \tag{10.43}$$

$$D^*F(x) = b^{*\mu}\nabla_\mu F - vq^{\mu\nu}\nabla_\mu q^\alpha_\nu\nabla_\alpha F. \tag{10.44}$$

We can then posit a stochastic geodesic equation

$$\tfrac{1}{2}(D^*D + DD^*)x^\mu = 0. \tag{10.45}$$

Working this out, we find that

$$v^\nu\nabla_\nu v^\mu - u^\nu\nabla_\nu u^\mu + vq^{\alpha\beta}\nabla_\alpha q^\gamma_\beta\nabla_\gamma u^\mu = 0. \tag{10.46}$$

If we plug in eqn (10.40) and set $v = h/m$ we find an evolution equation for v^μ:

$$v^\alpha\nabla_\alpha v^\mu = \frac{\hbar^2}{m^2}q^{\alpha\beta}[(\nabla_\beta LN\hat\rho)[\nabla_\alpha q^{\mu\nu}\nabla_\nu LN\hat\rho] - \nabla_\alpha q^\gamma_\beta\nabla_\gamma q^{\mu\nu}\nabla_\nu LN\hat\rho]. \tag{10.47}$$

This equation and eqn (10.42) together determine the evolution of $\hat\rho$ and v^μ, given a choice of initial data on a three-surface orthogonal to W^μ. As a check on the derivation we may ask what these equations give in the limit of weak fields and low velocities. In this approximation we may choose co-ordinates so that the metric has the form $g_{\mu\nu} = \text{diag}(1 + 2\Phi/c^2, -1, -1, -1)$. In addition we choose W_μ to be the unit time-like vector orthogonal to flat three-space. In the non-relativistic limit one can then choose $v_i = D_iS/m$. Making these choices one then finds that eqns (10.42) and (10.47) are equivalent to

$$i\hbar\frac{\partial\psi}{\partial t} = -\frac{\hbar^2}{2m}D^2\psi + m\Phi\psi, \tag{10.48}$$

where $\psi = \sqrt{\bar{\rho}} \exp iS/h$.† Thus, we have achieved our goal, which is the formulation of a theory (albeit one dependent on a preferred choice of slicing) which reduces to ordinary quantum mechanics in the presence of a Newtonian gravitational field in the weak field non-relativistic limit, and which reduces to a description of ordinary geodesic motion when Planck's constant is set to zero.

One thing we might like to know is whether, in the general case, eqn (10.47) reduces to an equation describing the evolution of a phase S so that, as in the flat space case this equation, together with eqn (10.42), describes the evolution of a complex wave function $\psi = \sqrt{\bar{\rho}} \exp iS/h$. That this is not the case can be seen as follows. If it were, then at least locally we could choose an S such that $v_\mu = \nabla_\mu S$. In this case the left-hand side of eqn (10.47) becomes $\frac{1}{2}\nabla_\mu[(\nabla_\nu S)(\nabla^\nu S)]$, and thus has vanishing curl. However, as the reader may verify, the curl of the right-hand side of eqn (10.47) does not, in general, vanish, unless $R^\lambda_{\mu\nu\sigma} = \nabla_\mu W^\alpha = 0$. Thus, in general, v^μ cannot be chosen orthogonal to any surfaces of constant phase, and the evolution described by eqns (10.42) and (10.47) cannot be reduced to the evolution of a complex wave function. However, as the evolution they generate conserves probability and is well defined they may be considered to be a sensible generalization of the Schrödinger equation to the case of a general background field.

One unsatisfactory feature which these equations possess is that the initial value problem is only well defined for data given on three surfaces orthogonal to W^μ. This is not surprising, as eqn (10.47) is a generalization of the non-relativistic Schrödinger equation, and even in the absence of gravity the initial value problem for the Schrödinger equation is only well defined in terms of a preferred time variable. This again underscores the fact that the present theory is only a warm-up exercise before looking for an extension of stochastic field theory to background gravitational fields.

One thing we would like to try is to see if there is a simple way to determine the preferred frame W^μ in terms of the variables of the theory. For example, the simplest possibility would be if we could determine W^μ by a condition that the particle motion is always Brownian in that frame in which its probability distribution is at rest. This would be possible if we could consistently set $v^\nu = fW^\mu$, where f is some function. This is, however, not in general possible, because, as W^μ has been assumed to be hypersurface orthogonal, in order to make this choice we need to be able to guarantee that if v^μ is chosen to be hypersurface orthogonal its

---

† This formulation then differs from one previously given by the author (Gravity Research Foundation essay, 1984 and EFI preprint) in which a small nonlinear term coupling $\dot{\Phi}/c^2$ to S was found. That formulation differed from the present one in that the $b \cdot W$ factors were missing from the stochastic differential eqns (10.33) and (10.34).

evolution according to eqn (10.47) must preserve the hypersurface orthogonality. However, as we have just shown, this is not in general the case.

We might try to drop the condition that W^μ be itself hypersurface orthogonal. However, in this case it is not clear whether there are in general any choices of surfaces for which eqns (10.42) and (10.47) have a well defined initial value formulation. Thus it seems that in this simple case there is no way to eliminate the need to give W^μ separately to define the theory in an arbitrary background gravitational field. One may hope that in a stochastic formulation of field theory it will be possible to determine the preferred frames dynamically in terms of the other variables of the theory.

However, in spite of these weaknesses, we may ask what we have learned from this formulation that may still be true in an extension to field theory. The most important thing is that while the equations define sensible time-evolution for v^μ and $\hat{\rho}$ in the presence of an arbitrary background field, consistent with probability conservation, this evolution can only be described by a linear equation on a linear space of states in the weak field non-relativistic limit. In the more general case, in which v^μ cannot be assumed to be hypersurface orthogonal, the equations are nonlinear and do not reduce to a single complex equation. Thus, in spite of the fact that probability conservation is maintained, the superposition principle does not apply in these more general circumstances. It seems likely that this will also be a feature of the stochastic field theory.

10.6. A non-local hidden variable theory

In the previous section we have seen that the equivalence between the stochastic formulation of quantum mechanics and the Hilbert space formulation of quantum mechanics may break down in the presence of gravitational fields. As the stochastic formulation seems to remain well defined this is good reason to try to take it seriously as a possible vehicle for extending quantum physics to domains in which gravity is important. At the same time the stochastic formulation has a few peculiar features, which, if we are to take it seriously as a possible fundamental theory, we would like to understand better. Chief among them is the high degree of correlation which must hold between the individual trajectories which comprise any ensemble which satisfies Nelson's three conditions, as we discussed above. Two things we would like to know are whether there in fact exist any Brownian ensembles of trajectory which satisfy Nelson's conditions, and if so, how to construct some examples of them.

It is clear that to answer this last question we should essentially have to construct a hidden variable theory, because we would be explicitly constructing an ensemble of classical trajectories (albeit not differentiable) which have the property that the evolution of their probability density and probability current density are governed by Schrödinger's equation. This is one important motivation for seeking to construct an example of a hidden variable theory.

Strictly speaking, it is not necessary to believe that there exists an underlying dynamics to make use of either the statistical interpretation or the stochastic formulation of quantum mechanics. However, as soon as one believes that these may contain more truth than the usual interpretation or the usual formulation, then it is hard to avoid curiosity about whether it is possible to make some description of an individual quantum system which is more complete than that which can be adduced from the wave function. Similarly, we want to know what sort of dynamics could give rise to ensembles which satisfy Nelson's conditions, and hence be responsible for the unusual properties exhibited by quantum fluctuations. Thus, if one is willing to entertain the possibility that there might be some truth in the kinds of arguments I discussed in Section 10.1, then it seems that one has at least to entertain the notion that quantum physics and gravity are tied up in a fundamental way, not at the level of quantum mechanics, but at the level where quantum mechanics is itself explained in terms of a more elementary theory.

Before proceeding, I would like to mention briefly some additional reasons which might motivate us to look for hidden variable theories. First, and most obviously, if quantum mechanics can be explained by a hidden variable theory then many of the puzzles concerning quantum mechanics can be resolved without doing violence to our common sense notions of realism, ontology and epistemology. For example, the measurement problem evaporates because quantum states simply describe ordinary statistical ensembles of systems.

Secondly, I believe that the history of physics shows that progress in physics is more often achieved through developments of physics rather than through developments in philosophy. If we have the possibility of dissolving the philosophically troubling implications of quantum mechanics by inventing new theories without these difficulties, then this seems to me to be preferable to struggling with radical solutions to the basic philosophical questions which are motivated by the fact that they make quantum mechanics more rational. Furthermore, once we challenge ourselves with the task of inventing a new theory which *explains* the peculiar features of the quantum world then we find ourselves engaged in trying to ask new questions about physics and invent new hypotheses about nature to answer them, and this is probably what, as physicists, we are more likely to succeed at doing. The philosophers have plenty of time to mull over the peculiarities of quantum mechanics, if they turn out to be unavoidable.

Now, we will pay a price for sticking with our common sense ideas about reality and our relation to it, and this is that physics will have to become explicitly non-local. However, the experimental evidence (Chapter 1; Freedman and Clauser (1972); Clauser (1976); Fry and Thomson (1976); Aspect *et al.* (1981, 1982*a*,*b*)) against the Bell inequalities (Bell 1965) is now so strong that we can state confidently, and independently of any specific theory, that nature is full of non-locality. All around us are occurring space-like separated events whose joint probability distributions, were we to measure them, would not factor into products of local probabilities. Thus, we have a choice only between non-locality hidden in the linear structure of the Hilbert spaces which describe many particle systems and non-locality explicitly built into the dynamics of a hidden variable theory. To go from the former to the latter seems a cheap price to pay to avoid having to alter radically the basic tenets of classical realism.

Of course, I do not hope, or wish, to convince those who have found an interpretation of quantum mechanics which they find satisfying, such as the adherents of the many-worlds interpretation. At this stage of ignorance it is undoubtably for the best if our different feelings impel us in different directions; I am merely stating mine. In any case, all of the above means nothing if the attempt to construct hidden variable theories does not lead to theories which are compelling, beautiful, and surprising on their own merits. Nothing remotely of this sort presently exists. However, what does exist are some examples of explicitly non-local hidden variable theories which show that, at least, theories of this type do exist. Furthermore these theories answer the questions about stochastic mechanics we mentioned above by giving explicit constructions of ensembles which, at least to a certain degree of approximation, satisfy Nelson's three conditions.

I would like briefly to describe the particular motivations which led to the construction of these theories. First of all, Bell's theorem seems to indicate the possibility that non-local correlations may be found between any pair of systems which have previously interacted. This indicates that, in order to explain these correlations, a hidden variable theory will have to have at least one hidden variable for every pair of particles that potentially could interact. This leads to the question of whether a hidden variable theory could be made by inventing new dynamical variables which are attributed to pairs of particles, in addition to the usual dynamical variables which describe properties of individual particles. The simplest possibility which might be tried is then to represent each dynamical variable of an N-particle system by an $N \times N$ matrix in which the diagonal elements are attributed to individual particles, while the off-diagonal elements are attributed to pairs of particles.

Let us suppose that we do assign a dynamical variable to every pair of particles, and that the dynamics will involve both these non-local variables and the local dynamical variables associated with individual particles. Then it follows that any local description of the physics must be statistical, even if the basic dynamics is deterministic. Even if we could invent experiment arrangements which controlled the variables associated with pairs of particles, each of which are nearby, the motion of the particles in our neighbourhood will be influenced by those variables assigned to pairs containing a nearby particle and a far away particle, and there is no way these variables could possibly be controlled in local experiments. These variables will have to be described probabilistically and, as there are many more particles which are far away than there are which are nearby, they will most likely swamp any contributions coming from the pairs of particles, both of which are local.

It is then intriguing to wonder if the statistical fluctuations seen in the local variables of quantum systems are due precisely to their couplings to these non-local variables. Perhaps the usual quantum mechanics, in which one has only local dynamical variables, results from doing ordinary statistical mechanics on the non-local variables, where all of the variables, local and non-local are coupled through some deterministic dynamics. Such a theory would be globally deterministic, but locally statistical.

There is, however, a difficulty with this that must be confronted. If there are many more non-local variables in which a particular particle is paired with particles far away than there are in which it is paired with nearby particles, then there are also many more non-local variables than there are local variables. If the whole system is treated according to statistical mechanics, then what is to prevent the system from reaching an equilibrium behaviour in which, by the equipartition theorem, the non-local variables completely dominate the energetics of the system such that the local physics is completely swamped. The system must be very special so that the non-local physics contributes what is, on a cosmological scale, a small perturbation to the local physics, rather than the reverse.

One way in which this problem might be resolved is the following. Suppose the non-local theory had the property that in the thermodynamic limit in which N, the number of particles, is taken to infinity, the physics became local and classical. However, suppose also that N, while very large, is finite. Then there will be statistical fluctuations in the evolution of these classical variables. These fluctuations will occur on scales which are of order of $1/\sqrt{N}$ the dimensions of the whole system. Now, it is interesting to note that the ratio of the Compton wavelength of a stable elementary particle (which determines the rough scale of quantum fluctuations) to the radius of the universe is approximately 10^{-40}, which is, roughly, $1/\sqrt{N}$, where N is the number of stable particles observed in the universe.

This relation, which is related to Dirac's large number hypothesis (Dirac 1937, 1938), may be only a coincidence, or it may be an important clue as to why there are quantum fluctuations. In any case it gives a clue as to how to arrange things in a non-local hidden variable theory so that, even though there are many particles, the non-local physics produces fluctuations in the local physics rather than swamping it altogether. In the example to be discussed below this is achieved by scaling the non-local variables by appropriate powers of $1/N$ so that in the thermodynamic limit their effects disappear. However, one might imagine that in some real theory this was a deep and important property of the limit. More ambitiously, we might imagine that the underlying theory is completely non-local in that it makes no reference to local properties in space–time, but that the description of local physics in a space–time of small dimension emerged in the thermodynamic limit in a way analogous to that in which thermodynamics emerges from the classical mechanics of systems with large numbers of degrees of freedom.

I shall now describe a class of hidden variable theories constructed to exhibit these ideas. I shall do this by listing the assumptions of the theory and then stating a theorem which gives a correspondence between systems satisfying these assumptions and systems satisfying Schrödinger's equation. There is not space to present the details of the proof, but this is given elsewhere (Smolin 1983).

The system to be described is a non-relativistic N particle system in two spatial dimensions where, as should be apparent from the above, N is large. I will discuss several extensions to other systems in the last section.

Our system will be described by an $N \times N$ complex matrix, $M_{ab}(t)$. The positions of the N particles are given in the complex plane by its eigenvalues $\lambda_a = x_a + ib_a$. We make the following hypotheses as to the evolution of this sytem.

(A) *Scaling hypothesis.* If we decompose M_{ab} into a diagonal and off-diagonal part:

$$M_{ab} = \delta_{ab}D_a + N_{ab}; \qquad N_{aa} = 0, \tag{10.49}$$

and

$$\frac{\overline{N_{ab}}}{\overline{D_a}} \approx \frac{\overline{\overline{N_{ab}}}}{\overline{\overline{D_a - D_b}}} \approx \frac{1}{N^q}, \tag{10.50}$$

where the double bar denotes here an average over the magnitudes of all of the elements. (A single bar will denote the complex conjugate in what follows.) q will be chosen so that the dynamics becomes purely local in the limit $N \to \infty$. In addition, this choice will permit us to make a perturbation expansion for the positions of the eigenvalues, λ_a, in terms of powers of $1/N$. To second order we have,

$$\lambda_a = D_a + \sum_{b \neq a} \frac{N_{ab} N_{ba}}{D_a - D_b} . \tag{10.51}$$

The second term describes a random walk in the complex plane of $N - 1$ steps. The sum then contributes a term of order $1/N^{2q-1/2}$, from which we infer that in order for classical mechanics to be restored in the $N \to \infty$ limit we must have $q > \frac{1}{4}$.

(B) *Rules about the off diagonal elements.* Each of the N_{ab} is in one of three states:
(1) Off: $N_{ab} = 0$.
(2) Dormant: $N_{ab} \neq 0$, but with some fixed or slowly varying value.
(3) Dynamical-varying on some short dynamical time scale.
For example, we might construct a model in which the dormant N_{ab} are at random points in a region of the complex plane while the dynamical N_{ab} move around in the region. There are two additional rules these variables must obey:
Basic rule: For a fixed a and b, only one of N_{ab} and N_{ba} may be dynamical at any one time.
Sufficiency rule: For a given a, let n_a be the number of N_{ab} which are not off. Then for all a, $n_a \gg 1$.

(C) *Underlying deterministic dynamics*: We assume that the evolution of the matrix D_{ab} is governed by a classical action principle of the form

$$L_M = \sum_a m_a \dot{D}_a \dot{D}_a + \tfrac{1}{2} m_N \sum_{a \neq b} \dot{N}_{ab} \dot{N}_{ab} - F(N_{ab}) - V(\lambda_a). \tag{10.52}$$

Here, $F(N_{ab})$ is some potential energy function which describes interactions between the elements of N_{ab}, as well as any forces needed to constraint them to regions of the appropriate size in the complex plane. There are also contributions to the potential energy which are functions only of the eigenvalues λ_a. These may be of the form

$$V(\lambda_a) = \sum_{a > b} V_2(|\lambda_a - \lambda_b|) + \sum_a V_{\text{external}}(\lambda_a), \tag{10.53}$$

where the two terms describe, respectively, two-body forces between the particles and the effects of external potentials on the particles. The m_a are just the masses of the particles, as can be seen by taking the $N \to \infty$ limit in which the N_{ab} go away, and we assume for simplicity that all of the off-diagonal elements have the same 'mass' m_N.

(D) *Mass fudge rule.* We assume that, for each a and b,

$$\frac{n_a}{n_b} = \frac{m_b}{m_a}.$$
(10.54)

This is motivated strictly by the fact that it guarantees Nelson's condition (3), eqn (10.25). It is in order to accommodate particles with different masses that some of the N_{ab} must be off.

(E) *Brownian character of the motion of the dynamical N_{ab}.* We assume that the dynamics and initial conditions are chosen such that an appropriate timescale Δt exists such that the motion of each dynamical N_{ab}, coarse grained over Δt, satisfies the conditions of Brownian motion eqns (10.1)–(10.6). For example, this can be achieved if there are short range elastic forces between the N_{ab} and if Δt is longer than the mean free time between collisions. The diffusion constant v of the dynamical N_{ab} is then computable in terms of the initial conditions and the parameters of the theory. The parameters must also be arranged so that Δt is shorter than $\hbar/E$, where E is the energy of the quantum system defined by the correspondence theorem below.

Another example of a system that satisfies this condition is, a variant of the model the Ehrenfests used in their explanation of Boltzmann's work (Ehrenfest and Ehrenfest 1959). We have two kinds of particles in a region of the complex plane. The dormant N_{ab} are fixed randomly at sites in the region. The dynamical N_{ab} move through the region with fixed velocity v, colliding elastically with the dormant N_{ab} and with the walls. The diffusion constant is $v = lv$, where l is the mean free path.

(F) *Assumptions concerning the statistical distributions of the N_{ab}.* We need to make some assumptions concerning how the ensemble of N_{ab} are distributed. These will be made in terms of the probability distribution which describes the ensemble, $\rho(N_{ab})$, which is defined so that it satisfies eqn (10.8). In order to ensure that the theory is symmetric under rotations in the complex plane we will assume that

$$\int dN_{ab}\rho(N_{ab})N_{ab} = 0$$
(10.55)

for all N_{ab}. We also need to assume that, for given a and b, the distributions of N_{ab} and N_{ba} are uncorrelated. Thus, we will require that

$$\int dN_{ab}\rho(N_{ab})N_{ab}N_{ba} = 0. \tag{10.56}$$

We need put no restrictions on the higher moments, except those concerning the average magnitudes of the N_{ab} from condition (A). For example,

$$\int dN_{ab}\rho(N_{ab})|N_{ab}|^2 \approx \frac{1}{N^{2q}}(D_a - D_b). \tag{10.57}$$

(G) *More scaling rules.* The motion of the dynamical N_{ab} may then be described by the formalism of stochastic processes described in Section 10.2. However, we can show that the coupling of the D_a to the fluctuations of the N_{ab}, while non-vanishing, through the terms (10.53), may be neglected for the present purposes. Thus, the D_a will be described by differentiable trajectories. We, however, do need to require that relations similar to eqn (10.50) hold between the velocities and accelerations of the D_a and the N_{ab}:

$$|V_{ab}| < \frac{1}{N^q}|D|\overline{\left|\frac{\partial^2 D}{\partial t^2}\right|\left|\frac{\partial D}{\partial t}\right|^{-1}} \tag{10.58}$$

$$\overline{D^*B_{ab} + DB_{ab}^*} < \frac{1}{N^q}\overline{\left|\frac{\partial^2 D}{\partial t^2}\right|}. \tag{10.59}$$

We are now ready to state a *correspondence theorem*:

Let $M(t)_{ab} = \delta_{ab}D(t)_a + N(t)_{ab}$ be an $N \times N$ complex matrix evolving in time such that conditions (A) to (D) and (G) are satisfied, and let us be given a statistical ensemble of such matrices, which satisfies conditions (E) and (F). Then,

(1) The eigenvalues of the matrix M_{ab}, λ_a, undergo Brownian motion (a special case of a theorem of Dyson (1962)), such that the diffusion constant for each λ_a, C_a, satisfies,

$$C_a m_a = \hbar + O(1/\sqrt{N}). \tag{10.60}$$

This *defines* a new constant $\hbar$ which is computable as a function of the diffusion constant for the off-diagonal elements, v;

(2) There exists a real function, $S(\lambda, \bar{\lambda})$, such that, through first order in $1/N^{2q-1/2}$,

$$m_a v_{a,x} = \frac{\partial S}{\partial x_a} \tag{10.61}$$

$$m_a v_{a,y} = \frac{\partial S}{\partial y_a}, \tag{10.62}$$

where $\lambda_a = x_a + iy_a$ and $v_a = v_{a,x} + iv_{a,y}$, with v_a the current velocity for λ_a.

(3) If we form the complex function

$$\psi(x_a, y_a) = \sqrt{\rho(x_a, y_a)}e^{iS/\hbar} \tag{10.63}$$

with $\rho(x_a, y_a)$ the joint probability distribution function for the λ_a, then ψ satisfies the equation

$$i\hbar\frac{\partial \psi}{\partial t} = -\sum_a \frac{\hbar^2}{2m_a}\left(\frac{\partial^2}{\partial x_a^2} + \frac{\partial^2}{\partial y_a^2}\right)\psi + \sum_{a>b} V(|\lambda_a - \lambda_b|)\psi$$

$$+ O(1/N^{2q-1/2}) + O(1/N^{1/2}). \tag{10.64}$$

We may note that the errors come both from the neglect of terms in the velocities and accelerations of the eigenvalues which are smaller by powers of $1/N^{2q-1/2}$ than the dominant terms coming from the interparticle forces and from the fluctuations in the value of Planck's constant from eqn (10.60).

While we do not have space here to give the proof of this theorem we may say one word about the method. The basic equations, which follow from our assumptions, relate expectation values of functions of the λ_a to expectation values of functions of the D_a and N_{ab}. For example, the probability distributions are related by

$$\rho_\lambda(\lambda) = \int (dN_{ab})(dD_a)\rho(N)\rho_D(D)\Pi_a\delta\left(\lambda_a - D_a - \sum_{b \neq a}\frac{N_{ab}N_{ba}}{D_a - D_b} + \ldots\right), \tag{10.65}$$

where ρ_λ and ρ_D, are, respectively, the probability distributions for λ and D. Similarly, given a quantity $F(D, N)$ as a function of D_a and N_{ab} we may find the expectation value of that quantity as a function of λ by

$$F(\lambda) = \int (dN_{ab})(dD_a)F(D, N)\rho(N)\rho_D(D)\Pi_a\delta$$

$$\times \left(\lambda_a - D_a - \sum_{b=a}\frac{N_{ab}N_{ba}}{D_a - D_b} + \ldots\right) \bigg/ \rho_\lambda(\lambda). \tag{10.66}$$

Using these relations one can show that the conditions (A) to (G) guarantee that Nelson's three conditions are satisfied to the orders indicated in the theorem.

I would like to close by mentioning, without details, three extensions of this result which are presently being pursued.

(1) An extension to a lattice quantum field theory for a complex scalar field in any number of space dimensions in which each eigenvalue λ_a gives the value of the field on one site of the lattice. In this case one discovers that the continuum limit is even more tricky than in the usual lattice

quantum field theory, because one must scale the parameters of hidden variable theory carefully with the lattice spacing so that h is finite in the limit. This is also interesting because one finds that, if the construction succeeds, Lorentz invariance will be recovered in the continuum limit in spite of the presence of non-local hidden variables defined with respect to some particular frame.

(2) An extension to a relativistic particle model in $3 + 1$ dimensions based on replacing the complex elements of M_{ab} by elements of the Hermitian quaternions. One can then try to put interactions in by adding to the action Wheeler–Feynman type terms. Singularities are then encountered because division by Hermitian quaternions corresponding to light-like intervals is not defined. These may be the usual singularities of quantum field theory in a new guise.

(3) A kind of a 'Machian' version of the theory presented here in which the off-diagonal elements are all constant and the universal stochastic fluctuations necessary to derive quantum mechanics arise from the transmittal, through the non-local terms of eqn (10.66), of thermal fluctuations due to the fact that some of the particles may be in hot regions (for example in stars.) One then finds that Planck's constant is only non-zero when some parts of the system are hot. While almost certainly wrong, this is an amusing idea to contemplate!

Acknowledgements

I would like to thank Lee Lindblom and Bob Wald for discussions during the course of the work described in Section 10.5. This work was supported in part by a grant, PHY 80-26043, from the National Science Foundation.

References

Ashtekar, A. and Geroch, R. (1974). *Rep. Prog. Phys.* **37**, 1211.
—— and Magnon-Ashtekar, C. R. (1955). *C. R. Acad. Sci., Paris* **286**, 875.
Aspect, A., Dalibard, J., and Roger, G. (1982*a*). *Phys. Rev. Lett.* **49**, 1804.
—— Grangier, P., and Roger, G. (1981). *Phys. Rev. Lett.* **47**, 460.
—— —— —— (1982*b*). *Phys. Rev. Lett.* **49**, 91.
Ballentine, L. E. (1970). *Rev. Mod. Phys.* **42**, 358.
Bell, J. S. (1965). *Physics* **1**, 195.
Bohm, D. (1952). *Phys. Rev.* **85**, 166, 180.
Clauser, J. F. (1976). *Phys. Rev. Lett.* **36**, 1223.
Davidson, M. (1981). *J. Math. Phys.* **22**, 2588.
DeWitt, B. S. (1975). *Phys. Rep.* **19C**, 297.
Dell, J. (1985). Yale University preprint.
Dirac, P. A. M. (1937). *Nature* **139**, 323.
—— (1938). *Proc. R. Soc. Lond.* A **165**, 199.

Dyson, F. J. (1962). *J. Math. Phys.* **3,** 1191.

Ehrenfest, P. and Ehrenfert, T. (1959). *The conceptual foundations of the statistical approach in mechanics.* Cornell University Press, Ithaca.

Einstein, A. (1949). In *Albert Einstein: Philosopher scientist* (ed. P. A. Schilpp). Library of Living Philosophers, Evanston, Ill.

Fényes, I. (1952). *Z. Phys.* **132,** 81.

Freedman, S. J. and Clauser, J. F. (1976). *Phys. Rev. Lett.* **36,** 1223.

Fry, E. S. and Thomson, R. C. (1976). *Phys. Rev. Lett.* **37,** 465.

Fulling, S. A. (1973). *Phys. Rev.* D **7,** 2850.

Garfinkle, D. and Wald, R. (1985). *Gen. Rel. Grav.* **17,** 461.

Guerra, F. and Loffredo, M. I. (1980). *Lett. Nuovo Cimento* **27,** 41.

Hawking, S. W. (1976). *Phys. Rev.* D **14,** 2460.

—— (1982). *Comm. Math. Phys.* **87,** 395.

—— (1984). *Nucl. Phys.* B **244,** 135.

Kay, B. (1977). Ph.D. Thesis, University of London.

—— (1978). *Comm. Math. Phys.* **62,** 55.

Kuchař, K. (1982). In *Quantum gravity II* (ed. C. J. Isham, R. Penrose, and D. W. Sciama). Oxford University Press, Oxford.

Nelson, E. (1966). *Phys. Rev.* **150,** 1079.

—— (1967). *Dynamical theories of Brownian motion.* Princeton University Press, Princeton.

—— (1979). Connection between Brownian motion and quantum mechanics. In *Lecture notes in physics*, vol 100. Springer, Berlin.

Robiscoe, R. and Shyn, T. (1970). *Phys. Rev. Lett.* **24,** 559.

Sciama, D. W. (1979). Thermal and quantum fluctuations in special and general relativity: an Einstein synthesis. In *Relativity, quanta and cosmology in the development of the scientific thought of Albert Einstein,* (ed. M. Pantaleo and F. de Finis). Johnson Reprint Corporation, New York.

——, Candlas, P., and Deutsch, D. (1981). *Adv. Phys.* **30,** 327.

Seiler, E. (1984). Stochastic qualization and gauge fixing in gauge theories. *Max Planck Institut preprint MPI-PAE/PTL 20/84.*

Smolin, L. (1982). On the nature of quantum fluctuations and their relation to gravitation and the principle of inertia. IAS preprint.

—— (1983). IAS preprint.

—— (1984*a*). *Gen. Rel. Grav.* **16,** 205.

—— (1984*b*). *Gen. Rel. Grav.* **17,** 417.

Triebwasser, S., Dayhoff, E. S., and Lamb, W. E., Jr. (1953). *Phys. Rev.* **89,** 98.

Unruh, W. G. (1976). *Phys. Rev.* D **14,** 870.

Wax, N. (ed.) (1954). *Selected papers in noise and stochastic processes.* Dover, New York.

11

Entropy, uncertainty, and nonlinearity

L. P. Hughston

11.1. Introduction

I propose that we should take a Bayesian attitude towards the wave function in quantum mechanics. In probabilistic terms this means, roughly speaking, that we are faced with the task of formulating 'best bets with the information given'. When we have exact knowledge of the Hamiltonian and exact knowledge of the relevant initial conditions and boundary conditions, then nothing special is gained in the adoption of the Bayesian stance. But suppose there is uncertainty—suppose we only have partial knowledge of the conditions under which a solution is sought. Then, providing we have suitable means at our disposal for the assignment of sensible *a priori* probabilities to the unknown data, we may proceed with Bayesian methods to construct an exact wave function for the system under consideration—only now we are obliged to interpret the wave function in a slightly more general way as providing us with information about the 'best bet' as to the particle's behaviour, rather than determining in any absolute sense an exact statistical distribution for an ensemble.

Needless to say in most realistic circumstances we do indeed lack complete information in advance about detail of the Hamiltonian, initial conditions, and so on; therefore we must enquire how in principle to formulate quantum mechanics under such conditions. Let us suppose $\psi(x, t)$ is the wave function for a particle in non-relativistic quantum theory. Then the position of the particle at time t can be regarded as a random variable $X(t)$ with probability density function $f(x, t) = \psi\bar{\psi}$. If the Hamiltonian, etc., are known, the Schrödinger equation determines $\psi(x, t)$, and hence $f(x, t)$, completely. On the other hand, if we lack knowledge about the relevant initial conditions or boundary conditions, the Schrödinger equation will not fix $f(x, t)$ uniquely; and if we do not know the Hamiltonian exactly we cannot even write down the Schrödinger equation, let alone solve it.

Thus we are faced with an example of the classical problem of the determination of *a priori* probabilities. What is the 'most reasonable' or 'least biased' choice for the distribution $f(x, t)$ when we are presented with data which are in themselves insufficient to determine $f(x, t)$ directly via the Schrödinger equation?

In what follows it is shown that the problem of *a priori* probabilities in quantum theory can be approached, with a degree of success, with the techniques of information theory; in particular, by use of the 'principle of maximum entropy' (Jaynes 1957; Rosenkrantz 1983), which can be regarded in some respects as a refinement of the Laplacian 'principle of indifference' (see, e.g., Keynes (1929), Ch. IV). The essential idea is as follows. If data insufficient to determine $f(x, t)$ are specified, then we supplement these data with the requirement that the *entropy* of the distribution be maximized, subject to the constraint of the data already given. In favourable circumstances this suffices to determine $f(x, t)$ uniquely, via the Schrödinger equation, thereby providing the 'least biased' solution to the problem.

11.2. Configuration entropy

Suppose the space under consideration is R^n, and let ξ be an arbitrary parameter with dimensions of length to the power $-n$. Then the *entropy of configuration* $S_\xi(t)$ associated with the random variable $X(t)$ is defined by

$$S_\xi(t) = -\int f \ln(f/\xi) \, d\tau, \qquad (11.1)$$

where $f(x, t)$ is the probability density function for $X(t)$, and $d\tau$ is the volume element on R^n.

Note that under a change of the parameter ξ we have $S_\xi - S_\eta = \ln(\xi/\eta)$. Thus the actual value of an entropy S_ξ is not necessarily meaningful, unless some particular significance can be attached to the given value of ξ; but the *difference* between two entropies, e.g. $S_\xi(t_1) - S_\xi(t_2)$ has an 'absolute' meaning, and can be thought of as a measure of the 'relative information content' of the two distributions under consideration: a gain in entropy corresponds to a loss in information. Roughly speaking, the more evenly spread out a wave function is, the higher is its entropy of configuration.

The entropy associated with other random variables can be defined similarly. Suppose, for example, we write

$$\phi(p, t) = (2\pi)^{-n/2} \int e^{-i\boldsymbol{p} \cdot \boldsymbol{x}} \psi(x, t) \, d\tau \qquad (11.2)$$

for the Fourier transform of the wave function at time t. Then $\tilde{f}(p, t) = \phi\bar{\phi}$ is the probability density function for the random variable $P(t)$, and the associated *entropy of momentum* at time t is

$$\tilde{S}_\xi(t) = -\int \tilde{f} \ln(\tilde{f}\tilde{\xi}) \, d\tilde{\tau}, \qquad (11.3)$$

where $d\tilde{\tau}$ is the volume element on momentum space. Since $\tilde{S}_\xi(t)$ is a measure of the spread of the distribution of $P(t)$, one might very well expect the existence of an inequality relating S_ξ and $\tilde{S}_\xi$. Such an inequality was conjectured by Everett in an extended version of his 1957 thesis (see Everett 1973), and independently by Hirschman (1957):

$$S_\xi + \tilde{S}_\xi \geq n(1 + \ln \pi). \qquad (11.4)$$

The proof of this relation was supplied by Bialynicki-Birula and Mycielski (1975a,b) and independently by Beckner (1975a,b); various generalizations have subsequently been noted by Deutsch (1983), Partovi (1983), and Bialynicki-Birula (1984). It will be observed that the inequality holds independently of the value of ξ. Moreover, inequality (11.4) is in a sense 'stronger' than the Heisenberg inequality,

$$V(X)V(P) \geq 1. \qquad (11.5)$$

where $V(X)$ denotes the variance of X. Indeed, as by Bialynicki-Birula and Mycielski (1975a), we have

$$\frac{2}{n} \xi^{2/n} V(P) \geq (e\pi)^{-1} \exp(2\tilde{S}/n) \geq e\pi \exp(-2S/n) \geq \frac{n}{2} \xi^{2/n} (V(X))^{-1}, \qquad (11.6)$$

from which both inequalities (11.4) and (11.5) follow.

Now, suppose we are presented with a problem for which the appropriate initial data would consist of the specification of the density function $f(x, t)$ at time $t = 0$. Suppose, moreover, that we are not told $f(x, 0)$, but rather are merely given the moments $E(X^r)$ at $t = 0$ for, say, several values of r. What does one do under such circumstances?

We must make the most 'reasonable' or, equivalently, the 'least biased' choice for $f(x, 0)$ possible: therefore we demand that the 'information content' of the distribution of the random variable X be minimized at $t = 0$, subject to the constraints of the moments given.

Suppose for example we are given the first two moments, i.e. $E[X]$ and $V[X] = E[X^2] - (E[X])^2$. Then, for some value of ξ, we demand that

$$\delta \int [-f \ln(f/\xi) + \lambda xf + \mu(x - \alpha)^2 f + vf] \, d\tau = 0, \qquad (11.7)$$

where $\lambda, \mu,$ and v are Lagrange multipliers, to be determined by the constraints

$$\int xf \, d\tau = \alpha, \qquad \int (x - \alpha)^2 f \, d\tau = \sigma^2, \qquad \int f \, d\tau = 1. \qquad (11.8)$$

Carrying out the variation in eqn (11.7) and imposing the constraints (11.8) we obtain

$$f(x) = \frac{1}{\sqrt{2\pi}\,\sigma} \exp[-(x - \alpha)^2/2\sigma^2], \qquad (11.9)$$

where $\sigma^2 = V[X]$. Thus, perhaps not too surprisingly, the maximum entropy distribution for given expectation and variance is $N(\alpha, \sigma^2)$. Though but little information has been specified, nevertheless $f(x)$ is determined. Accordingly to our Bayesian point of view, $f(x)$ does *not* in this circumstance necessarily represent an 'approximation' to some unknown 'true' distribution: rather, it *is* the true distribution, in so far as such a notion makes sense at all. (For further discussion of the maximum entropy approach to moment problems see Mead and Papanicolaou (1984)).

11.3. The problem of the undetermined potential

Let us consider a dynamical situation governed by the Schrödinger equation for a single particle:

$$i\frac{\partial}{\partial t}\psi(x, t) = -\frac{1}{2m}\Delta\psi(x, t) + U(x, t)\psi(x, t). \qquad (11.10)$$

Imagine a circumstance where the potential $U(x, t)$ is not known exactly: all we are told is, say, its expectation value $\omega(t)$, defined by

$$\omega(t) = \int U(x, t)f(x, t) \, d\tau, \qquad (11.11)$$

where $f(x, t) = \psi(x, t)\bar{\psi}(x, t)$. Thus we regard the potential as a function $U(X, t)$ of the random variable $X(t)$, and we are given its expectation

$$\omega(t) = E[U(X, t)], \qquad (11.12)$$

for each value of t.

With this information alone at our disposal we wish to determine $\psi(x, t)$. The problem looks hopeless! Given initial data $\psi(x, 0)$ at $t = 0$ we evidently cannot use the Schrödinger to evolve it, since we do not know the Hamiltonian.

Nevertheless I would argue that, providing we take a Bayesian stance, we can indeed determine the wave function exactly for such a system.

The principle of maximum entropy amounts to the 'morally reasonable' maxim that, as best as can be managed, we should not assume any information other than what we are given. Thus the potential $U(X, t)$ is to be chosen in such a way that the entropy $S_\xi(t)$ associated with $X(t)$ is maximized, for each value of t, subject to the constraint (11.12). This leads us to the variational problem

$$\delta \int [-f \ln(f/\xi) + \lambda f U(x, t) + \mu f] \, d\tau = 0, \qquad (11.13)$$

together with the constraints (11.11) and $\int f \, d\tau = 1$. Carrying out the variation in eqn (11.13) we obtain

$$\lambda U(x, t) = \ln(f/\xi) + 1 - \mu. \qquad (11.14)$$

We now absorb the term $1 - \mu$ into the definition of ξ with the understanding that ξ is no longer definite, but is to be determined, eventually, by the normalization condition on f. Then, with this new value for ξ, eqn (11.14) reads

$$\lambda U(x, t) = \ln(f/\xi). \qquad (11.15)$$

Providing $E(U) \neq 0$ we can take the expectation of each side of this relation to obtain

$$\lambda = -S_\xi(t)/\omega(t), \qquad (11.16)$$

whence:

$$U(x, t) = -\frac{\omega(t)}{S_\xi(t)} \ln(f/\xi). \qquad (11.17)$$

Note that by virtue of eqn (11.1) the 'Bayesian' potential (11.17) satisfies the constraint (11.11), as desired. If $E(U) = 0$, then λ can be determined instead by examination of $E(U^2)$ or a suitable higher moment.

11.4. Nonlinearities in the Schrödinger equation

Inserting expression (11.17) for $U(x, t)$ into the Schrödinger eqn (11.10) we obtain

$$i \frac{\partial}{\partial t} \psi = -\frac{1}{2m} \Delta \psi - \beta \psi \ln(\psi \bar{\psi}/\xi), \qquad (11.18)$$

where $\beta(t) = \omega(t)/S_\xi(t)$. Thus we see that *the effect of statistical uncertainty in the form of the potential is to introduce a nonlinear term into Schrödinger's equation.*

It should be stressed that the nonlinearities here do not arise directly from any physical process, but rather represent an accurate impression of the original uncertainty in the specification of the physical problem.

It is worth noting that the modified Schrödinger eqn (11.18) is essentially identical in form to a nonlinear wave equation proposed by Bialynicki-Birula and Mycielski (1975*b*, 1976, 1979), who argue in favour of its numerous attractive properties. In their theory β is regarded as a *fundamental constant*, and is not necessarily related to the entropy of configuration as it is here. Perhaps it should be emphasized that eqn (11.18) is not to be viewed in any sense as being incompatible with or an alternative to the Schrödinger equation, but rather in statistical terms represents the particular form the Schrödinger equation takes, in this problem, when methods of Bayesian inference are employed to determine the potential in the situation where the actual information supplied is meagre.

As has been pointed out by the authors noted above, eqn (11.18) admits stable stationary solutions. Let us examine a solution in the case $n = 1$ in order to get some notion as to the values the various parameters of the theory can take on.

We assume that $\psi(x, t)$ is in an energy eigenstate with $\psi(x, t) = e^{-iEt}\psi(x)$. We also assume that $E(X) = \alpha$, and, for simplicity, that the expectation of the momentum vanishes. We find that $\psi(x) = f^{1/2}$ with $f(x)$ given by expression (11.9). The variance turns out to be

$$\sigma^2 = \frac{1}{2m(E - \omega)}. \tag{11.19}$$

Note that since $m \gg E$, the spread of the distribution is necessarily much broader than the Compton wavelength of the particle. The other main parameters of the theory turn out to be given by

$$\beta = 2(E - \omega), \tag{11.20}$$

$$S = \frac{1}{2}\frac{\omega}{E - \omega}, \tag{11.21}$$

$$\xi = \frac{1}{2\sigma}\exp\left(\frac{E + \omega}{E - \omega}\right), \tag{11.22}$$

and the potential itself turns out to be

$$U(x) = E - 2\omega + 8m(E - \omega)^2(x - \alpha)^2. \tag{11.23}$$

11.5. Fluctuations

In conclusion we consider a slightly more complicated problem involving an undetermined potential. Here we shall assume that the precise form of $U(x, t)$ in eqn (3.1) is not known, but that $U(x, t)$ can be written in the form

$$U(x, t) = A(x, t) + B(x, t), \tag{11.24}$$

where $A(x, t)$ is a known potential, and $B(x, t)$, which is to represent fluctuations, is unknown apart from its mean

$$\omega(t) = E[B(X, t)] \tag{11.25}$$

and its variance

$$\sigma^2[B] = E[B^2] - (E[B])^2. \tag{11.26}$$

This leads us to the variational problem

$$\delta \int [-f \ln(f/\xi) + \lambda f B(x, t) + \mu f (B - \omega)^2 + v f] \, d\tau = 0. \tag{11.27}$$

Carrying out the variation we get

$$-\ln(f/\xi) - 1 + \lambda B + \mu(B - \omega)^2 + v = 0. \tag{11.28}$$

With an appropriate renormalization of the parameter ξ, which will now be determined by the condition $\psi(x, t)$ is normalized, as in the previous example, we get

$$B(x, t) = \frac{\sigma[B]}{\sigma[\ln^{1/2}(\xi/f)]} \{\ln^{1/2}(\xi/f) - E[\ln^{1/2}(\xi/f)]\} + E[B], \tag{11.29}$$

where $E[B]$ and $\sigma[B]$ are the given data. If we assume the fluctuations are small, then eqn (11.29) can be used as a basis for perturbation theory, where the probability density function $f(x, t)$ appearing in the expression for $B(x, t)$ is associated with the *unperturbed* wave function which is constructed by solving the Schrödinger equation with the potential $A(x, t)$.

Acknowledgements

The author wishes to express his gratitude to I. Bialynicki-Birula, D. Deutsch, K. C. Hannabuss, R. Penrose, and W. Saslaw for helpful remarks in connection with this work.

References

Beckner, W. (1975a). *Proc. Natl. Acad. Sci. USA* **72**, 638.
—— (1975b). *Ann. Math.* **102**, 159.
Bialynicki-Birula, I. (1984). *Phys. Lett.* **103A**, 253.
—— and Mycielski, J. (1975a). *Commun. Math. Phys.* **44**, 129.
—— —— (1975b). *Bull. Acad. Polon. Sci. Cl. III.* **23**, 461.
—— —— (1976) *Ann. Phys. N.Y.* **100**, 62.

—— —— (1979). *Physica Scripta* **20,** 539.

Deutsch, D. (1983). *Phys. Rev. Lett.* **50,** 631.

Everett III, H. (1973). In *The many-worlds interpretation of quantum mechanics* (ed. B. S. DeWitt and N. Graham). Princeton University Press, Princeton.

Hirschman, I. I. (1957). *Am. J. Math.* **79,** 152.

Jaynes, E. T. (1957). *Phys. Rev.* **106,** 620.

Keynes, J. M. (1929). *A treatise on probability.* Macmillan, London.

Mead, L. R. and Papanicolaou, N. (1984). *J. Math. Phys.* **25,** 2404.

Partovi, M. H. (1983). *Phys. Rev. Lett.* **50,** 1883.

Rosenkrantz, R. D. (ed.) (1983). *E. T. Jaynes: papers on probability, statistics and statistical physics.* Reidel, Dordrecht.

12

Events and processes in the quantum world

Abner Shimony

12.1. Prospective

The concern of this paper will be different from that of most studies of quantum gravity. There will be no discussion of the problem of quantizing the gravitational field equations, hence nothing about the modifications which quantization requires of space–time structure at distances of the order of the Planck length ($\sim 10^{-33}$ cm). The emphasis will be rather upon the implications of quantum mechanics for certain general properties of events and processes which occur in the theatre of space–time, specifically with *non-locality* and *nonlinearity*.

Quantum mechanics is undoubtedly a non-local theory when it treats correlated spatially separated systems. When one examines closely the character of this non-locality, however, one does not find reasons for modifying the causal structure of space–time as described by special or general relativity theory, but rather reasons for refining the concept of an event. The occurrence of definite outcomes of measurements implies that there are processes in nature governed by nonlinear laws. Nonlinearity is very peculiar from the standpoint of quantum mechanics, since it is contrary to the linearity of the time-dependent Schrödinger equation, which is commonly assumed to govern the dynamics of any isolated physical system. Hence it is necessary to inquire whether a rational treatment of the occurrence of outcomes is possible without some modifications of current quantum mechanics. Whatever decision is reached on this question, however, it does not appear to bear directly on space–time structure, but rather on the general character of processes.

One of the attractions of emphasizing events and processes, rather than space–time structure, is the availability of relevant experimental evidence at the atomic level, in contrast to the frustrating inaccessibility of evidence at the level of the Planck length. Rejoicing over this fact must be tempered, however, by the formidable conceptual difficulties of interpreting experimental results. The main purpose of this chapter is to present these difficulties to an audience of experts on the interrelations of quantum mechanics and space–time theory, in the hope of eliciting new ideas which will free us from our current impasse.

12.2. Some features of the quantum-mechanical world-view

When the formalism of quantum mechanics is supplemented by a few principles of interpretation, there are striking consequences concerning the nature of the physical world. These consequences by themselves do not constitute a philosophy—neither a theory of knowledge, nor a metaphysics, nor a methodology, and obviously not a value theory—but they are so fundamental that a comprehensive philosophy has an obligation to take them into account. Accordingly, it seems appropriate to refer collectively to these consequences as 'the quantum mechanical world-view'.

One consequence is the possibility of a pure state in which a physical variable has an *indefinite* value. Pure states are represented by normalized vectors of an appropriate Hilbert space $\mathcal{H}$, or more accurately, they are in one-to-one correspondence with the rays or one-dimensional subspaces of $\mathcal{H}$. If $\mathscr{F}$ is a physical variable (usually called an 'observable', but this location should be avoided if one accepts the principle of avoiding anthropocentrism in interpreting the formalism) which has the value f_1 in the state represented by the normalized vector u_1 and the value $f_2 (\neq f_1)$ in the state represented by the normalized vector u_2, then in the state represented by the vector u,

$$u = c_1 u_1 + c_2 u_2, \qquad c_1 \neq 0, c_2 \neq 0, \qquad |c_1|^2 + |c_2|^2 = 1, \qquad (12.1)$$

the physical variable $\mathscr{F}$ does not have a definite value. Furthermore, if we add to the formalism the principle of interpretation that the pure state represents a maximal specification of the system itself, rather than a compendium of someone's knowledge of the system, then *the indefiniteness of $\mathscr{F}$ is objective and and not a matter of incomplete knowledge*. In this way, the superposition principle, which is exemplified in eqn (12.1) has a striking philosophical implication. (It should be added that even if superselection principles hold, restricting the domain of applicability of the superposition principle, there will still remain an abundance of instances of eqn (12.1), and its philosophical implication remains intact.)

The coefficients c_1 and c_2 in eqn (12.1) determine the probabilities of the outcomes f_1, f_2 respectively, upon condition that a measurement of the physical variable $\mathscr{F}$ is performed. Specifically, $|c_1|^2$ is the probability of the outcome f_1, and $|c_2|^2$ is the probability of the outcome f_2. Given the principle of interpretation that a pure quantum state represents a maximal specification of the system itself, we cannot regard these probabilities as epistemic—i.e., as the reasonable degrees of belief concerning the values of a quantity which has a definite though unknown value prior to measurement. The probabilities of classical statistical mechanics can reasonably be understood in this way, but not the quantum mechanical probabilities. The latter must be regarded as ontic probabilities for the objectively indefinite physical variable to take on one or another value when it becomes definite; they are implicit in the physical state of affairs rather than in someone's knowledge.

The combination of indefiniteness of value with definite probabilities of possible outcomes can be compactly referred to as *potentiality*, a term suggested by Heisenberg (1958, p. 185). When a physical variable which initially is merely potential acquires a definite value, it can be said to be *actualized*. So far, the only processes we have mentioned in which potentialities are actualized are measurements, but in a non-anthropocentric view of physical theory the measurement process is only a special case of the interaction of systems, of special interest to scientists because knowledge is thereby obtained, but not fundamental from the standpoint of physical reality itself. It remains an open problem, however (discussed further in Section 12.5), to determine under what circumstances the actualization of potentialities generically takes place.

The pure quantum states of a composite system $1 + 2$ are represented by normalized vectors in the tensor product Hilbert space

$$\mathcal{H} = \mathcal{H}_1 \otimes \mathcal{H}_2 \tag{12.2}$$

where $\mathcal{H}_i$ is the Hilbert space associated with component i ($i = 1, 2$). If u_1, u_2 are orthogonal normalized vectors in $\mathcal{H}_1$ and v_1 and v_2 are orthogonal normalized vectors in $\mathcal{H}_2$, then both $u_1 \otimes v_1$ and $u_2 \otimes v_2$ are normalized vectors in $\mathcal{H}$, representing possible pure states of $1 + 2$. Application of the superposition principle yields immediately that

$$\Psi = \frac{1}{\sqrt{2}} (u_1 \otimes v_1 + u_2 \otimes v_2) \tag{12.3}$$

is a normalized vector of $\mathcal{H}$ and represents a pure state of $1 + 2$. But a little algebraic analysis shows that

$$\forall w \in \mathcal{H}_1, \quad z \in \mathcal{H}_2, \quad w \otimes z \neq \Psi. \tag{12.4}$$

In Schrödinger's locution, Ψ is an 'entangled' state—one in which neither system 1 by itself nor system 2 by itself can be said to be in a pure state; neither can be fully specified without reference to the other. It must be emphasized that the concept of entanglement is inseparable from the role of potentiality in quantum mechanics. Suppose that u_1 and u_2 are related to the physical variable $\mathcal{F}$ as above, and v_1 and v_2 represent states in which a physical variable $\mathcal{G}$ of 2 have respective values g_1 and g_2 ($g_1 \neq g_2$). When the state of $1 + 2$ is represented by Ψ, then both $\mathcal{F}$ and $\mathcal{G}$ are merely potential, but in an interlocked manner: $\mathcal{F}$ and $\mathcal{G}$ can be actualized so as to have respective values f_1 and g_1, or so as to have respective values f_2 and g_2, and in fact the probability of each of these joint outcomes is $\frac{1}{2}$; but they cannot be actualized either as f_1, g_2 or as f_2, g_1.

If 1 and 2 are spatially well separated, then entanglement implies a peculiar kind of quantum non-locality, the character of which is the primary concern of Section 12.4.

It may be objected at this point that we have proceeded recklessly in our interpretation of the quantum mechanical formalism. By supplementing the formalism with the principle that a pure quantum state is a maximal specification of the system itself, and by insisting upon the strict avoidance of anthropocentrism, have we not inserted an unhealthy dose of metaphysics into a scientific theory? There are two evident avenues for escaping from the metaphysical consequences which have been catalogued above. One is a hidden variables theory, which maintains that a complete specification of a physical system requires the supplementation of the quantum state, the supplementary information being commonly known as 'hidden variables'. If this avenue is taken, then quantum indefiniteness can be understood epistemically, as incomplete knowledge of the details concerning the system. Quantum mechanical probabilities can then be interpreted in the same way as the probabilities of classical statistical mechanics when some physical constraints are imposed (e.g. equilibrium with specified reservoirs) but some details are unknown. Finally, quantum mechanical entanglement can also be understood epistemically, as the information that one system supplies about the other in virtue of correlations established by their mode of joint preparation. *Prima facie*, this avenue of escaping from the metaphysical peculiarities inferred above is very attractive indeed: it is non-anthropocentric, it is 'realistic' concerning the ontological status of physical systems, and it continues an intellectual tradition which was immensely successful in classical physics. We can easily understand why this avenue was recommended by Einstein and de Broglie, in spite of their full realization of the power of quantum mechanics as a statistical theory and in spite of their own great contributions to it. The significance of Bell's theorem and the experiments based upon it is that this very attractive avenue of interpretation must be abandoned, unless modifications are made which deprive it of its attractiveness. In Section 12.3 a summary will be given of the impasse which Bell's theorem has exhibited in the program of hidden variables theories.

The other avenue for escaping from the peculiar metaphysics of quantum mechanics is to relax the taboo against anthropocentrism. If *any* physical theory—and quantum mechanics in particular—is regarded as nothing more than an instrument for summarizing human experience and anticipating more experience in the future, then the peculiar features of quantum mechanics cannot be attributed to the things in themselves. There is a long philosophical tradition behind this point of view, from Berkeley through Kant to Mach, and many of the founding fathers of quantum mechanics accepted it either deliberately, as a thoughtfully considered epistemological commitment, or opportunistically, as a way of avoiding painful metaphysical difficulties. There is a large literature on the cogency of positivistic and instrumentalistic interpretations of scientific theories, and I shall refer to Suppe's (1974) survey in order to avoid reviewing familiar arguments. As a surrogate for an argument I shall mention a paraphrase of a statement of Einstein which Wigner gave in an unpublished lecture (Varenna, 1970): 'Do you really believe that the *sun* is nothing more than our perceptions?'

12.3. Bell's theorem and local hidden variables theories

The program of a hidden variables interpretation of quantum mechanics is logically viable, provided that sufficient latitude is allowed in the interpretation. Specifically, it is essential that the interpretation be 'contextual', in the sense that the value which is assigned by a complete state λ to a physical variable $\mathcal{F}$ must depend also upon the 'context' of the measurement of $\mathcal{F}$ (see Bell 1966; Gudder 1970; Shimony 1984*a*). Contextualism permits the program of hidden variables interpretations to escape the well known impossibility theorems of Gleason (1957), of Kochen and Specker (1967), and of Bell himself (1966).

Logical viability is not good enough, however. An acceptable hidden variables theory must also satisfy physical desiderata. In particular, unless revolutionary evidence is presented in favour of changing the relativistic conception of space–time structure, an acceptable hidden variables theory must conform to this structure. So far the profound differences between the special and the general theories of relativity have not impinged significantly upon the analysis of hidden variables theories. The demand upon hidden variables theories which has played the central role is common to the special and general theories: namely, there is no direct causal connection between points with space-like separation. This thesis, which will be called simply 'locality', does not actually occur as a premise of Bell's theorem (1964), but it did provide the heuristics for his crucial premise. For convenience of analysis I shall present neither Bell's original theorem nor any of his later variants, but rather shall give an argument inspired by his work. The argument consists of a modification of a theorem of Clauser and Horne (1974), a modification of a theorem of Jarrett (1984), and some informal discussion.

Theorem 1. Let $\langle \Lambda, \Sigma(\Lambda), \rho \rangle$ be a classical probability space, i.e. Λ is a non-empty set, $\Sigma(\Lambda)$ a σ-algebra of subsets of Λ, and ρ a probability measure on $\Sigma(\Lambda)$. Let a and b be parameters, for each value of which and for each $\lambda \in \Lambda$ there are random variables $p_\lambda^1(a)$, $p_\lambda^2(b)$, and $p_\lambda(a, b)$ with values in $[0, 1]$. Suppose further that the factorizability condition

$$p_\lambda(a, b) = p_\lambda^1(a)p_\lambda^2(b) \tag{12.5}$$

is satisfied. Then

$$-1 \leqq p(a', b') + p(a', b'') + p(a'', b') - p(a'', b'') - p^1(a') - p^2(b') \leqq 0,$$
$$\tag{12.6}$$

where

$$p^1(a) = \int_\Lambda p_\lambda^1(a)\, d\rho, \tag{12.7a}$$

$$p^2(b) = \int_\Lambda p_\lambda^2(b)\, d\rho, \tag{12.7b}$$

$$p(a, b) = \int_\Lambda p_\lambda(a, b)\, dr. \tag{12.7c}$$

Although Theorem 1 is somewhat more abstract than that stated by Clauser and Horne (1974), it is proved in essentially the same way.

There are various possible interpretations of the terms of Theorem 1 (see Shimony (1984a), Section 4.) In the most interesting interpretation, λ is the complete state of a pair of particles propagating in different directions from a common origin and subjected to tests chosen at will by experimenters, with a and b being parameters characterizing the tests to which the first and second particles respectively are subjected. For example, if particles 1 and 2 are photons, then a may be the angle which the transmission axis of a polarization analyser for 1 makes with a fixed axis perpendicular to the propagation direction; and similarly for b. Among the values of the parameter a we shall include ∞, which means that particle 1 is not subjected to a polarization test—i.e. no polarization analyser is placed in its path; a similar interpretation is given to $b = \infty$. We shall restrict our attention to situations in which the only possible outcomes when $a \neq \infty$ are $x_a = \pm 1$, and the only possible outcomes when $b \neq \infty$ are $x_b = \pm 1$. For example, $x_a = 1$ could mean that particle 1 passes through its analyser when the parameter has the value a, and $x_a = -1$ could mean that particle 1 impinges upon the analyser with this value of the parameter but fails to pass through it; and $x_b = \pm 1$ is similarly interpreted. When $a = \infty$, i.e. particle 1 is not subjected to a polarization test, then the trivial outcome is designated by $x_a = \infty$; and likewise if

$b = \infty$. It will be assumed that for each $\lambda \in \Lambda$ and each choice of the parameters a, b there is a probability function $p_\lambda(x_a, x_b \mid a, b)$ which satisfies the following conditions:

$$0 \leq p_\lambda(x_a, x_b \mid a, b) \leq 1 \quad \text{for} \quad \begin{matrix} x_a = \pm, \infty \\ x_b = \pm, \infty; \end{matrix} \tag{12.8a}$$

$$\text{if } a \neq \infty, b \neq \infty, \text{ then } \sum_{X=\pm1} \sum_{Y=\pm1} p_\lambda(X, Y \mid a, b) = 1; \tag{12.8b}$$

$$\text{if } a \neq \infty, \text{ then } \sum_{X=\pm1} p_\lambda(X, \infty \mid a, \infty) = 1; \tag{12.8c}$$

$$\text{if } b \neq \infty, \sum_{Y=\pm1} p_\lambda(\infty, Y \mid \infty, b) = 1; \tag{12.8d}$$

$$p_\lambda(\infty, \infty \mid \infty, \infty) = 1. \tag{12.8e}$$

It will be useful to define some marginal and conditional probability functions in terms of $p_\lambda(x_a, x_b \mid a, b)$:

$$p_\lambda^1(x_a \mid a, b) = \sum_X p_\lambda(x_a, X \mid a, b), \tag{12.9a}$$

where $X = \pm1$ if $b \neq \infty$, $X = \infty$ if $b = \infty$;

$$p_\lambda^2(x_b \mid a, b) = \sum_{X=\pm1} p_\lambda(X, x_b \mid a, b) \tag{12.9b}$$

where $X = \pm1$ if $a \neq \infty$, $X = \infty$ if $a = \infty$;

$$p_\lambda^1(x_a \mid a, b, x_b) = \frac{p_\lambda(x_a, x_b \mid a, b)}{p_\lambda^2(x_b \mid a, b)} ; \tag{12.10a}$$

$$p_\lambda^2(x_b \mid a, b, x_a) = \frac{p_\lambda(x_a, x_b \mid a, b)}{p_\lambda^1(x_a \mid a, b)} . \tag{12.10b}$$

Two kinds of independence conditions can now be compactly defined:

Parameter Independence: For all λ, a, b,

$$p_\lambda^1(x_a \mid a, b) = p_\lambda^1(x_a \mid a, \infty), \tag{12.11a}$$

$$p_\lambda^2(x_b \mid a, b) = p_\lambda^2(x_b \mid \infty, b). \tag{12.11b}$$

(According to this condition the probability of the outcome concerning a single particle of the pair is independent of the choice of the parameter of the test performed upon the other particle.)

Outcome Independence:

$$p_\lambda^1(x_a \mid a, b) = p_\lambda^1(x_a \mid a, b, x_b). \tag{12.12a}$$

$$p_\lambda^2(x_b \mid a, b) = p_\lambda^2(x_b \mid a, b, x_a). \tag{12.12b}$$

We can now prove a theorem which is a modification of the

decomposition theorem of Jarrett (1984), with his 'Completeness Condition' replaced by our condition of Outcome Independence.

Theorem 2. If the conditions of Parameter and Outcome Independence are both satisfied, then

$$p_\lambda(x_a, x_b \mid a, b) = p_\lambda(x_a, \infty \mid a, \infty) \cdot p_\lambda(\infty, x_b \mid \infty, b). \tag{12.13}$$

Proof.

$$\begin{aligned}
p_\lambda(x_a, x_b \mid a, b) &= p_\lambda^1(x_a \mid a, b, x_b) \cdot p_\lambda^2(x_b \mid a, b) \\
&= \frac{p_\lambda^2(x_b \mid a, b, x_a) p_\lambda^1(x_a \mid a, b)}{p_\lambda^2(x_b \mid a, b)} p_\lambda^2(x_b \mid a, b) \\
&= p_\lambda^1(x_a \mid a, b) p_\lambda^2(x_b \mid a, b) = p_\lambda^1(x_a \mid a, \infty) p_\lambda^2(x_b \mid \infty, b) \\
&= p_\lambda(x_a, \infty \mid a, \infty) p_\lambda(\infty, x_b \mid \infty, b).
\end{aligned}$$

If the conditions of Parameter and Outcome Independence are satisfied, then the conclusion of Theorem 2 constitutes the crucial premise of Theorem 1, once some notational identifications are made. It suffices to define

$$p_\lambda(a, b) = p_\lambda(1, 1 \mid a, b), \tag{12.14a}$$

$$p_\lambda^1(a) = p_\lambda(1, \infty \mid a, \infty), \tag{12.14b}$$

$$p_\lambda^2(b) = p_\lambda(\infty, 1 \mid \infty, b). \tag{12.14c}$$

With these identifications, inequality (12.6) follows. What is most remarkable is the generality of the inequality: it holds regardless of the character of the probability space $\langle \Lambda, \Sigma(\Lambda), \rho \rangle$. Inequality (12.6) does, however, depend upon the factorizability condition (12.5), which in itself is neither obvious nor obligatory, but Theorem 2 shows that there are conditions which imply factorizability. Furthermore, if the event which consists of the choice of parameter *a* together with the consequent test performed on particle 1 has space-like separation from the event which consists of the choice of parameter *b* together with the test performed on particle 2, then relativistic locality ensures that these two events are not directly connected causally. Consequently, it is plausible that both Parameter and Outcome Independence should hold. This crucial final step in the argument will, however, be re-examined in Section 12.4, since the concept of locality is subtle.

If ideal apparatus is assumed, it is easy to give examples of quantum states of a particle pair for which inequality (12.6) is violated by the quantum mechanical predictions. For example, let the quantum polarization state of a pair of photons propagating respectively in the $\hat{z}$ and $-\hat{z}$

directions be

$$\Psi = \frac{1}{\sqrt{2}}[u_x(1) \otimes u_x(2) + u_y(1) \otimes u_y(2)], \qquad (12.15)$$

where $u_x(i)$, $u_y(i)$ represent states of polarization of photon i ($i = 1, 2$) along the $\hat{x}$ and $\hat{y}$ axes respectively. The quantum mechanical prediction based upon Ψ for the joint passage of 1 and 2 through analysers with transmission axes making angles a, b respectively with the x-axis is easily shown to be

$$p_\Psi(a, b) = \tfrac{1}{2}\cos^2(a - b), \qquad (12.16)$$

and the probability of the passage of a single photon is

$$p_\Psi^1(a) = p_\Psi^2(b) = \tfrac{1}{2}. \qquad (12.17)$$

If a', a'', b', b'' are respectively chosen to be $22\tfrac{1}{2}°$, $0°$, $45°$, $67\tfrac{1}{2}°$, then

$$p_\Psi(a', b') = p_\Psi(a', b'') = p_\Psi(a'', b') = 0.4268,$$

and

$$p_\Psi(a'', b'') = 0.0732.$$

Hence

$$p_\Psi(a', b') + p_\Psi(a', b'') + p_\Psi(a'', b') - p_\Psi(a'', b'')$$
$$- p_\Psi^1(a') - p_\Psi^2(b') = 0.2072, \quad (12.18)$$

in violation of inequality (12.6). Hence no hidden variables theory satisfying relativistic locality (as construed above) can agree with all the predictions of quantum mechanics. This result is one of the forms of Bell's theorem.

Even when the assumption of ideal apparatus is dropped it is possible to exhibit situations in which the quantum mechanical predictions violate inequality (12.6). As a result, experimental confrontations between quantum mechanics and all hidden variables theories satisfying the factorizability condition (12.5) are possible. The confrontation is sharpened in the experiment of Aspect *et al.* (1982), in which the polarization analysis events on the two photons have space-like separation, which by the argument above leads almost inevitably to the factorizability condition. This spectacular experiment yielded results in agreement with quantum mechanics and in disagreement with inequality (12.6). As in all actual experiments, a determined advocate of a hypothesis can find loopholes to explain the disconfirming evidence, but in this case the loopholes are unattractive and unpromising. The evidence is overwhelming against the family of local hidden variables theories.

12.4. Aspects of non-locality

The experimental violation of inequality (12.6) indicates that a re-examination of Parameter and Outcome Independence is needed. Is each of them required by relativistic locality? Which of them is violated by quantum mechanics?

It is easy to see that a violation of Parameter Independence would be the basis of a violation of relativistic locality, as Jarrett (1984) pointed out. Suppose that for some specified a, b

$$p_\lambda^1(x_b \mid a, b) \neq p_\lambda^1(x_b \mid \infty, b), \tag{12.19}$$

and let an ensemble of pairs of particles be prepared, each pair in the complete state λ, propagating within a short time interval from a common origin. Let Experimenter I interact with particles 1 of the ensemble in a region R_1 and Experimenter II with particles 2 of the ensemble in a region R_2, and suppose that R_1 and R_2 have space-like separation. The protocol is also laid down that Experimenter I will either choose the value a for each of the analysers upon which the particles 1 impinge or else the value ∞ (i.e. the analysers will be removed from the paths of the 1 particles). Experimenter II chooses the value b for the analysers upon which the particles 2 impinge and observes the resulting statistics. Because of Inequality (12.19) the statistics observed by Experimenter II will depend upon the choice made by Experimenter I, and by making the ensemble large enough the information about the decision can be transmitted from I to II with a probability as close to unity as desired. Furthermore, the whole process can be automated and conscious observers dispensed with, so that an explosion can be triggered in R_2 (with a probability as close to unity as desired) by the automated selection in R_1 of one of the two parameter values a or ∞. The only way to deny that a violation of relativistic locality occurs under the conditions just supposed is to insist upon certainty as a criterion for transmission of a message, and that demand does not seem reasonable.

The quantum mechanical predictions concerning ensembles of pairs of particles do not violate Parameter Independence, provided that non-locality is not explicitly built into the interaction Hamiltonian of the particle pair. Specifically, it is impossible to capitalize upon the entanglement of the quantum state of a two-particle system for the purpose of sending a message to an observer of one of the particles by performing an operation upon the other particle. This is a theorem which has been proven in great generality by Eberhard (1978), Ghirardi *et al.* (1980), and Page (1982), and I shall not give a proof here.

There seems to be no way to use a violation of Outcome Independence for the purpose of sending a message faster than light. Consider again the two Experimenters I and II observing respectively the 1 and 2 particles of an ensemble of pairs, and suppose that for specified values a, b of the adjustable parameters

$$p^2_\lambda(1 \mid a, b) \neq p^2_\lambda(1 \mid a, b, 1), \tag{12.20}$$

from which it follows that

$$p^2_\lambda(1 \mid a, b, 1) \neq p^2_\lambda(1 \mid a, b, -1). \tag{12.21}$$

Experimenter I may try to capitalize upon inequality (12.21) to send a message to Experimenter II by blocking the transmission of all 2 particles which have partners (i.e., 1 particles) for which the outcome is j ($j = 1$, -1). Since the statistics of $x_b = 1$ differ, in view of inequality (12.21), according to the choice of j, a message can be transmitted with a probability as close to unity as desired by sufficiently increasing the size of the ensemble. But this message cannot be transmitted any faster than the flight of the 2 particles themselves, and therefore it cannot be superluminal. I see no way of varying the experimental arrangement so as to achieve superluminal transmission. Consequently, a violation of Outcome Independence does not constitute a counter-example to relativistic locality in the same sense as a violation of Parameter Independence. Shimony (1984b) called the kind of non-locality implicit in a violation of Outcome Independence 'uncontrollable non-locality'. I now think that this terminology is misleading, since inequality (12.21) does permit a controlled message to be transmitted, though not superluminally.

Situations can be readily exhibited in which the quantum mechanical predictions violate Outcome Independence. Consider again the Ψ of eqn (12.15), and suppose that it represents the complete polarization state of the composite system $1 + 2$. Let the parameters a and b be chosen equal, for specificity 0 (in other words, both polarization analysers are oriented with transmission axes along $\hat{x}$). Then, if ideal analysers are assumed,

$$p^1_\Psi(1 \mid 0, 0) = \tfrac{1}{2}, \tag{12.22}$$
$$p^1_\Psi(1 \mid 0, 0, 1) = 1, \tag{12.23}$$

thus violating Outcome Independence.

We have now answered both of the questions with which Section 12.4 commenced, and the answers mesh in a most interesting manner. Quantum mechanics violates Outcome Independence, which cannot be used to send a superluminal message; and it conforms to Parameter Independence, a violation of which would permit superluminal communication. We may summarize by adapting a well known political slogan: there is 'peaceful co-existence' between quantum mechanics and relativity theory. This result, incidentally, throws light upon a question that must have occurred to some readers: if quantum mechanics is non-local in some sense, why should we feel obliged to restrict our attention to local hidden variables theories? Why not escape from most of the peculiar metaphysics of quantum mechanics by paying the price of accepting a small part of this peculiarity, namely non-locality? There are, in fact, some serious proposals, notably by Vigier and collaborators (Vigier 1982) for escaping from Bell's Theorem by building non-locality into their hidden variables theories, specifically, by postulating shock waves of superluminal velocity in a hypothetical 'quantum ether'. It is far from clear, however, that there is peaceful co-existence between relativity theory and such hidden variables theories. (See Shimony (1984*b*) for a challenge on this point.) Consequently, local hidden variables theories may be possible only if relativity theory is replaced by a radically different theory of space–time structure, whereas the non-locality of quantum mechanics may leave relativistic space–time structure intact, but only change our conception of an event in space–time.

In spite of the discriminations made above, the idea of peaceful co-existence between relativistic space–time structure and quantum mechanics remains puzzling, and a further examination is needed of the concept of an event. Let us return once more to the state of photon pair $1 + 2$ represented by Ψ of eqn (12.15), which we shall again assume to constitute a complete specification of the polarization properties of the pair. Let $A_x(i)$ be the physical variable which takes on the value 1 if photon i ($i = 1, 2$) is polarized along the direction $\hat{x}$ and the value -1 if it is polarized along the direction $\hat{y}$ (which is perpendicular both to x and to directions $\hat{z}$ and $-\hat{z}$ of propagation of the two photons); and let $A_{x'}(i)$ have an analogous meaning concerning another direction $\hat{x}'$ in the plane perpendicular to $\hat{z}$. We note—as we have not done so far—that Ψ is cylindrically symmetrical and hence can be written not only in the form of eqn (12.15), but also as

$$\Psi = \frac{1}{\sqrt{2}} [u_{x'}(1) \otimes u_{x'}(2) + u_{y'}(1) \otimes u_{y'}(2)]. \qquad (12.24)$$

Now let us suppose that $A_x(1)$ and $A_{x'}(2)$ are measured with ideal apparatus, and that the acts of measurement have space-like separation. Furthermore, let Σ be an inertial frame of reference in which the measurement of $A_x(1)$ occurs earlier than that of $A_{x'}(2)$, and Σ' an inertial frame in which the reverse time order holds. It is reasonable (though, as we shall see, not obligatory) to describe the sequence of relevant events from the standpoint of Σ to be the following: $t_0 < t_1 < t_2$

and

at t_0: $A_x(1)$, $A_x(2)$, $A_{x'}(1)$, $A_{x'}(2)$ are all indefinite;

at t_1: $A_x(1)$ and $A_x(2)$ are definite, but not $A_{x'}(1)$ and $A_{x'}(2)$ (because the first measurement reduces the superposition Ψ, yielding either $u_x(1) \otimes u_x(2)$ or $u_y(1) \otimes u_y(2)$);

at t_2: $A_x(1)$ and $A_{x'}(2)$ are definite, but $A_x(2)$ and $A_{x'}(1)$ are indefinite (because the second measurement reduces $u_y(2)$ so as to yield either $u_{x'}(2)$ or $u_{y'}(2)$ but leaves $u_x(1)$ intact).

The sequence of events from the standpoint of Σ' is the following (with reasons suppressed, since they are analogous to the ones given for Σ): $t_0' < t_1' < t_2'$ and

at t_0': $A_x(1)$, $A_x(2)$, $A_{x'}(1)$, $A_{x'}(2)$ are all indefinite;

at t_1': $A_{x'}(2)$ and $A_{x'}(1)$ are definite, but $A_x(2)$ and $A_x(1)$ are indefinite;

at t_2': $A_{x'}(2)$ and $A_x(1)$ are definite, but $A_{x'}(1)$ and $A_x(2)$ are indefinite.

The criterion for definiteness of a physical variable A at a given time is that the quantum state at that time is an eigenstate of A, as commonly assumed in the interpretation of the quantum mechanical formalism. In addition, a plausible premise has been tacitly employed, even though it is not essential to the interpretation of the formalism: namely, that the definiteness or indefiniteness of a physical variable constitutes an event.

When the sequences from the standpoints of Σ and Σ' are compared there are some agreements and some disagreements. There is agreement concerning the initial events (at t_0 and t_0' respectively), and also concerning the final events (at t_2 and t_2' respectively). There is disagreement concerning the intermediate events. Consequently, the two accounts of *processes* from initial to final sets of events are in disaccord. But it is important to note that the process is a theoretical construction. $A_x(2)$ is not measured by an observer in Σ but only inferred from the measured value of $A_x(1)$ in virtue of the correlation expressed by Ψ; and likewise $A_{x'}(1)$ is inferred rather than measured directly by an observer in Σ'. As to causality, there is also some agreement and some disagreement between the descriptions in Σ and Σ'. There is no pair of events such that an observer in Σ identifies one as the cause and the other as the effect, while an observer in Σ' makes the opposite identification. However, an observer in Σ identifies the measurement of $A_x(1)$ as the cause of the transition of $A_x(2)$ from being indefinite to being definite, whereas an observer in Σ' does not recognize this causal sequence; and likewise an observer in Σ' identifies the measurement of $A_{x'}(2)$ as the cause of the transition of $A_{x'}(1)$ from being indefinite to being definite, whereas an observer in Σ does not do so. The thesis of peaceful co-existence presupposes a conceptually coherent reconciliation of the descriptions from the standpoints of Σ and Σ'. Even more desirable, in the spirit of the geometrical formulation of space–time theory, would be a co-ordinate-free account.

I do not believe that we are able, in our present state of understanding of the foundations of quantum mechanics, to meet these desiderata. The reason is that the puzzling events in question consist of transitions from indefiniteness to definiteness of physical variables, in other words, of the actualization of potentialities. Although the concept of potentiality is crucial to the world-view of quantum mechanics (as argued in Section 12.2), the theory is notoriously uninformative about the process of actualization. More will be said about this problem in Section 12.5, though without a firm conclusion. For the present, however, I wish to present a conjecture which has some coherence even in the absence of a full account of the actualization of potentialities.

The conjecture is that *the reduction of a superposition, however that is achieved, does not* ipso facto *constitute the actualization of a physical variable of which the reduced state is an eigenvalue*. When this conjecture is applied to the previously considered sequence of events in Σ the consequence is that $A_x(2)$ is not definite at t_1; and when it is applied to the sequence of events in Σ' the consequence is that $A_{x'}(1)$ is not definite at t_1'. These statements are made in full cognizance of the fact that at t_1 the state of the composite system $1 + 2$, from the standpoint of Σ, is represented either by $u_x(1) \otimes u_x(2)$ or by $u_y(1) \otimes u_y(2)$, and in either state the probability that $A_x(2)$ has the value 1 is either unity or zero. That is to say, an observer who is aware of the result of the measurement made upon $A_x(1)$ would know with certainty what would be found if $A_x(2)$ were measured—provided, of course, that there were no intervening disturbance of system 2, such as the measurement of $A_{x'}(2)$. My conjecture, however, says that certainty with regard to what would be found concerning $A_x(2)$ is not tantamount to the actualization of $A_x(2)$, or equivalently, to the occurrence of the event that $A_x(2)$ is definite. According to the conjecture, the actualization of a potentiality must not be conceived as a limiting case of probability—as probability 1 or 0. Instead, actuality and potentiality are radically different modalities of reality. These modalities are obviously related, since the probabilities associated with a quantum mechanical potentiality govern the propensities of actualization, and in the extreme case of probabilities 1 and 0 they positively exclude one or the other possible outcome. Nevertheless, excluding one outcome does not constitute actualization of the other outcome.

The conjecture just presented removes the discrepancy between the description of intermediate states from the standpoints of Σ and Σ'. According to the conjecture, from neither standpoint is $A_x(2)$ or $A_{x'}(1)$ definite. Of course, there is a discrepancy of the time ordering of the measurements of $A_x(1)$ and $A_{x'}(2)$, but this is the familiar frame-dependence of temporal ordering of events with space-like separation, and quantum mechanics introduces nothing new on this point.

My conjecture seems to be at odds with the well known criterion for the existence of an element of physical reality formulated by Einstein *et al.* (1935): 'If, without in any way disturbing a system, we can predict with certainty (i.e., with probability equal to unity) the value of a physical quantity, then there exists an element of physical reality corresponding to this physical quantity'. The metaphysical content of the criterion of Einstein *et al.* is minimal. Beyond assuming that there are systems and properties that have existence independent of human experience, they seem to be making no metaphysical commitment. The primary import of their criterion is epistemological: it gives a sufficient condition for asserting that the value of physical variable is objectively definite. My conjecture, however, has a considerable metaphysical content, for it distinguishes between two modalities of reality, and it modifies the criterion of Einstein *et al.* by abstaining from the identification of the 'existence of an element of physical reality' with existence in the modality of actuality. It is well known that Bohr (1935) offered an answer to the argument of Einstein *et al.* for the incompleteness of quantum mechanics. However, my conjecture is only vaguely cognate to Bohr's answer. Bohr's fundamental dichotomy is between results of experiments, describable in the language of classical physics, and the quantum mechanical wave function, which he took to be an algorithm or intellectual instrument for calculating the probabilities of experimental outcomes. By contrast, the conjecture presented above is quite metaphysical and attributes a mode of reality to the quantum state, and a different one to actualizations, and it makes no commitment that the latter occur only in the context of deliberate experimentation by human beings.

12.5. Nonlinearity and the actualization of potentialities

The quantum mechanical law of dynamical development for an isolated system is the time-dependent Schrödinger equation, which can be written in integrated form as

$$\psi(t) = U(t)\psi(0), \tag{12.25}$$

where

$$U(t) = e^{-iHt/\hbar} \tag{12.26}$$

for an appropriate self-adjoint operator H. For each t, $U(t)$ is a unitary operator on the Hilbert space to which the vectors $\psi(t)$ belong. For our present purposes, the most important characteristic of $U(t)$ is linearity, which has the effect of propagating superpositions. When eqn (12.25) is applied to a composite system consisting of a microscopic system together with a measuring apparatus, it has the consequence—under suitable initial conditions—of leading to a final state in which the indexical apparatus variable (i.e. the one from which the value of the microscopic variable of interest is inferred) is indefinite: a pointer needle does not point in any definite direction. This peculiarity of the measurement process shows that quantum mechanics does not have, in any obvious way, the conceptual tools for explaining how potentialities are actualized. There is a large and sophisticated literature on attempts to solve the problem of the actualization of potentialities strictly within the formalism of quantum mechanics, but in my opinion none of these attempts succeed. I shall not give even a brief review of this literature, partly because of shortage of space and partly because of the availability of excellent surveys (e.g. d'Espagnat 1976).

The problem of actualizing potentialities is, of course, completely circumvented by a hidden variables theory, and indeed this is one of the major motivations for investigating such theories. But we have already summarized the strongest evidence against hidden variables, and therefore must seek elsewhere. A very attractive program, which makes a minimal modification of quantum mechanics and avoids anthropocentric interpretations, is to keep the entire time-independent formalism of quantum mechanics (states, superpositions, physical variables, probabilities of outcomes, etc.) but to substitute a nonlinear differential equation for the linear time-dependent Schrödinger equation. It should be stated at the outset that there are obstacles to this program in the form of theorems (beautifully surveyed by Simon 1976) which derive the time-dependent Schrödinger equation from the time-independent part of the quantum mechanical formalism, provided that very mild and reasonable suppositions are made about temporal evolution. These obstacles are not insuperable, however, for these mild and reasonable suppositions may in fact be false, and alternatives to them are worth exploring.

A number of desiderata for a nonlinear quantum dynamics can be reasonably stated:

(i) Suppose that we have a situation in which phenomenologically it is known that actualization of potentialities occurs (e.g. a measurement situation), and suppose that initially the state of the system is represented by $\sum c_n y_n$, where each u_n is an eigenvector of the physical variable which is actualized. Then the nonlinear equation must govern a process

$$\sum c_n u_n \rightarrow u_k$$

for some k, with no residual contributions from u_i with $i \neq k$, in a finite time interval.

(ii) In such a situation, the probability of the transition to u_k is $|c_k|^2$.

(iii) The dynamical law should be general, not restricted to the measurement situation.

(iv) It must be possible to show in certain cases that the transition described in (i) is very rapid.

(v) There must be approximate agreement with the standard linear dynamics of quantum mechanics in the case of micro-systems.

Comments:

Desideratum (i) retains the point of view that the quantum state is a complete description of the physical system—a thesis to which we have returned in view of the difficulties exhibited for hidden variables theories. Implicit, however, in desideratum (i) is a view of actualization: namely, that it is the limiting case of potentiality, in which the final state is an eigenstate of the physical variable of interest, or alternatively, in which the probability of one value of the physical variable is unity and that of all others is zero. Desideratum (i) is therefore in direct conflict with the conjecture concerning the relation between potentiality and actuality proposed for relativistic reasons in Section 12.4. The promise of success of a nonlinear variant of the Schrödinger equation which fulfils desideratum (i) would therefore seriously undercut my conjecture.

Desideratum (ii) is essential in order to regard the measuring process as a special case of the interaction between physical systems, which in turn is a necessary condition for the avoidance of anthropocentrism.

Desideratum (iv) is imposed by respect for phenomenology, for it is known that there are measurement processes in which the time interval between an initial state described by a superposition and a final definite result is very short indeed. Coincidence counting technology indicates that the time interval may be of the order of a few tens of picoseconds.

Desiderata (iii) and (v) are imposed by the excellent agreement of experimental results with quantum mechanical predictions in many experiments involving microscopic systems.

A serious investigation of a nonlinear variant of the time-dependent Schrödinger equation has been reported by Pearle (Chapter 7 , see also Pearle 1969). Quite apart from considerations of agreement with experiment, Pearle's nonlinear equation is heuristically interesting. It is successful in achieving desiderata (i) and (ii), though I am sceptical that it, or any moderate variant of it, can achieve desiderata (iii) and (v).

The equation of Bialynicki-Birula and Mycielski (1976) was not designed for the purpose of solving the problem of the actualization of potentialities, and in fact it fails to fulfil the crucial desideratum (i). Nevertheless, it is noteworthy because of investigations which bear upon desideratum (v). The equation is

$$\frac{\partial \psi}{\partial t} = H\psi - b \log|a\psi|^2 \, \psi, \qquad (12.27)$$

where H is the usual quantum mechanical Hamiltonian operator, a is a constant introduced for dimensional reasons and of no great physical importance, and b is a non-negative quantity with dimension of energy which determines the magnitude of the nonlinearity. Neutron optics experiments have been performed for the purpose of setting upper bounds to b, and the results are very striking. A neutron interferometer experiment of Shull *et al.* (1980) showed that b has an upper bound of about 3×10^{-13} eV, and a neutron Fresnel diffraction experiment of Gähler *et al.* (1981) set a bound of about 3×10^{-15} eV. These are, to my knowledge, by far the best experimental confirmations of the linearity of the temporal evolution of the quantum state. Although these experiments tested only the Bialynicki-Birula–Mycielski equation, and not nonlinear equations generally, they indicate that nonlinear terms are likely to be required by experiment to be very small indeed.

It is premature to pronounce judgment upon the program of nonlinear variants of the Schrödinger equation. Nevertheless, my personal view is pessimistic. In particular, there is great tension between desiderata (iii) and (v). In order that the reduction of the superposition described in desideratum (i) should proceed very rapidly, the nonlinearity must be large, at least in certain measurement situations; but in order to account for the great success of the time-dependent Schrödinger equation in the experiments of Shull *et al.* and Gähler *et al.* the nonlinearity must be small. The only way out of this impasse is to hypothesize that the magnitude of the nonlinearity depends crucially upon either the mass or the number of degrees of freedom of the system, or upon both; for the confirmation of the time-dependent Schrödinger equation has been achieved with microscopic systems (neutrons), whereas the rapid reduction of superpositions is required for a measuring apparatus, which is macroscopic. So far, no one has proposed a concrete implementation of this proposal. Consequently, I remain pessimistic that the reduction of a superposition, i.e. the actualization of a potentiality, can be correctly described by a continuous process in space–time, governed by a nonlinear differential equation.

12.6. Conclusions

Work during the last two decades has led us to a high plateau concerning the interpretation of quantum mechanics, the topography of which was described in Sections 12.2 and 12.3, and the first part of Section 12.4. These results can be stated with great confidence. By contrast, the statements in the latter part of the paper were made diffidently. It is only a conjecture that the actualization of a potentiality must be sharply distinguished from probabilities unity and zero. And it is only a conjecture that the reduction of superpositions is inexplicable by substituting some appropriate nonlinear differential equation for the time-dependent Schrödinger equation.

Nevertheless, it should be emphasized that these two conjectures mesh quite well. As noted in the first comment on the desiderata, the success of the program of a nonlinear variant of the Schrödinger equation would lend plausibility to the view that actualization is the limiting case of potentiality—the achievement of the extreme values of probability. Hence our brief survey of the poor prospects of a nonlinear differential equation governing the temporal evolution of the quantum state agrees well with the conjecture concerning actualization presented in Section 12.4.

Are we able to make any more positive statement about the conditions under which potentialities are actualized? I shall risk one more conjecture: that the actualization of a potentiality is discontinuous and stochastic. One obvious motivation for this conjecture is that desideratum (iv) is easily satisfied, for the stochastic process can be accomplished with arbitrarily great speed. Someone may comment at this point that after much agonizing I have returned to a rather conventional interpretation of quantum mechanics, for in textbook presentations and in all versions of the Copenhagen interpretation the outcome of a measurement is stochastic and discontinuous. I wish, however, to maintain a distance from this conventional interpretation, until and unless overwhelming considerations drive me to accept it. My reason, once again, is to avoid anthropocentrism. The actualization of potentialities seems to be such a fundamental transition in the physical world that I cannot seriously believe that it occurs only when experiments are performed by human beings. We simply are not that important in the scheme of things, and we should not try to undo the Copernican revolution. Consequently, we must be attentive to non-anthropocentric accounts of the actualization of potentialities, even if the formulations up to now are quite unsatisfactory. At this Conference Andrew Frenkel has presented a theory due primarily to Károlyházy, in which the reduction of superpositions is linked to stochastic processes governing the space–time metric. Another possibility that is not exhausted is that the superposition principle fails for some kinds of macroscopic systems. Even if the fascinating efforts (described by Anthony Leggett in Chapter 3) to exhibit macroscopic quantum coherence are successful, there still may be parts of the macroscopic domain which serve as the locus of actualization. Finally, there is mind, as London and Bauer pointed out in their work on measurement (1939), and as Wigner has been lecturing for three decades (e.g. 1961). If only human minds were the locus of actualizations of potentialities, then anthropocentrism would re-appear, but if there are non-human minds— as we have good reason to believe—might they not be as efficacious in the reduction of superpositions as our own?

My final conclusion is that the high plateau which we have reached in the interpretation of quantum mechanics is a darkling plain.

Appendix: the Everett interpretation

Although I have not referred to the Everett interpretation (1957) in the body of this Chapter, I briefly mentioned it in my original lecture, and then most of the discussion following the lecture was devoted to it. This is not the first time that I have observed a discussion of the Everett interpretation expanding to fill the available vessel.

Part of the appeal of the Everett interpretation is the metaphor of a trunk with many branches. The ramification occurs whenever, according to more orthodox interpretations, a superposition is reduced, but instead of a choice among many branches the Everett interpretation retains them all. The metaphor tacitly presupposes, however, a preferred basis, in which the vector representing the state of the system is expressed: the vector itself is the trunk, and the projections upon the basis vectors are the branches. But objectively there is no preferred basis. There is a continuum of possible bases, all on the same footing. But when one speaks in this more accurate way about the mathematics of the quantum state, the quasi-familiarity of the original metaphor is lost, and with it the appeal.

There have been some valiant attempts to re-establish the branching metaphor by singling out a basis—notably the 'pointer basis' of Zurek (1981) and the 'environmentally determined basis' of Deutsch (1981). I shall not attempt to explore the technicalities of these efforts, but shall remark upon a slippery feature which they share. Only if the entity which is responsible for the singling out of the basis—the pointer of the apparatus in the one case, the environment in the other—is endowed with a solid set of properties does the project even begin to have intuitive appeal. But if the fundamental ontological fact is merely that the universe is in a quantum state, this solidity evaporates. Only in some bases (relatively few, it would seem, whatever that precisely means) are there branches in which the pointer exists and in which the environment has its salient properties, and even for these bases most branches do not endow the apparatus or the environment with the requisite properties for singling out a basis. There is, consequently, a danger of a regress in the process of basis determination.

Even if a single branch system is accepted, the Everett interpretation appeals to our imaginations only because we endow each branch with the same kind and the same degree of actuality which we habitually (due to our subjectivity or our limitations) attribute to only one branch. Along the branch that I (subjectively) experience I may blunder, but along another branch, with equal actuality, I triumph gloriously. The Everett interpretation can be used in this way to mitigate sorrows, but this use is two-edged, for it equally well implies the speciousness of happiness. Moral questions apart, this procedure of attributing the same actuality to each branch results from the fundamental proposal of Everett to elimi-nate the distinction between potentiality and actuality. And, of course, if all branches are on the same footing, it is true that this distinction vanishes. But what is to be conflated with what? Making all branches equally actual appeals to our imaginations, nourished as they have been by drama and novels. But might it not be at least as appropriate, or more so, to make all branches equally potential, equally in limbo, equally remote from existential definiteness? Might not the elimination of ac-tuality in favour of potentiality be demanded by the occurrence of interferences among branches, as provided by the quantum formalism?

The Everett interpretation is sometimes defended by an analogy to the Copernican theory: the true physical situation seems to be in contradic-tion with appearances because naive people fail to take into account their own physical states. I believe, however, that this analogy is faulty, and its weakness is instructive. The reconciliation of the apparent turning of the celestial sphere with the correct physical kinematics is achieved by an analysis of visual appearances in terms of the actual relative motions of the bodies observed, including that of the observer. But the Everett interpretation lacks an ingredient needed to implement this analogy: namely, a principle in terms of which one branch of the subject is endowed with subjective immediacy. Without such a principle there is only potentiality, as sketched in the preceding paragraph. But with such a principle the interpretation would lose its purity, and indeed there would be a reduction of a superposition based upon an interaction of the physical world with a mind.

My final objection to the Everett interpretation does not rest upon any anomalies or conceptual difficulties, but simply upon the consideration that it is an immense extrapolation of the linear dynamics of quantum mechanics from the microscopic domain which supplies experimental confirmation for this dynamics. We are so far from having evidence for the validity of this dynamics for macroscopic systems, for organisms, for creatures endowed with mentality, and for the space–time field that the extrapolation of this dynamics to the entire cosmos should be recognized as sheer conjecture.

Some of the points which I have just made are amplified, and others are given, in Bell (1978) and Stein (1984).

References

Aspect, A., Dalibard, J., and Roger, G. (1982). *Phys. Rev. Lett.* **49**, 1804.

Bell, J. S. (1964). *Physics* **1**, 195.

—— (1966). *Rev. Mod. Phys.* **38**, 447.

—— (1978). *Epistemol. Lett.* **20**, 1.

Bialynicki-Birula, I. and Mycielski, J. (1976). *Ann. Phys. N.Y.* **100**, 62.

Bohr, N. (1935). *Phys. Rev.* **48**, 696.

Clauser, J. F. and Horne, M. A. (1974). *Phys. Rev.* D **10**, 526.

Deutsch, D. (1981). Quantum theory as a universal physical theory. University of Texas at Austin preprint.

Eberhard, P. (1978). *Nuovo Cimento* **46B**, 392.

Einstein, A., Podolsky, B., and Rosen, N. (1935). *Phys. Rev.* **47**, 777.

d'Espagnat, B. (1976). *Conceptual foundations of quantum mechanics* 2nd edn. Benjamin, Reading, Massachusetts.

Everett, H. (1957). *Rev. Mod. Phys.* **29**, 454.

Gähler, R., Klein, A. G., and Zeilinger, A. (1981). *Phys. Rev.* A **23**, 1611.

Ghirardi, G. C., Rimini, A., and Weber, T. (1980). *Lett. Nuovo Cimento* **27**, 263.

Gleason, A. (1957). *J. Math. Mech.* **6**, 885.

Gudder, S. (1970). *J. Math. Phys.* **11**, 431.

Heisenberg, W. (1958). *Physics and philosophy*. Harper and Row, New York.

Jarrett, J. (1984). *Noûs* **18**, 569.

Kochen, S. and Specker, E. P. (1967). *J. Math. Mech.* **17**, 59.

London, F. and Bauer, E. (1939). *La théorie de l'observation en mécanique quantique* (Actualités scientifiques et industrielles 775). Hermann, Paris. (Engl. transl. in *Quantum theory and measurement* (ed. J. A. Wheeler and W. Zurek) (1983). Princeton University Press, Princeton.

Page, D. (1982). *Phys. Lett.* **91A**, 57.

Pearle, P. (1969). *Phys. Rev.* D **13**, 857.

Shimony, A. (1984*a*). *Br. J. Phil. Sci.* **35**, 25.

—— (1984*b*). In *Proc. Int. Symp. on Foundations of Quantum Mechanics in the Light of New Technology*, (eds. S. Kamefuchi, H. Ezawa, Y. Murayama, M. Namiki, S. Nomura, Y. Ohnuki, and T. Yajima). Physical Society of Japan, Tokyo. pp. 25–30.

Shull, C. G., Atwood, D. K., Arthur, J., and Horne, M. A. (1980). *Phys. Rev. Lett.* **44**, 765.

Simon, B. (1976). In *Studies in mathematical physics: essays in honor of Valentine Bargmann* (ed. E. Lieb, B. Simon, and A. Wightman). Princeton University Press, Princeton. pp. 327–49.

Stein, H. (1984). *Noûs* **18**, 635.

Suppe, F. (ed.) (1974). *The structure of scientific theories*. University of Illinois Press, Urbana.

Vigier, J.-P. (1982). *Astr. Nachr.* **303**, 55.

Wigner, E. P. (1961). In *The scientist speculates* (ed. I. J. Good). Heinemann, London.

Zurek, H. (1981). *Phys. Rev.* D **24**, 1516.

13

The many-worlds interpretation of quantum mechanics in quantum cosmology

Frank J. Tipler

At the beginning of my talk, I asked the audience 'Who believes in the many-worlds interpretation of quantum mechanics?'. Out of a total of perhaps 60 people in the audience, about 10 hands went up: the only hands I remember belonged to David Deutsch and Bryce DeWitt. There was a look of profound scepticism on the face of Roger Penrose (who had told me five months previously in New Orleans that 'I cannot believe the many-worlds interpretation makes physical sense'.) Bob Wald frowned and slowly shook his head.

I then asked, 'Who does not believe in the many-worlds interpretation?'. About 30 hands went up (including those of Roger Penrose and Bob Wald); clearly Bryce, David (and I) were in a minority at this meeting. Finally I asked, 'Who is neutral on the many-worlds interpretation?'. The remaining 20 hands went up. I shall do my best in this short paper to persuade both the sceptics and those have not yet formed an opinion as to the validity of the many-worlds interpretation (MWI) that this interpretation is philosophically more beautiful than competing interpretations, and that it can be used in quantum cosmology as a powerful tool not only to interpret the wave function of the universe, but also to give us some information about the equation which this wave function obeys. I shall present my arguments only in outline; the details will be given elsewhere (Tipler 1986; Barrow and Tipler 1986).

Most sceptics, I've found, have a mistaken idea of what the MWI really means, so it behoves me to review it briefly. The MWI is a theory of measurement, so it is concerned with describing how the universe looks to us *qua* human beings. At some stage during *any* measurement, the information is digitalized, and this is true even for the measurement of continuous variables, for example position or momentum. A typical position measurement—of an α-particle nucleus, say—would be carried out by letting the α-particle pass through an array of atoms such as those of a photographic plate. The array cannot make position measurements of unlimited accuracy; at best, the accuracy would be limited by the size of the atom. Even if we were to improve the accuracy of the position measurement at this level of the measuring process, the position measurement would in the end be digitalized when it is transmitted to human beings, for the data corresponding to an arbitrarily precise position measurement would in general exceed the storage capacity of a human brain. Hence we can model any measurement by a measurement of a discrete variable.

Let us suppose the discrete variable we wish to measure has two states $|\uparrow\rangle$ and $|\downarrow\rangle$ (which may represent an electron with spin up and spin down, respectively). In the MWI picture—and indeed, in the picture of any quantum theory of measurement which includes the measuring apparatus in the physical analysis—a measurement consists of correlating the states of system to be measured with states of the measuring apparatus. Since we have two possible system states, it is convenient to allow the measuring apparatus designed to measure them to have three states, $|n\rangle$, $|u\rangle$, $|d\rangle$, where $|n\rangle$ represents a neutral state which the measuring apparatus will be in prior to the measurement.

If the system just happens to be in an eigenstate of the variable to be measured; i.e., either in the state $|\uparrow\rangle$ or in the state $|\downarrow\rangle$, the measuring apparatus had better correctly record this fact if it is to deserve the name 'measuring apparatus'. That is, the measurement process must be represented by a unitary operator M with the properties

$$M|\uparrow\rangle|n\rangle = |\uparrow\rangle|u\rangle \tag{13.1a}$$

$$M|\downarrow\rangle|n\rangle = |\downarrow\rangle|d\rangle, \tag{13.1b}$$

where I have assumed that the apparatus state $|u\rangle$ is to record the system as being in state $|\uparrow\rangle$ if in fact it is in that state, and $|d\rangle$ to record the system as being in state $|\downarrow\rangle$ if in fact it is in that state. If the measuring operator has an effect on the apparatus states other than eqns (13.1), the apparatus it represents is defective and we junk it and get an apparatus which does work; i.e. which has the effect (13.1) on the eigenstates of the 'universe' which is the system and apparatus regarded as a single quantum system. It will be noted that I have also assumed that my measurement apparatus has no effect on the system eigenstates when a measurement is made. Such a measurement has been termed a 'von Neumann measurement' by DeWitt (1973); it constitutes the minimal interference a measurement can have on a system.

Since all dynamics in quantum mechanics occurs via *linear* unitary operators, and since in quantum mechanics it is assumed that any state is a ray in some Hilbert space, which means it can be represented as a linear combination of states which span the Hilbert space (in our example $|\uparrow\rangle$ and $|\downarrow\rangle$), it follows that eqns (13.1) also define the effect of a measurement M on *any* state $|\psi\rangle = \alpha\,|\uparrow\rangle + \beta\,|\downarrow\rangle$, where α and β are constants:

$$M\,|\psi\rangle\,|n\rangle = M(\alpha\,|\uparrow\rangle + \beta\,|\downarrow\rangle)\,|n\rangle = M(\alpha\,|\uparrow\rangle\,|n\rangle + \beta\,|\downarrow\rangle\,|n\rangle)$$
$$= \alpha\,|\uparrow\rangle\,|u\rangle + \rho\,|\downarrow\rangle\,|d\rangle. \tag{13.2}$$

If we regard human beings and more generally macroscopic objects as quantum mechanical objects—if for example $|u\rangle$ and $|d\rangle$ are regarded as mental states of human beings—then we are led inexorably by the last expression in eqn (13.2) to the conclusion that an observation by a human being on some system which is not in an eigenstate of the observable will result in the human observer being 'split' by the observation: in one 'world' the system will be in state $|\uparrow\rangle$ and the observer will measure it as being in state $|\uparrow\rangle$ (that is the observer will be in state $|u\rangle$), and in the other 'world' the system will be in state $|\downarrow\rangle$ and the observer will measure it as being in state $|\downarrow\rangle$. Acceptance of this counter-intuitive (!!) conclusion is the many-worlds interpretation.

However, many presentations of the MWI have made it appear more counter-intuitive than it really is. For example, many accounts assert that 'the entire universe is split by a measurement'. This is not true. Only the observed/observer system splits; only that restricted portion of the universe acted on by by the measurement operator M splits. If we let $|\text{Cosmos}\rangle$ be the parts of the universe which is not acted on by M—and this will be almost all of it—then the state of the entire Universe before the measurement is $|\psi\rangle\,|n\rangle\,|\text{Cosmos}\rangle$, and the effect of the measurement is

$$M\,|\psi\rangle\,|n\rangle\,|\text{Cosmos}\rangle = \alpha\,|\uparrow\rangle\,|u\rangle\,|\text{Cosmos}\rangle + \beta\,|\downarrow\rangle\,|d\rangle\,|\text{Cosmos}\rangle$$
$$= (\alpha\,|\uparrow\rangle\,|u\rangle + \beta\,|\downarrow\rangle\,|d\rangle)\,|\text{Cosmos}\rangle, \tag{13.3}$$

which shows that $|\text{Cosmos}\rangle$ is *not* split. If a human being is the measuring apparatus, then any interaction with the rest of the universe will split the human being if—but only if—the interaction is such that it could result in different states of the human organism which are distinguishable by the human sensory system. The splitting is denoted by correlations between the human and the various subsets of the Universe with which he interacts.

Since the information stored by human beings is finite, the set of all possible measurements can split a human being into only a finite number of pieces. A very rough estimate of an upper bound for the number of pieces can be obtained by assuming that each atom in a human being can be in two states distinguishable by the human consciousness. As the number of atoms in a human being is about 10^{26}, the upper bound to the number of pieces is 2 raised to the 10^{26} power. Admittedly a very large number, but still finite.

Because in a von Neumann measurement a more drastic change occurs in the measuring apparatus than in the system, it is more appropriate to regard the *observer* as splitting rather than the universe. Furthermore, the observer splits because he is designed to split. A measuring apparatus is defined to be a physical system which undergoes a physical change sufficiently large to be easily detectable by humans when interacting with the system to be measured. The measuring apparatus also must be capable of going into several macroscopically distinguishable states as a consequence of the measurement interaction. In the above example, if the system is in state $|\uparrow\rangle$, the apparatus records this fact by undergoing a drastic change from the state $|n\rangle$ to the state $|u\rangle$, and if the system is in state $|\downarrow\rangle$, the apparatus must undergo the drastic change from $|n\rangle$ to $|d\rangle$. From these fundamental requirements, the apparatus *must* split if M is a linear operator.

Those physicists who reject the MWI are of course aware of this. The main reason they prefer to think that the measurement process is *not* correctly represented by the linear operators that do correctly describe non-measurement dynamics has been succinctly stated by Abner Shimony: '. . . the continuous evolution of the total quantum state [eqn (13.2) above] is obtained by Everett at the price of an extreme violation of Ockham's principle, the entities being entire universes' (Shimony 1963). As we have seen above, however, the splitting of the 'universe' in the MWI really refers only to the system/apparatus rather than to the universe in the usual sense of the system/apparatus and everything else that exists. It is nevertheless true that there are more 'worlds'—more existent entities—in the universe in the MWI than in an interpretation where $|n\rangle$ goes to either $|u\rangle$ or $|d\rangle$ but not both. In fact, in a private conversation, John A. Wheeler told me that the main reason for his current rejection of what was once called the 'Everett–Wheeler theory', was his distaste for the enlarged ontology it definitely implies.

I personally think the enlarged ontology of the MWI is really a very strong argument for *accepting* rather than rejecting it. One of the most fundamental problems in cosmology is the problem of the Universal Initial Conditions: in particular, why do we happen to live in a universe with density parameter Ω very close to 1? By 'very close' I mean within two orders of magnitude. As is well known, this requires enormous fine tuning at the Planck time in a non-inflationary universe. In any classical cosmology, the actual universe would be just one of an infinite number of possible universes, and we would have no ultimate reason for picking one over another; we would be unable to say why the one unique point in initial data space was chosen for actual physical existence rather than another. This is a mathematical consequence of the Einstein field equations having an initial data space with a non-trivial topology. Furthermore, every hidden variable theory known to me has an initial data space which consists of more than a single point, and I suspect it would be exceedingly difficult if not impossible to construct a hidden variable theory whose initial data space was only a single point, since such a theory could not be applied to laboratory phenomena where the initial conditions do vary. Thus the Universal Initial Conditions would have to be accepted as an *a priori* postulate in any classical cosmological theory, including a hidden variable cosmological theory.

The enlarged ontology of the MWI automatically obviates the problem of the Universal Initial Conditions: the Universe consists of *all* logically possible universes. (When capitalized, the word 'Universe' will mean everything that exists; when not capitalized, 'universe' will refer to a single Everett branch whose evolutionary history is approximately given by the classical evolution equations with our particular branch being modelled very closely by the usual Friedmann universe.)

The ontological enlargement required by the MWI is precisely analogous to the spatial enlargement of the universe which was an implication of the Copernican theory. Indeed, philosophers in Galileo's time used Ockham's principle to support the Ptolemaic and Tychonic systems *against* the Copernican system. For example, the philosopher Giovanni Agucchi (Drake 1978, p. 212) argued in a letter to Galileo that one of the three most powerful arguments against the Copernican system was the existence of the vast useless interstellar void which the Copernican system required. Thus at the very least it behoves us to be very cautious in accepting Ockham's-principle objections against the MWI. The Copernican system was accepted in spite of the ontological enlargement it insisted on because of the greater theoretical elegance of the heliocentric system over its competitors. Similarly, I would contend that the MWI should be accepted because it is the most elegant interpretation of the quantum mechanical formalism.

If all logically possible universes are present, why are we in one with $\Omega = 1$? This question can be answered only by invoking the probability interpretation of $|\langle R | \Psi \rangle|^2$, where $|\Psi\rangle$ is the wave function of the Universe. For simplicity we will evaluate the Universal wave function in a co-ordinate space of one variable, the radius of the universe R. In other words, in what follows we will consider only what the MWI will say about interpreting a quantized Robertson–Walker universe.

Up to now our discussion of the MWI has not made any reference to probability, to probability amplitudes, or indeed to scalar products in Hilbert space. The analysis of observer splitting requires only the assumption that the observer states $|n\rangle$, $|u\rangle$, and $|d\rangle$ be 'distinguishable'; it is not necessary to assume that this distinguishability arise from the vanishing of $\langle u \,|\, d\rangle$, etc. But questions involving the likelihood of a given classical universe must use probability amplitudes and scalar products.

As DeWitt (1973) has shown in his analysis of how the probability interpretation of the wave function arises in the MWI, it is not physically meaningful to talk about the probability of the various worlds into which a quantum system splits if the system can be measured only once. Rather, if it is desired to measure the probability of the various worlds, it is necessary to measure the splitting of an ensemble of systems all of which are in the identical quantum state. For the system with the two distinguishable states $|\uparrow\rangle$ and $|\downarrow\rangle$ and in the state $|\psi\rangle = \alpha\,|\uparrow\rangle + \beta\,|\downarrow\rangle$, it is necessary to make m separate measurements on the system $|\psi\rangle\,|\psi\rangle\,|\psi\rangle\ldots|\psi\rangle$, which is the tensor product of m systems $|\psi\rangle$, each in identical quantum states. The kth measurement of the system $|\psi\rangle\ldots|\psi\rangle$ is represented by the measurement operator M_k, which causes the state of the kth member of $|\psi\rangle\ldots|\psi\rangle$ to be recorded in the memory of an apparatus with m memory slots. The initial state of the apparatus memory can be represented as $|n, n, \ldots, n\rangle$, with m n's. The effect of the kth measurement is analogous to eqn (13.2):

$$M_k\,|\psi\rangle\ldots(\alpha\,|\uparrow\rangle + \beta\,|\downarrow\rangle)\ldots|\psi\rangle\,n, \ldots, n\rangle$$
$$= \alpha\,|\psi\rangle\,|\psi\rangle\ldots|\uparrow\rangle\,|\psi\rangle\ldots|\psi\rangle\,|n, n, \ldots, n, u, n, \ldots n\rangle$$
$$+ \beta\,|\psi\rangle\,|\psi\rangle\ldots|\downarrow\rangle\,|\psi\rangle\ldots|\psi\rangle\,|n, n, \ldots, n, d, n, \ldots, n\rangle, \qquad (13.4)$$

so only the kth slot in the apparatus memory is changed. The complete set of m measurements of the state of $|\psi\rangle\ldots|\psi\rangle$ is represented by the product of operators $M_m M_{m-1}\ldots M_2 M_1$ acting on $|\psi\rangle\ldots|\psi\rangle\,|n, \ldots, n\rangle$. The result is to split the universe (the m systems and the apparatus) into a linear superposition of states in which all possible sequences of u's and d's appear in the apparatus memory. Each sequence of u's and d's in the apparatus memory represents an Everett world in which that sequence of u's and d's were measured. The standard interpretation of $|\langle\uparrow\,|\,\psi\rangle^2/ |\langle\psi\,|\,\psi\rangle|^2$ is that it is the probability that $|\uparrow\rangle$ will be measured when the system is in the state $|\psi\rangle$. It has been shown by Finkelstein (1963), Hartle (1968), and Graham (see DeWitt 1973) that as $m \to \infty$, the Everett worlds in which the percentage of u's in the apparatus memory sequence is different from $|\langle\uparrow\,|\,\psi\rangle^2/|\langle\psi\,|\,\psi\rangle|^2$ is of measure of zero in the Hilbert space. The standard probability interpretation is obtained if it is assumed that measure zero events in the most fundamental physical space (Hilbert space in quantum mechanics) correspond to events of zero probability. It is not possible to obtain the standard probability interpretation without this probability assumption (or some variant of it), and this is often used as a criticism of the MWI. However, it is a matter of logic that it is impossible for probabilities to appear in any theory without introducing probability into the theory in the form of probability axioms. Even the hidden variable theories have to introduce the notion of probability in this way in order to recover the standard probability interpretation of quantum mechanics (Tipler 1984). The only question is what probability axioms are the most natural. I submit that 'measure zero in physical space means zero probability' is the most natural, as it introduces probability into quantum mechanics in the same way that probability enters statistical mechanics. Once the formula $|\langle \uparrow | \psi \rangle|^2/|\langle \psi | \psi \rangle|^2$ is interpreted as the probability that the state $|\uparrow\rangle$ will be measured, it is natural to normalize the system wave function to unity: $|\langle \psi | \psi \rangle|^2 = 1$. Only if it is possible to measure the ratios of the state $|\uparrow\rangle$ to the state $|\downarrow\rangle$ by measuring the state of an ensemble of systems does it make physical sense to normalize the system wave function.

However well this may work for quantum subsystems of the Universe, there is a difficulty which arises from the very uniqueness of the Universe: there is only one Universe, and hence only one wave function $|\Psi\rangle$ which represents it. Thus it is not possible to construct an ensemble of quantum Universes $|\Psi\rangle \ldots |\Psi\rangle$ on which to carry out measurements of probabilities. Therefore, if it is the radius of the Universe that is being measured, the *first* measurement of the universal radius will split the Universe into universes which collectively have all possible radii. Subsequent measurements of R will yield results which are consistent with the first splitting; that is, in each universe the observer will see an essentially classical evolution no matter what the shape of the Universal wave function. The physics of a quantum Universe according to the MWI is pictured in Fig. 13.1. The important point is that an observer in a branch universe is incapable of determining the overall shape of $|\langle R | \Psi \rangle|^2$, for this determination would require measurements on an ensemble of universes. It follows that an observer is incapable of measuring $|\langle R | \Psi \rangle|^2/\langle \Psi | \Psi \rangle|^2 \, dR$, so there is no physical reason for normalizing the Universal wave function.

Nevertheless, even non-normalizable wave functions can be regarded as coding probability information. The natural assumption to make—the assumption which is the most natural generalization of the probability assumption in non-cosmological quantum mechanics where we *can* normalize wave functions—is that at any given time, the relative probability of finding ourselves in a universe with a radius between R_1 and R_2 instead of between R_3 and R_4 is given by

$$\left[\int_{R_1}^{R_2} |\langle R | \Psi \rangle|^2 \, dR \right] \div \left[\int_{R_1}^{R_2} |\langle R | \Psi \rangle|^2 \, dR + \int_{R_3}^{R_4} |\langle R | \Psi \rangle|^2 \, dR \right]. \quad (13.5)$$

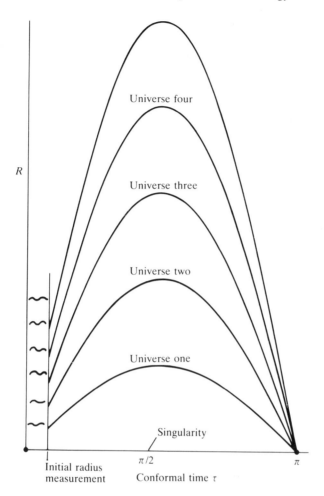

Fig. 13.1. The branching of a quantum universe. Before the first interaction occurs that can encode a scale measurement, the Universe, represented before this interaction occurs as a series of wavy lines, has no radius. After the first scaled interactions occur, the Universe has been split by the interactions into a large number of branches, in each of which an essentially classical evolution is seen. These branches are represented here by the sine curves, each of which goes through the final singularity at conformal time $t = \pi$. The collection of all sine curves represents all the classical radiation gas Friedmann models. Each curve is defined by $R_{\max}$, the radius of the universe at maximum expansion. In the quantum Universe, *all* the classical universes are present, one classical universe defining a single branch.

This assumption, is a probability assumption, and it allows us to calculate the probability of finding ourselves in one class of universes instead of another.

Let us suppose that the co-ordinate space domain of the Universal wave function $\langle R \mid \Psi \rangle$ is $(0, +\infty)$, because negative radii are difficult to interpret (see, however, Brill (1975)). Furthermore, let us suppose that the Universe is closed, which for quantum Robertson–Walker universes means that R will be interpreted as the scale factor in a metric of the form

$$\mathrm{d}s^2 = R^2(t)[-\mathrm{d}t^2 + \mathrm{d}\chi^2 + \sin^2\!\chi(\mathrm{d}\theta^2 + \sin^2\!\theta \, \mathrm{d}\phi^2)]. \quad (13.6)$$

The probability that we will find ourselves in a branch universe of radius R or less is obtained from eqn (13.5) by setting $R_1 = 0$, $R_2 = R$, $R_3 = R$, and $R_4 = +\infty$. If the Universal wave function $\langle R \mid \Psi \rangle$ is non-normalizable, but $|\langle R \mid \Psi \rangle|^2$ is bounded above as $R \to 0$, this probability is zero. That is, it is overwhelmingly likely that we will find ourselves in a branch universe which is closed, but arbitrarily closed to the spatially flat Robertson–Walker universe with $\Omega = 1$.

I claim that this is the most natural resolution of the Flatness Problem in cosmology. In contrast to the inflationary universe solution, it does not involve fine-tuning physical constants. Instead, it requires only that we adopt the MWI and assume that the Universal wave function is non-normalizable. But as we have seen, there is no possible physical measurement which could require us to normalize the Universal wave function. It is a generally accepted principle of physics that what is not forbidden by physical law is compulsory, so it seems likely that the Universal wave function is in fact not normalizable.

One often sees in the literature assertions that 'the most likely radius for the universe according to quantum mechanics is the Planck length'. Such assertions are not true. They arise from attempting to impose classical concepts on a quantum world: the only length scale that can appear in quantum gravity is the Planck length, so it is inferred that the radius corresponding to the peak of the wave function is the Planck length. But the wave function is envisaged to be strongly peaked because strongly peaked wave functions are used in elementary texts as models of classical objects, and it is of course true that the evolution of the actual universe is presently close to that predicted by the classical Einstein equations.

However, if one analyses not the time evolution of the spatial position of a particle's wave function but rather the time evolution of the *observations* of the particle's position, one discovers that wave functions which are spread out over all space can nevertheless yield observations which closely approximate classical motion. For example, an α-particle with a definite momentum p has a wave function $\langle R \mid \psi \rangle = \exp[ipR]$, which implies $|\langle R \mid \psi \rangle|^2 = $ constant, and yet Heisenberg (1930) and Mott (1929) (reprinted in Wheeler and Zurek (1983)) showed long ago that the motion of such a particle would appear classical to very high accuracy. What happens is that the first measurement of position splits the observing system into all possible initial positions and the time evolution in each of the corresponding worlds is essentially classical. (See Tipler (1986) for a detailed discussion of this.) But this is just what we pictured as happening in quantum cosmology in Fig. 13.1. The radius of a particular branch universe is not determined by any intrinsic scales (such as the Planck length) in the Universal Hamiltonian, but rather by the initial measurement of the radius of the Universe.

If the observed cosmological length scale—the observed radius of the universe—arises not from an intrinsic scale in the Hamiltonian but rather from position measurements, this suggests that the fundamental Hamiltonian may not have any fundamental length scale; i.e. it suggests the fundamental Hamiltonian is that of a radiation field. Furthermore, the philosophy underlying MWI—and the Copernican cosmological principle as applied to the set of all branch universes—suggests that *a priori* we should regard the various universe branches as equally likely (which would mean $|\langle R \mid \Psi \rangle|^2 = $ constant for the Universal wave function). But this would be impossible if there were an intrinsic length scale in the Hamiltonian itself. Thus the MWI suggests that the Universal Hamiltonian is conformally invariant; the world-view of the MWI suggests restrictions on the equation which the Universal wave function obeys.

In arguing for the MWI, I am arguing for a change in world-view as sweeping as the change in world-view ushered in by the Copernican Revolution. In 1610 there were three interpretations of the observed planetary motion, the Ptolemaic, the Tychonic, and the Copernican, all of which were empirically equivalent and entirely viable (Kuhn 1959), and two of which—the Tychonic and the Copernican—were actually mathematically equivalent if applied to circular orbits (Kuhn 1959). The Ptolemaic system was made the most implausible by Galileo's observations with the telescope which he announced in 1610, just as hidden variable theories were rendered implausible in the opinion of many by Professor Aspect's beautiful experiments. What finally convinced Galileo of the truth of the Copernican system as opposed to the Tychonic system was the fact that astronomers who would not regard the Earth's motion as real were under a great handicap in understanding the motions they observed, regardless of the formal mathematical equivalence of the Copernican and Tychonic systems (Drake 1980, p. 54). This was also the major factor in convincing other physicists and astronomers of the truth of the Copernican system (Kuhn 1959). Furthermore, the Tychonic system was dynamically ridiculous and almost impossible to apply to problems besides the particular problems of planetary orbits it had been designed to analyse. Similarly, the wave function collapse postulated by the Copenhagen interpretation is dynamically ridiculous, and this interpretation is difficult if not impossible to apply in quantum cosmology. Physicists who think in terms of the Copenhagen and hidden variable interpretations are handicapped in thinking about quantum cosmology. Indeed, I know of no general relativistic quantum cosmology which does *not* use the MWI, at least tacitly.

I suggest that the many-worlds interpretation will eventually replace the Copenhagen and hidden variable interpretations like the Copernican system replaced the Ptolemaic and the Tychonic.

Acknowledgement

This work was supported in part by the National Science Foundation under grant number PHY-8409672.

References

Barrow, J. D. and Tipler, F. J. (1985). *The anthropic cosmological principle*. Oxford University Press, Oxford.
Brill, D. R. (1975). In *Quantum theory and the structures of time and space* (ed. L. Castell, M. Driesch, and C. F. von Weizsacker). Carl Hanser, Munich.
DeWitt, B. S. (1973). In *The many-worlds interpretation of quantum mechanics* (ed. B. S. DeWitt and N. Graham). Princeton University Press, Princeton.
Drake, S. (1978). *Galileo at work*. University of Chicago Press, Chicago.
—— (1980). *Galileo*. Hin & Wang, New York.
Finkelstein, D. (1963). *Trans. N.Y. Acad. Sci.* **25**, 621.
Hartle, J. D. (1968). *Am. J. Phys.* **36**, 704.
Heisenberg, W. (1930). *The physical principles of quantum theory*. University of Chicago Press, Chicago.
Kuhn, T. S. (1959). *The Copernican revolution*. Vintage, New York.
Mott, N. F. (1929). *Proc. R. Soc. Lond.* A **126**, 76.
Shimony, A. (1963). *Am. J. Phys.* **36**, 755.
Tipler, F. J. (1984). *Phys. Lett.* A **103**, 188.
—— (1986). Interpreting the wave function of the Universe, *Phys. Reports* (in press).
Wheeler, J. A. and Zurek, W. H. (1983). *Quantum theory and measurement*. Princeton University Press, Princeton.

14

Three connections between Everett's interpretation and experiment

David Deutsch

It has come to my attention that there are still some conference participants who harbour residual doubts about the Everett interpretation (see Everett 1957; DeWitt 1973) of quantum theory. I thought it might be helpful to leave aside for a moment the theoretical arguments *pro* and *contra* (such as they are) and look at a seldom-mentioned but important aspect of Everett's interpretation, namely its connection with experiment.

There is a widespread belief that the conventional (i.e. wave function collapse) view and Everett's are alternative *interpretations* of the same quantum formalism, and are identical in their experimental predictions. Everett himself believed this. But he was mistaken. The conventional 'interpretation', unlike Everett's, is actually more than just an interpretation of the quantum formalism. It postulates a *modification of the formalism*, whose nature is unspecified but whose effect is somehow to introduce non-unitary evolution into quantum dynamics. More specifically, it asserts that certain physical systems (human brains and the like) violate the quantum principle of superposition. In fact, the characteristic assertion made by *all* realistic 'interpretations' other than Everett's is that

$$\text{`superpositions of distinct states of consciousness} \atop \text{do not occur in nature'.} \qquad (14.1)$$

Such an assertion is not just a matter of airy-fairy metaphysics. It is, in principle at least, an experimentally testable statement (see Experiment 3 below, and Deutsch (1985*a*)).

The three experiments I am going to describe span a huge range of practical feasibility. As you might expect, the most persuasive one (Experiment 3, described below) is the least practical at the present state of technology. Experiment 3 is nothing less than a direct, unambiguous experimental test of the Everett quantum theory against any theory that has the wave function collapsing. On the other hand Experiment 1, although its status as a crucial test is more questionable, has the advantage of being perfectly feasible today. Experiment 2 is intermediate, both in feasibility and in persuasive power, between Experiments 1 and 3.

With your assistance I propose to perform Experiment 1 right now.

14.1. Experiment 1 (first part)

Let the quantum state of the universe be $|\Psi\rangle$. In the Heisenberg picture $|\Psi\rangle$ is of course constant in time. As Page and Wooters (1983) have recently stressed, the physical effects of time 'passing' are consequences of the fact that $|\Psi\rangle$ is not an eigenstate of an appropriate physical time observable $\hat{t}$. Thus if $|t\rangle |\lambda\rangle$ is the simultaneous eigenstate of $\hat{t}$ and a suitable observable $\hat{\lambda}$ representing the rest of the universe, with eigenvalues t and λ,

$$|\Psi\rangle = \int dt \int d\lambda c(t, \lambda) |t\rangle |\lambda\rangle . \qquad (14.2)$$

In the 'classical approximation' the amplitude $c(\lambda, t)$ has support only near $\lambda = \bar{\lambda}(t)$, where $\bar{\lambda}(t)$ represents a 'classical history' of the universe. For the purposes of this experiment it will suffice to confine our attention to the state of the contents of the room you are in, including yourself, approximating the dynamical evolution as both isolated and 'classical', i.e.

$$|\Psi\rangle = \int dt c(t) |t\rangle |\bar{\lambda}(t)\rangle. \qquad (14.3)$$

Now perform two experimental observations: First, measure the present time (i.e. the value of a suitable 'time-observable' such as the appearance of your watch face), and call it t_1. And then note that you are *conscious* at the time when you are observing the value t_1. In the approximation (14.3) we may summarize your observations so far as

$$\bar{\lambda}(t_1) = \text{'I am conscious, observing } t_1\text{'}. \qquad (14.4)$$

We shall now leave Experiment 1 running for a while.

14.2. Experiment 2: quantum parallelism

Imagine a computer from whose quantum components, including the processor and the memory, dissipation and external perturbations have been eliminated sufficiently for interference to be observable between its computational states. (For explicit descriptions of computer models with this property see e.g. Benioff (1982), Albert (1983), Feynman (1982), and Deutsch (1985*b*)).

Let the computer be programmed in such a way that it executes an algorithm $\pi(f)$ computing the function

$$f: \{0, 1\} \rightarrow \{0, 1\}. \tag{14.5}$$

That is, $\pi(f)$ takes a single bit as parameter and returns a single bit as its result. The fact that the domain and range of f are so small does not of course mean that f is easy to compute from $\pi(f)$. In fact let us suppose that $\pi(f)$ is an algorithm of great complexity whose effect is not known and whose execution time T is very large. The effect of the dynamical evolution of the computer during the computation of either $f(0)$ or $f(1)$ is summarized, in the Heisenberg picture, by

$$|\Psi\rangle = |a, 0; t = 0\rangle = |a, f(a); t = T\rangle, \quad (a \in \{0, 1\}). \tag{14.6}$$

The first label in the kets in eqn (14.6) represents the state of a suitable 'input' observable $\hat{I}(t)$ and the second the state of a suitable 'output' observable $\hat{O}(t)$. Equation (14.6) states that if the input of the computer is prepared with the value a, then at a time T later the output contains, with unit probability, the value $f(a)$.

Now suppose that what is of interest to the programmer of this computer is not specifically $f(0)$ or $f(1)$, but a certain one-bit function G of them both:

$$G(f(0), f(1)) \in \{0, 1\}. \tag{14.7}$$

Specifically, let G be the operation

$$\begin{aligned} G(a, b) &\equiv a + b \quad (modulo\ 2) \\ &\equiv a \oplus b. \end{aligned} \tag{14.8}$$

If the computer were a classical machine there would be no way of evaluating $f(0) \oplus f(1)$ without computing *both* $f(0)$ and $f(1)$ *via* $\pi(f)$. Therefore the total time required would be at least $2T$. This could be reduced to about T if *two* classical processors were used in parallel. However, we shall see that because eqn (14.6) describes a quantum system, an additional range of options is open.

Let the computer be prepared at $t = 0$ in a state which is not an eigenstate of the input observable $\hat{I}(0)$, specifically in the linear combination

$$2^{-1/2}(|0, 0; t = 0\rangle + |1, 0; t = 0\rangle) \tag{14.9}$$

of eigenstates. According to eqn (14.6), the state of the computer in terms of eigenstates of $\hat{I}(T)$ and $\hat{O}(T)$ will be

$$2^{-1/2} \sum_{a=0}^{1} |a, f(a); t = T\rangle . \tag{14.10}$$

Thus at time T the state 'contains' the results, $f(0)$ and $f(1)$, of *both* computations—yet only one processor was used.

Of course there is a catch: It is easy to show that no measurement performed at time T on the state (14.10) without a knowledge of $f(0)$ or $f(1)$ can give an observer a knowledge of both $f(0)$ and $f(1)$. Remembering that we are interested only in $f(0) \oplus f(1)$, let us lower our sights: Is it possible to obtain a knowledge of just this combination from state (14.10) (again without executing $\pi(f)$)? Again the answer is 'no'.

What is possible, however, is *quantum parallelism* (Deutsch 1985*b*), a trick which allows us to compute certain parallelizable functions, including our G, with a single quantum processor more rapidly than with the corresponding classical processor—but with probability less than one. In the following example $G(f(0), f(1))$ is computed with probability $\frac{1}{2}$ in a time only a little greater than T:

At time T, let a non-degenerate observable $\hat{R}(T)$, whose eigenstates are

$$|zero\rangle \equiv \tfrac{1}{2}(|0, 0; t = T\rangle - |0, 1; t = T\rangle + |1, 0; t = T\rangle - |1, 1; t = T\rangle)$$

$$|one\rangle \equiv \tfrac{1}{2}(|0, 0; t = T\rangle - |0, 1; t = T\rangle - |1, 0; t = T\rangle + |1, 1; t = T\rangle)$$

$$|fail\rangle \equiv \tfrac{1}{2}(|0, 0; t = T\rangle + |0, 1; t = T\rangle + |1, 0; t = T\rangle + |1, 1; t = T\rangle)$$

$$|error\rangle \equiv \tfrac{1}{2}(|0, 0; t = T\rangle + |0, 1; t = T\rangle - |1, 0; t = T\rangle - |1, 1; t = T\rangle)$$

$$\tag{14.11}$$

be measured. Such an observable exists, since the states (14.11) form an orthonormal set. Furthermore, the measurement can be performed in a fixed, short time. By considering the scalar products of these states with all four possible states of the form (14.10) (corresponding to the four possible graphs of the function f), it is easily shown that

(i) If the outcome of the measurement is '*zero*' (i.e. the eigenvalue corresponding to the state $|zero\rangle$) then the value of $G(f(0), f(1))$ is *necessarily* zero.

(ii) Similarly, if the outcome is '*one*' then $G(f(0), f(1))$ must be 1.

(iii) If the outcome is '*fail*', nothing can be inferred about the value of $G(f(0), f(1))$, nor can further measurements not involving the algorithm $\pi(f)$ gain any information about it.

(iv) The outcome is never '*error*' if the measurements are all made with perfect precision. (An arbitrarily high precision is possible in principle).

(v) The probability of the outcome '*fail*' is $\frac{1}{2}$, whatever the graph of f may be.

With probability $\frac{1}{2}$ all the information in state (14.10) about the graph of f has been destroyed; but with equal probability, $\frac{1}{2}$, the value of $G(f(0), f(1))$ has become known to the observer. Note that in the latter event the observer knows *with certainty* that he has obtained the true value of $G(f(0), f(1))$. We are not *guessing* G here (as classical 'stochastic algorithms' do), we are trading execution speed for 'execution probability' in a way that is not possible classically, even with a classical stochastic computer.

What is the physical explanation of why quantum parallelism works? The description of the process according to Everett's scheme is simple: Two different algorithms, computing $f(0)$ and $f(1)$, are executed simultaneously by different instances of the same processor in different Everett branches. After a time T, the quantity $f(0)$ has been evaluated in one branch, and $f(1)$ in the other. At this point an outside observer *could* measure $\hat{O}(T)$, but the result would be to transfer the split to himself. In one branch he would obtain $f(0)$, and in the other $f(1)$. Subjectively, there would be a probability $\frac{1}{2}$ of obtaining the result of each computation while making the other inaccessible. To make use of quantum parallelism, the observer measures not $\hat{O}(T)$ but $\hat{R}(t)$, an observable which does not commute with $\hat{O}(T)$ or $\hat{I}(T)$. The effect is to transfer information (namely $f(0) \oplus f(1)$) depending non-trivially on the whole graph of f, into a single branch—at the cost of erasing all information about the graph of f from the other branch.

The 'explanation' in terms of non-Everett 'interpretations' is also simple—and vacuous:

> 'The probabilities of the various *outcomes* are as in (i) to (v) above, but further explanation is impossible. Describing the process internal to the quantum computer as an objective sequence of phenomena is impossible.' (14.12)

That is, of course, unless the complexity (or whatever) of the algorithm $\pi(f)$ is such that the computer itself counts as an 'observer'. In that case, because of (1), non-Everett 'interpretations' predict that quantum parallelism will not work, and the final result will be random (cf. Experiment 3).

I do not claim that when quantum parallelism is eventually observed experimentally, the conventional 'interpretation' will be 'refuted' empirically. This experiment is not a crucial test as Experiment 3 is. Nevertheless, the routine use of quantum parallelism in future commercial quantum computers will already make the conventional interpretation untenable for a different reason: It will be useless. What will a practical quantum programmer make of the statement (14.12) that his description of his program as proceeding in a sequence of parallel steps is 'meaningless', that only the final result exists physically? He will know from everyday experience that the execution of the steps is real in every way that counts to him: all his ingenuity is devoted to planning precisely which operations are to be executed in which order and in which Everett branch. To use N-fold quantum parallelism he must write N independent programs. And if he makes a mistake in his coding for any one of them, the whole program will not work. If told that the Everett branches do not exist in reality, he might well ask—and I issue this as a challenge to Everett's opponents—'On the occasions when a quantum processor delivers the results of N processor-days of computation within a day of being programmed, *where was the computation done?*'.

14.3. Experiment 3: the quantum observer

Everett's opponents (at least those who do not reject realism) are obliged to agree on statement (14.1) in order to avoid Everett's famous prediction of multiple co-existing consciousnesses for a single observer. (Avoiding this is their motivation for opposing Everett in the first place.) According to Everett, the same quantum formalism describes the dynamical evolution of all physical systems, whether they possess consciousness, or any other property, or not. The non-Everett 'interpretations' require certain systems, always including 'conscious' ones, to evolve according to different dynamics.

It is therefore necessary, in order to perform a crucial test of Everett's quantum theory against all the others, to determine experimentally whether or not the superposition principle holds for states of distinct consciousness.

This experiment is performed on a quantum system with four subsystems. The first three have the simplest kinematics possible in quantum theory—they are two-state systems. The fourth is a quantum computer, on which an 'artificial intelligence' program runs. It is a quantum observer, and plays a dual role, as both the observer and the observed object of the experiment.

The state of the combined system is a unit vector in the Hilbert space $\mathcal{H}$ scanned by the basis states

$$|s_1, s_2, s_3, S\rangle, \tag{14.13}$$

where s_1, s_2, and s_3 take values in the set $\{\downarrow, \uparrow\}$, and S takes values in some enormously large set whose structure we shall fortunately not have to know.

For definiteness imagine that subsystem 1 is the $\updownarrow$-component of the spin of a single silver atom. At the beginning of the experiment this atom is prepared in the state

$$|\rightarrow\rangle \equiv 2^{-1/2}(|\uparrow\rangle + |\downarrow\rangle). \tag{14.14}$$

Systems 2 and 3 constitute a 'sense organ' for the quantum observer. They are prepared in their 'receptive' state, $|\downarrow, \downarrow\rangle$. The silver atom passes through a Stern–Gerlach apparatus; that is, through an inhomogeneous magnetic field which causes the atom's motion to come to depend on the $\updownarrow$-component of its spin. The trajectory corresponding to the state $|\uparrow\rangle$ passes close to system 2, and the interaction is such that if a silver atom passes along that trajectory the state of system 2 is 'flipped', i.e. its effect is

$$|\uparrow, \downarrow, \downarrow\rangle \rightarrow |\uparrow, \uparrow, \downarrow\rangle \tag{14.15}$$

Similarly the trajectory corresponding to the silver atom state $|\downarrow\rangle$ passes close to system 3, and interacts with it according to

$$|\downarrow, \downarrow, \downarrow\rangle \rightarrow |\downarrow, \downarrow, \uparrow\rangle \tag{14.16}$$

The quantum observer (system 4) is capable of examining the sense organ. Among other things, it (he/she?) can measure a (system 2 + system 3) observable $\hat{E}$ with eigenstates

$$\begin{aligned}
|receptive\rangle &\equiv |\downarrow, \downarrow\rangle, \\
|spin\text{-}up\rangle &\equiv |\uparrow, \downarrow\rangle, \\
|spin\text{-}down\rangle &\equiv |\downarrow, \uparrow\rangle, \\
|error\rangle &\equiv |\uparrow, \uparrow\rangle.
\end{aligned} \tag{14.17}$$

Obtaining the values '*spin-up*' or '*spin-down*' for $\hat{E}$ will initiate in the observer a subjective experience of 'having consciously observed' the silver atom's $\updownarrow$ spin component to be $\uparrow$ or $\downarrow$ respectively. Obtaining the value '*receptive*' will initiate the experience of 'not observing' the spin (yet). And as in experiment 2, the value '*error*' indicates an experimental error (or 'optical illusion'). It can be made to occur with an arbitrarily small frequency, according to how accurately non-perturbing the measurements are.

Having subjectively experienced one of these four states of consciousness, the observer continues the experiment by making a public, irreversible record, not of which one of the four it experienced, but only of *whether it has observed a definite spin value or not*. For example, it may type, with ink on paper, the words

> 'I hereby confirm that I am at this moment contemplating in my mind the memory of having observed one and only one of the two possible values, i.e. $\uparrow$ or $\downarrow$, of the $\updownarrow$-component of the spin of the silver atom. I know which of the two it was, but I shall not reveal which.' $\qquad(14.18)$

What is involved physically in an introspective contemplation of the kind reported by the observer in statement (14.18)? In detail, we do not know, of course. But in general terms, it must include a sequence of measurements of some sort, performed by subsystems of system 4 on each other. These measurements are links in a measurement chain which began with the measurement of $\hat{E}$ by system 4.

We lose no generality by requiring that this measurement of $\hat{E}$ take place in two steps: first, the observable

$$\hat{P} \equiv |\downarrow, \downarrow\rangle\langle\downarrow, \downarrow| + |\uparrow, \uparrow\rangle\langle\uparrow, \uparrow| \qquad (14.19)$$

is measured, and then

$$\hat{Q} \equiv |\downarrow, \downarrow\rangle\langle\downarrow, \downarrow| - |\downarrow, \uparrow\rangle\langle\downarrow, \uparrow| + |\uparrow, \downarrow\rangle\langle\uparrow, \downarrow| - |\uparrow, \uparrow\rangle\langle\uparrow, \uparrow| \qquad (14.20)$$

$\hat{P}$ and $\hat{Q}$ commute with each other and form a complete set of observables for systems 2 and 3. They also commute with $\hat{E}$. Therefore the declaration (14.18), saying that the observer's experience has been that of observing one or other of the values $\downarrow$ or $\uparrow$ for $\hat{E}$, can be rephrased in terms of $\hat{P}$ and $\hat{Q}$:

> 'I hereby confirm that I am at this moment contemplating in my mind the results of my observations of both $\hat{P}$ and $\hat{Q}$. The value of $\hat{P}$ was 1. I shall not reveal the value of $\hat{Q}$.' (14.21)

The dynamical evolution which has taken place so far in the experiment took place in five sequential steps. In summary:

(1) Measurement of the $\updownarrow$-component of the spin of the silver atom by systems 2 and 3.
(2) Measurement of $\hat{P}$ by a subsystem of system 4.
(3) Measurement of $\hat{Q}$ by another subsystem of system 4.
(4) 'Introspection' (i.e. isolated self-interaction) by system 4.
(5) 'Irreversible' recording of the observed value of $\hat{P}$ in some outside observable (ink and paper, the collective memory of the scientific community, or what you will).

The state after step 5, labelled by eigenvalues of the $\updownarrow$-spin component, $\hat{Q}$, $\hat{P}$, the quantum observer's consciousness, and the outside observables in which the value of $\hat{P}$ was recorded, has the form

$$2^{-1/2}(|\uparrow, Q = 1, P = 1, \text{'thinking that } Q = 1 \text{ and } P = 1\text{'}, \text{'}P = 1\text{'}\rangle$$
$$(14.22)$$
$$+ |\downarrow, Q = -1, P = 1, \text{'thinking that } Q = -1 \text{ and } P = 1\text{'}, \text{'}P = 1\text{'}\rangle).$$

The next step is to undo, by reversing the dynamical evolution, steps 4, 3, 2, and 1. This is possible in principle because these steps involve only the quantum computer, which can effect any desired unitary transformation upon the state of a subsystem of itself (Deutsch 1985*b*), and the silver atom which is a microscopic system. The 'irreversible' step 5, which involves the outside world, is not undone. Then finally the $\leftrightarrow$-component of the spin of the silver atom is measured.

At last we are in a position to detect the experimental difference between Everett's view and others: If any of the 'interpretations' incorporating statement (14.1) is true, then the measured value will be $\rightarrow$ or $\leftarrow$ with equal probability (the whole experiment must be performed sufficiently often for the results to be statistically significant). If Everett is right, the result will invariably be $\rightarrow$. This is because, according to statement (14.1) the state must have collapsed at some time during step 4 to one of the two eigenstates of the $\updownarrow$-component of the spin of the silver atom, and neither of these states contains information distinguishing $\rightarrow$ from $\leftarrow$ in any way. If Everett is right, then the state after steps 4 and 5 will still be a linear superposition giving equal weight to the $\downarrow$ and the $\uparrow$ values of the $\updownarrow$-spin, which has become correlated with $\hat{Q}$ but not with $\hat{P}$. The value of $\hat{Q}$, unlike $\hat{P}$, has not been irreversibly measured, so when the dynamical evolution is partially reversed interference between different eigenstates of the $\updownarrow$-spin will reconstitute the original eigenstate of the $\leftrightarrow$-spin.

14.4. Experiment 1 (second part)

You will now be able to confirm three facts about the present values of the observables you measured in the first part of Experiment 1, namely:

(1) The time (i.e. the state of your watch) now, t_2, is different from what it was then (t_1).
(2) You are still conscious.
(3) Your state of consciousness is different (I hope!) from the one you observed earlier.

But this implies that in the superposition (14.3).

$$\bar{\lambda}(t_2) = \text{'}I \text{ } am \text{ } conscious, \text{ } observing \text{ } t_2\text{'}. \tag{14.23}$$

Now eqns (14.3), (14.4), and (14.23) imply that you have detected experimentally a linear superposition of distinct states of human consciousness. This is contrary to statement (14.1).

14.5. Discussion

Of course, Experiment 1 is a trick. In a sense all it shows is that conventional (non-Everett) 'interpretations' are very awkward indeed to express in the Heisenberg picture, (for my honest attempt see e.g. Deutsch 1985*a*), and quite impossible to reconcile with Page and Wooters' (1983) apparently reasonable requirement that in quantum cosmology time must be an observable. (In all non-Everett 'interpretations' even the Heisenberg state is not constant, and must be a function of parameter time.)

As I have said, Experiment 1 is not an experimental test of Everett as Experiment 3 is. Nevertheless as a thought experiment it teaches us something about the Everett ontology: *that it is not, after all, so different from what we were used to in pre-quantum physics*. We have seen (eqn (14.2)) that in quantum cosmology the multiplicity of 'universes existing at different times' is just a special case of the Everett multiplicity of universes existing with different values of a more complete set of observables. While, admittedly, alternative formulations are viable *in this case*, we have in performing experiment 1 used successfully a formulation of quantum theory in which different states of consciousness *do* co-exist in the state vector, and we have had no trouble in explaining our experimental observations in terms of it. We are so used to the phenomenon of 'time passing' that we do not normally regard it as an indictment of the whole of physics that all values of *t* occur on an equal footing, that the different instants they refer to are all equally real, but that 'we are only aware of one of them' (*at a time*!). We accept indirect evidence of the physical existence of the past and future, and would scorn an 'interpretation of relativity' in which the manifold 'collapsed' to a single space-like hypersurface.

Experiment 2 exposes the barrenness of the conventional 'interpretations' when applied to a physical process in which quantum interference effects are important. Some members of the audience objected that Experiment 2 is simply the two-slit interference experiment 'writ large'. In a sense they are right, in that even the two slit interference experiment, understood as including its complement of refinements such as 'delayed choice', cannot be interpreted realistically except by Everett. But the quantum computer sharpens considerably the horns of the dilemma faced by Everett's opponents. It is one thing to deny that the electron had a definite location at one slit or the other; after all this differs by only a hair's breadth from Everett's interpretation, that the electron had *two* definite locations simultaneously. It is quite another thing to deny that a complex sequential process, in the course of which previously unknown knowledge was brought into existence by means of a definite, but parallel, computational algorithm, ever took place. *Ex nihilo nihil*.

If Experiment 3 were performed and gave a result contradicting statement (14.1), it would provide evidence of the physical existence of multi-valued observables, including the observables involved in consciousness.

The quantum observer in Experiment 3 *experiences* the splitting and re-merging of its own consciousness in the same sense that it experiences its own past: by observing physical evidence for which there is no alternative realistic interpretation.

References

Albert, D. Z. (1983). *Phys. Lett.* **98A,** 249.
Benioff, P. A. (1982). *Int. J. Theor. Phys.* **21,** 177.
Deutsch, D. (1985*a*). *Int. J. Theor. Phys.* **24,** 1.
—— (1985*b*). *Proc. R. Soc. Lond.* **A400,** 97.
DeWitt, B. S. (1973). In *The many-worlds interpretation of quantum mechanics*, (ed. B. S. DeWitt and N. Graham). Princeton University Press, Princeton.
Everett, H. (1957). *Rev. Mod. Phys.* **29,** 454.
Feynman, R. P. (1982). *Int. J. Theor. Phys.* **21,** 467.
Page, D. M. and Wooters, W. K. (1983). *Phys. Rev.* D **27**.

15

Transition amplitudes versus transition probabilities and a reduplication of space–time

Iwo Bialynicki-Birula

15.1. Introduction

The rise of quantum physics, very surprisingly, has exerted little influence on our views about space and time. With the exception of quantum gravity, all theories assume that quantum processes occur in the space–time of classical physics, and even in quantum gravity the modifications are presumably restricted to regions of space–time of the size of the Planck length.

In this chapter I argue that the very structure of all quantum theories suggests a revision of the classical notions of space and time. I will present evidence that two copies of space–time, rather than one, are the proper arena for all quantum processes. At the heart of this observation lies the very well known fact that every set of equations and formulae in quantum theory, from which all the transition amplitudes are determined, may always be written in two equivalent forms, differing by complex conjugation. We obtain one set from the other by reversing the sign of the imaginary unit i. This is clearly seen already in the simplest case of the non-relativistic quantum mechanics. The sign of i may be chosen according to the following two conventions:

$$
\begin{array}{ll}
\text{I. } \textit{Standard convention} & \text{II. } \textit{Opposite convention} \\
i\hbar\,\partial_t\Psi = H\Psi, & -i\hbar\,\partial_t\Psi = H\Psi, \\
[x, p] = i\hbar, & [x, p] = -i\hbar, \qquad (15.1) \\
i\hbar\dot{F} = [F, H], & -i\hbar\dot{F} = [F, H].
\end{array}
$$

Convention I originated with matrix mechanics (Born and Jordan 1925), while convention II was used originally by Schrödinger (1926, 1927) in his first papers on wave mechanics. The clash of the signs was bound to occur, since in both formulations of quantum mechanics, considered independently from each other, it is natural to describe the time evolution through the harmonic factors

$$e^{2\pi i v t} \quad \text{or} \quad e^{iEt/\hbar}. \tag{15.2}$$

When the two formulations were compared, it became obvious that one of the signs had to be reversed to achieve consistency. Of course, both formulations, I and II, are physically completely equivalent; they give identical transition probabilities. The transition probabilities are defined as squared moduli of transition amplitudes,

$$P = A^* \cdot A, \tag{15.3}$$

so that a change of convention results merely in an interchange of A and A^*.

The interchange of the two formulations can also be effected by the sign reversal of the Planck constant $\hbar$. This is the first indication that the existence of the two formulations may have something to do with space–time properties of quantum mechanics. After all, the Planck constant is a dimensional quantity that depends on the units of length and time. In particular, it changes its sign when the unit of time is reversed. As a result of this, the interchange of the formulations I and II is essentially equivalent to the Wigner time reversal (Bialynicki-Birula 1957).

The association of the transition amplitude A with the process evolving forward in time and of the complex conjugate amplitude A^* with the time-reversed process enables one to view the transition probability P as describing a process that evolves around a loop in time. Such loops in time were introduced long ago (Schwinger 1960, 1961) as a mathematical device that helps to evaluate expectation values. I believe that these loops in time should be taken more realistically and I shall present here some tentative ideas on this subject.

15.2. Probabilities and two-sided diagrams

Let me begin with non-relativistic quantum mechanics. The transition probability for a particle to travel from x_0, t_0 to x, t can be expressed as a product of two path integrals (Feynman and Hibbs 1965):

$$p(xt; x_0 t_0) = \int Dx_1 \int Dx_2 \exp \frac{i}{\hbar} (S_1 - S_2). \tag{15.4}$$

This formula is explicitly symmetric under complex conjugation and under the change of the sign of $\hbar$. As a result, expression (15.4) is

Fig. 15.1

invariant under the interchange of the two conventions. However, if expression (15.4) is to be considered as more fundamental than the single path integral representing the transition amplitude, then both trajectories x_1 and x_2 must be treated on an equal footing. This can be achieved in a natural way if we introduce two copies of space–time and we sum over all trajectories lying in both copies. At the end points both copies are identified (Fig. 15.1). This identification is presumably related to the classical nature of the measuring devices that determine the initial and final co-ordinates of the particle. I shall say more about it later. The perturbation expansion of the transition probability $p(xt; x_0t_0)$ further explains my rationale for the reduplication of space–time in quantum mechanics. Upon expanding the transition probability into a power series in the potential V, we can represent each term by a simple two-sided diagram (Fig. 15.2). The total probability is the sum of contributions from all possible combinations of elementary scattering processes occurring in both copies of space–time.

In order to clarify a little the role of the classical description in the context of the reduplicated space–time, I will perform a change of variables from x_1 and x_2 to x and $\bar{x}$ (Bialynicki-Birula 1971). In terms of

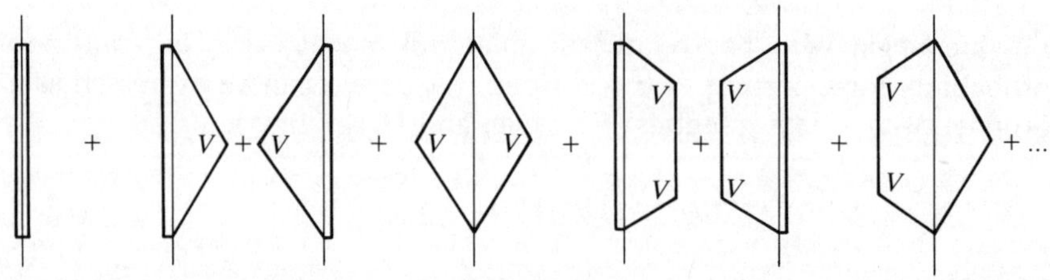

Fig. 15.2

the new variables, the formula (15.4) reads

$$p(xt; x_0t_0) = \int Dx \int D\bar{x} \exp\left[-i \int_{t_0}^{t} dt(m\ddot{x} + \nabla V(x)) \cdot \bar{x} + O(\hbar^2)\right]$$

$$= \int Dx\, \delta(m\ddot{x} + \nabla V) + O(\hbar^2), \tag{15.5}$$

where

$$x_1 = x + \hbar\bar{x}/2, \qquad x_2 = x - \hbar\bar{x}/2. \tag{15.6}$$

Formulae (15.5) and (15.6) show clearly that in the classical limit, when $\hbar \to 0$, the two copies of space–time collapse into one, and only one trajectory, namely the one that is determined by the Newton equations, contributes to the transition probability.

The same concept of a reduplicated space–time can be used also in relativistic theories. As a matter of fact, the description based directly on transition probabilities that makes use of two-sided diagrams of the type shown in Fig. 15.2 has several advantages in quantum electrodynamics and in other gauge theories. This description was introduced in the past (Bialynicki-Birula 1970; Bialynicki-Birula and Bialynicka-Birula 1975) to simplify the study of gauge invariance and renormalization. The two-sided diagrams representing the lowest order expressions for the transition probability in the electron–electron (Møller), electron–positron (Bhabha), and electron–photon (Compton) scattering are shown in Fig. 15.3. Two-sided diagrams are similar to the vacuum polarization diagrams—they never have open lines. It is this very feature of the two-sided diagrams that leads to significant simplifications, when the corresponding expressions for the transition probabilities are written down. In particular, it has been possible to eliminate all wave function renormalization constants, which are a source of serious complications in gauge theories, due to their gauge dependence. It is also known (Bloch and Nordsieck 1937; Jauch and Rohrlich 1954; Lee and Nauenberg 1964) that the transition probabilities, but not the transition amplitudes, are free from infra-red divergences in theories with massless particles.

I have wondered for some time whether through all these cancellations and simplifications nature is not trying to tell us something quite important. Perhaps the two-sided diagrams representing, as they are, directly observable quantities, are closer to reality. If so, we ought to depart from the conventional approach to quantum scattering and treat all processes as *really* going around a loop in time in the manner represented by two-sided diagrams. The proper arena for quantum processes would then be not one copy of Minkowski space–time, but rather two copies of it—one copy for each side of the diagram. Once two copies of space–time are introduced, we must face some burning questions: how are these two copies joined together; what is the nature of

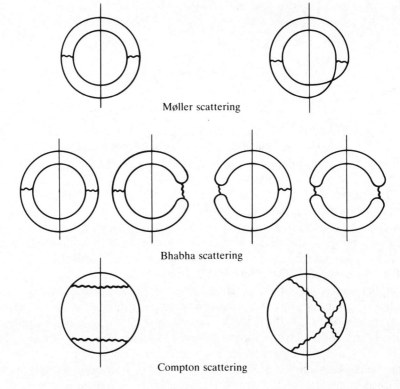

Møller scattering

Bhabha scattering

Compton scattering

Fig. 15.3

this join; does this description apply equally well to particles? I must admit that the discussions that I had in Oxford during the conference have shattered some of my naive hopes for a simple solution to these problems. However, I believe that the main idea has not been proven to be worthless.

15.3. Transition probabilities on scri

In the full relativistic theory, in contradistinction to quantum mechanics, the two copies of space–time should be joined together along the future and the past infinity. This conclusion can be reached from the analysis of the most promising relativistic theories—gauge field theories. Transition probabilities for finite times in such theories are not gauge invariant and hence devoid of physical meaning. This is because the propagators of particles carrying charges, Abelian or non-Abelian, are gauge dependent; only their mass-shell values are gauge invariant. In addition, the whole

renormalization procedure is adapted (mass-shell renormalization) to the
S-matrix, i.e. to infinite time-lapse calculations. The limits of the propagators, when $t \to \pm\infty$, yield the same transition amplitudes as those
obtained more conventionally by Fourier transforming the truncated
propagators and taking the mass-shell values. This equivalence enables
one to interpret the two-sided diagrams as representing processes occurring in space–time. Originally (Bialynicki-Birula 1970) these diagrams
were introduced to describe transition probabilities in momentum space,
as they appear in the S-matrix calculations when the states are labelled
with the momentum vectors of the initial and final particles. Since the
space–time interpretation of the S-matrix plays an essential role in my
analysis, I shall describe in some detail the space–time description and
the momentum description of relativistic scattering processes.

The truncated propagators, whose Fourier transforms determine the
transition amplitudes in relativistic field theory, are related to the full
propagators through the relations.

$$G(x_1, \ldots, x_n) = (-\mathrm{i})^n \int \mathrm{d}^4 y_1 \ldots \mathrm{d}^4 y_n G(x_1, y_1) \ldots G(x_n, y_n) F(y_1, \ldots, y_n),$$
(15.7)

where I have taken, for simplicity, a theory of only one type of scalar
particles. Instead of taking the Fourier transform of F and evaluating it
on the mass-shell, one may take the limits of the full propagator G, when
its arguments x_i are removed to infinity along the world lines of particles
moving freely with velocities v_1. This follows from the fact that the
(renormalized) one-particle propagator $G(x, y)$ behaves for large times
as the free Feynman propagator Δ_F and from the asymptotic form of the
free propagator:

$$\int \mathrm{d}^4 y \, \Delta_F(\boldsymbol{v}t - \boldsymbol{r}', t - t') F(\boldsymbol{r}', t'; \ldots)_{t \to \pm\infty} \sim$$

$$- (\sqrt{m}/2)(-2\pi \mathrm{i} |t| \sqrt{1 - v^2})^{-3/2} \mathrm{e}^{-\mathrm{i}m |t| \sqrt{1-v^2}} \int \mathrm{d}^4 y \, \mathrm{e}^{\pm \mathrm{i}p \cdot y} F(y, \ldots), \quad (15.8)$$

where p denotes the four-momentum of the particle,

$$(p^\mu) = m(1 - \boldsymbol{v}^2)^{-1/2}(1, \boldsymbol{v}),$$
(15.9)

and $\boldsymbol{r}'$ and t' are the space and time components of the four-vector y.

For massless particles, the situation is quite different. Due to the
absence of dispersion in the propagation of the waves describing these
particles, the asymptotic behaviour of the massless propagators is
characterized by the first power of $|t|$ in the denominator. This factor is
common in the massive and in the massless cases. In the massive case, we

obtain an additional factor of $|t|^{1/2}$ in the denominator due to the spreading of the wave packet in the direction of the propagation. The counterpart of formula (15.8) in the massless case has the form:

$$\int d^4 y D_F(\boldsymbol{n}t - \boldsymbol{r}', t + t_0 - t') F(\boldsymbol{r}', t'; \dots)_{t \to \pm\infty} \sim (-8\pi^2 i\, |t|)^{-1}$$

$$\times \int_0^\infty d\omega e^{\mp i\omega t} \int d^4 y e^{\pm ik \cdot y} F(y, \dots) \quad (15.10)$$

where k is the four-momentum vector of the massless particle,

$$(k^\mu) = \omega(1, \boldsymbol{n}). \quad (15.11)$$

Thus, the full information about the scattering of massless particles is contained in the limits of the propagators, when $t \to \pm\infty$, if they are known for all the values of the initial time t_0. In other words, one must take the limit, when the advanced time $v = t + r$ tends to $+\infty$, or the retarded time $u = t - r$ tends to $-\infty$ while the retarded time (in the first case), or the advanced time (in the second case) is held fixed. Upon comparing the formulae (15.8) and (15.10), we notice one additional important difference: the limit of the propagator in the massless case does not give directly the scattering amplitude of a particle with a given four-momentum, but its Fourier transform with respect to energy. Still, the limits (15.8) and (15.10) of the propagators with respect to all their arguments determine all the elements of the S-matrix. Of course, prior to taking the limits $t \to \pm\infty$ we must remove the kinematical factors in the expressions (15.8) and (15.10): the appropriate powers of $|t|$ and the oscillating factor describing the change of the phase of the wave function along its free trajectory.

Our analysis has shown that the transition amplitudes in relativistic quantum field theory are determined by the appropriately scaled values of the propagators on the boundary of the Minkowski space–time. Therefore, it seems natural to use in their analysis the concepts introduced by Penrose (1964, 1968) in his 'conformal treatment of infinity'. Such a conformal treatment, however, is not suitable to the description of massive particles (cf. also Binegar *et al.* (1982) for a somewhat different approach). The description of those particles is clearly pathological; their trajectories all begin at one point I^- (past time-like infinity) and they all end at one point I^+ (future time-like infinity). As a result, the limits of the (scaled) propagators at the points I^- and I^+ are ill-defined: they depend on the path along which these points are approached. I shall restrict myself in my analysis to the massless case and make only some speculative remarks about the massive case at the end.

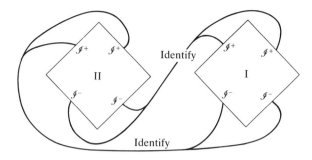

Fig. 15.4

Transition amplitudes for massless particles are determined by the (scaled) values of the propagators with their arguments lying in the future null infinity $\mathscr{I}^+$ and the past null infinity $\mathscr{I}^-$. As I have argued above, in order to accommodate two-sided diagrams we need two copies of the Minkowski space–time joined together along their common boundary, in such a way that closed loops in time are possible. In the conformal framework of Penrose there is essentially only one way of doing it: the $\mathscr{I}^+$ of the first copy is to be identified with the $\mathscr{I}^-$ of the second copy and vice versa (Fig. 15.4).

The two copies of Minkowski space–time with their boundaries identified in such a manner form a hypertorus. The hypertorus is divided into two halves by a distinguished light cone made of $\mathscr{I}^+$ and $\mathscr{I}^-$—the scri.

The basic tenets of relativistic quantum theory of massless particles may be stated as follows. Quantum processes take place on the whole hypertorus. The corresponding transition probabilities pick up contributions from trajectories lying in both halves of the hypertorus. The only difference between the contributions from the first and from the second half is that the imaginary unit i is replaced by $-i$. All transition probabilities 'live' on the scri, where the two copies of the space–time meet each other. These statements follow from the standard formulation, so that the physical content of the theory remains unchanged. However, the transition probabilities appear here, in a natural manner, as functions of the retarded and advanced times that label the points on the scri, rather than as functions of the energies of the incoming and outgoing particles. Otherwise, 'everything is just translated into an alternative, but equivalent, mathematical framework. But the new framework suggests different directions in which to proceed, from the ones which might seem natural in a more conventional theory'. These words were written (Penrose 1975) in connection with the twistor program, but I believe that they are also applicable here. Additional reasons for quoting Penrose stem from my expectations that the twistor formalism should prove very useful in the analysis of the transition probabilities on the scri. After all, twistors are a natural tool when the conformal properties play a role, and one is led, *nolens volens*, to the conformal compactification in order to describe two copies of the Minkowski space–time in 'a compact way'.

At the end I should say something about the mass, which is a source of some embarrassment in every conformal theory. The standard treatment of mass does not look very promising. In the conformally compactified space–time all trajectories of massive particles converge at one point, which in turn leads to a pathological behaviour of the transition probabilities treated as functions on the scri. These functions must be highly singular at I^+ and I^- in order to give different values at these points for all different directions along which these points are approached.

A possible way out of these difficulties is to assume that at the very fundamental level massive particles always move with the speed of light. The result of the mass term is a series of random 'kicks', which cause discontinuous changes in the direction of the particle velocity (*Zitterbewegung*). That the Dirac equation admits such an interpretation is shown by Penrose and Rindler (1984) (zigs and zags) and by others in connection with the Feynman path-integral approach to the Dirac equation. In the most recent publication on this subject (Gaveau *et al.* 1984) the authors write: 'An electron propagates at the speed of light and with a certain chirality (like a two-component neutrino), except that at random times it flips both direction of propagation (by 180°) and handedness: the rate of such flips is precisely the mass, m (in units $\hbar = c = 1$)'. If this analogy with Brownian motion is assumed to be valid also for interacting particles, then we could imagine that a random kick, combined with a kick caused by the interaction, might send the particle across the scri to the other copy of space–time. It is also possible that quantum tunnelling plays a role here.

More hard facts, however, will have to be ascertained before we would know whether these speculations can be taken seriously.

Acknowledgements

I would like to thank Z. Bialynicka-Birula, A. Bialynicki-Birula, C. Fronsdal, J. Kijowski, A. Trautman and S. Woronowicz for many stimulating conservations. Discussions during the Oxford Meeting, most notably those with R. Penrose, have contributed significantly to the evolution of my ideas.

References

Bialynicki-Birula, I. (1957). *Bull. Acad. Polon. Sci. Cl. III.* **5**, 805.
—— (1970). *Phys. Rev.* D **2**, 2877.

—— (1971). *Ann. Phys. N.Y.* **67**, 252.

—— Bialynicka-Birula, Z. (1975). *Quantum electrodynamics*. Pergamon, Oxford.

Binegar, B., Fronsdal, C., Flato, M., and Salano, S. (1982). *Czech. Jour. Phys.* B **32**, 439.

Bloch, F. and Nordsieck, A. (1937). *Phys. Rev.* **52**, 54.

Born, M. and Jordan, P. (1925). *Z. Phys.* **34**, 858.

Feynman, R. P. and Hibbs, A. R. (1965). *Quantum mechanics and path integrals* McGraw-Hill, New York.

Gaveau, B., Jacobson, T., Kac, M., and Schulman, L. S. (1984). *Phys. Rev. Lett.* **53**, 419.

Jauch, J. M. and Rohrlich, F. (1954). *Helv. Phys. Acta* **27**, 613.

Lee, T. D. and Nauenberg, M. (1964). *Phys. Rev.* B **133**, 1549.

Penrose, R. (1964). In *Relativity, groups and topology* (ed. C. M. DeWitt and B. DeWitt). Gordon and Breach, New York.

—— (1968). In *Battelle rencontres 1967* (ed. C. M. DeWitt and J. A. Wheeler). Benjamin, New York.

—— (1975). In *Quantum theory and the structure of time and space* (ed. L. Castell, M. Drieschner, and C. F. von Weizsäcker). Carl Hauser, Munich.

Penrose, R. and Rindler, W. (1984). *Spinors and space–time* Vol. 1. Cambridge University Press, Cambridge.

Schrödinger, E. (1926). *Ann. Phys.* **81**, 129.

—— (1927). *Ann. Phys.* **82**, 186.

Schwinger, J. (1960). In *Brandeis University summer institute in theoretical physics, lecture notes*. Brandeis University.

—— (1961). *J. Math. Phys.* **2**, 407.

16

Leibnizian time, Machian dynamics, and quantum gravity

J. B. Barbour

16.1. Introduction

Historically, dynamics was bedevilled from its beginning by the invisibility of space and time. Newton (1686) championed the view that space and time, although invisible, do exist and provide the arena within which motion occurs. Leibniz (1716) argued that there is no such thing as absolute space but only the relative configurations of simultaneously existing bodies and that time is merely the succession of such instantaneous configurations and not something that flows quite independently of the bodies in the universe and their motion. What Leibniz was advocating was that dynamics should be based exclusively on observable elements; it should not contain elements that are not in principle directly observable. This, of course, was Mach's standpoint too (Mach 1872), which Einstein (1916) adopted wholeheartedly when developing general relativity.

Quantum mechanics (QM) inherited the kinematic structure of Newtonian dynamics. Its most fundamental operators—those of momentum, angular momentum, and energy—correspond, respectively, to displacement and rotation in Newton's invisible space and displacement in invisible time. The aim of this paper is to question the extent to which the existing framework of QM is appropriate for quantization of Einstein's theory of general relativity (GR). It will be pointed out that the use of absolute space and time in Newtonian dynamics leads to a characteristic failure of predictive power. A simple theory of dynamics (gauge-invariant dynamics) will then be discussed that uses only directly observable quantities and does not suffer from this failure of predictive power. It will then be shown that GR has essentially the same structure as gauge-invariant dynamics. Finally, it is argued that for this reason it may not be appropriate to attempt to quantize GR within the existing framework of QM.

In his introduction to the discussion meeting, Roger Penrose pointed out that hitherto the attempts to quantize GR have tended to regard QM as sacrosanct and have sought to make GR fit QM. He wondered whether, in view of the failure of this approach, the time had not come to look at the possibility of modifying QM. This paper is a contribution in that direction.

16.2. Predictive power of classical dynamics

The points raised in this paper can be illustrated by a *Gedanken-experiment*: suppose two successive 'snapshots' are taken of a universe of n material points (with known masses m_i, $i = 1, \ldots, n$) moving in Euclidean space under gravity in accordance with Newton's laws. The snapshots, which show only the relative distances between the bodies, differ intrinsically by a small amount, but the separation in time between them is not known. Is it possible to predict the future evolution of the system uniquely?

It is not. Some essential information is missing. From the relative distances in the two snapshots, it is impossible to deduce either the angular momentum or the kinetic energy of the system, but these two quantities exert a profound influence on the subsequent motion. Poincaré (1905) found such a failure of predictive power very curious. The equations of dynamics are of second order in time, so that initial positions and velocities are required in the initial-value problem. But in Newtonian dynamics, these quantities must be specified in absolute space and time. The purely relative—and observable—quantities do not quite suffice to determine the absolute quantities. As a result, quite different futures can evolve from apparently identical initial conditions (cf. Barbour 1982).

The next section presents a dynamical theory without this curious defect.

16.3. Gauge-invariant dynamics

Because the magnetic field is always observed to satisfy div $B = 0$, it can be represented in terms of a vector potential A, $B = \text{curl}\,A$, with, however, A determined only up to the gauge transformation

$$A \rightarrow A + \nabla\psi, \qquad \psi = \psi(x). \tag{16.1}$$

Similarly, the $n(n - 1)/2$ observable relative distances between n ($n \geq 5$) bodies in Euclidean space satisfy algebraic relations which reduce the number of independent quantities to $3n - 6$. These relations make it possible to represent the observable data by means of Cartesian coordinates r_i. Like the vector potential A, these Cartesian 'potentials' are determined only up to the six-parameter 'gauge transformation'

$$r_i \to \mathscr{E} r_i + g, \tag{16.2}$$

where $\mathscr{E}$ is an orthogonal 3×3 matrix and g is a vector.

Imagine a continuous sequence of snapshots of the n bodies as they move relative to each other. Each snapshot represents an instant of Leibnizian time. Label the sequence by an arbitrary monotonically increasing label τ. Since the gauge of the potentials in each snapshot is independent of the gauges in the preceding and following snapshots, $\mathscr{E}$ and g in transformation (16.2) become arbitrary functions of τ:

$$r_i(\tau) \to \mathscr{E}(\tau) r_i(\tau) + g(\tau). \tag{16.3}$$

Note that the gauge group (16.3) is uniquely determined by the nature of the (algebraic) relations satisfied by the (observable) relative distances and the Leibnizian concept of time, according to which the label τ is itself determined only up to

$$\tau \to \tau', \quad \tau' = \tau'(\tau), \quad d\tau'/d\tau > 0, \tag{16.4}$$

since no time label τ is distinguished *a priori*.

If the complete universe is considered, it is not difficult to construct a dynamical theory invariant under transformations (16.3) and (16.4), which may be called the *Leibniz group* (Barbour and Bertotti 1977). Only the essential features will be given here; the details can be found in Barbour and Bertotti (1982).

A gauge-invariant action is constructed. Let $\{r_i\}$ be an initial configuration and $\{r_i'\}$ differ from it infinitesimally: $r_i' = r_i + \delta r_i$. Now, if the gauge of, say, r_i' is changed, δr_i will change too. Let

$$\delta T = \sum_{i=1}^{n} m_i (\delta r_i / \delta \tau)^2, \tag{16.5}$$

and let $\delta \hat{T}$ be the minimum of eqn (16.5) over all δr_i obtained by applying transformation (16.3) to the representation of the second configuration. Then $\delta \hat{T}$ is uniquely determined by observable quantities and is gauge invariant.

To obtain the gauge-invariant analog of Newtonian gravitational dynamics, let

$$V = \sum_{i<j} m_i m_j / r_{ij},$$

and consider the variational principle

$$\delta I = 0, \quad I = \int (V\hat{T})^{1/2} \, d\tau, \tag{16.6}$$

where the product form and the square root are taken in order to ensure that the integrand of eqn (16.6) is homogeneous of degree one and therefore invariant with respect to (16.4).

It is easy to show that the physically distinct solutions of eqn (16.6) are identical to the solutions of Newtonian gravitational dynamics for which the considered n-body universe has angular momentum about the centre of mass equal to zero and, in addition, zero total energy. These results are consequences of the invariance of eqn (16.6) with respect to transformations (16.3) and (16.4), respectively, or, as Bertotti and I put it, implementation of the First and Second Mach's Principles (Barbour and Bertotti 1982).

16.4. Two different concepts of motion

In the Newtonian concept of motion, each body moves primarily in the arena of absolute space and time, and only secondarily with respect to other bodies, the interactions with which give rise to deviations from the basic uniform rectilinear inertial motion. In the Leibnizian view, there is just a sequence of snapshots. Space and time are to be *constructed* from the observable data in the snapshots, which can be regarded as lying initially in a random heap.

Except in uninteresting degenerate cases, the presumed continuity of the changes of the relative configurations will permit a unique ordering of the sequence. After this, it is possible to *stack horizontally*. This is essentially the operation that occurs in the minimization of eqn (16.5). Take the first snapshot, place the second on top of it, and move the second around on the top of the first until eqn (16.5) is minimized. This will happen when: (1) the centres of mass coincide, (2) there is no overall rotation. This is what causes the total angular momentum to vanish. The third snapshot can then be stacked relative to the second, the fourth relative to the third, and so forth. The complete sequence is thus ordered and stacked horizontally.

To stack *vertically*, i.e., to determine a separation in 'time' between successive snapshots, we use the equations of motion deduced from eqn (16.6):

$$\frac{d}{d\tau}\left(\frac{V^{1/2}}{\hat{T}^{1/2}}\frac{dr_i}{d\tau}\right) = \frac{1}{2}\frac{\hat{T}^{1/2}}{V^{1/2}}\frac{\partial V}{\partial r_i} \tag{16.7}$$

(see Barbour and Bertotti (1982); the horizontal stacking is assumed already performed). Up to its origin, there is a unique choice of the label τ, which is still arbitrary in eqn (16.7), that casts eqn (16.7) into an especially simple form. Namely, choose τ such that

$$\hat{T}^{1/2} = V^{1/2}. \tag{16.8}$$

Then: (1) with respect to τ, the bodies now follow Newton's laws exactly. By virtue of the horizontal stacking, bodies subject to no forces move along straight lines; by virtue of the vertical stacking, they now move uniformly as well. As Poincaré (1905) asserted, 'time' is chosen so as to put the equations of dynamics into simplest form. (2) The determination of the distinguished time simultaneously defines an 'energy' of the system, which however is exactly zero (V is minus the usual Newtonian potential energy).

To summarize: to formulate gauge-invariant dynamics, it is not necessary to presuppose absolute space and time. Instead, they are constructed after the dynamical problem has been solved using observable relative data. It is necessary to consider the entire universe, which, after horizontal and vertical stacking, is necessarily found to have vanishing energy and angular momentum as a consequence of the gauge-invariant dynamics. However, well separated subsystems of the universe can perfectly well have non-vanishing values of these parameters. When applied to the complete universe, gauge-invariant dynamics is full rational in the sense required by Poincaré, i.e. seemingly identical initial data cannot give rise to quite different dynamical evolutions.

Kretschmann (1917) argued that Einstein's principle of general covariance has only methodological but not physical significance. For example, Newtonian dynamics can also be made invariant with respect to transformations (16.3) and (16.4) and, like gauge-invariant dynamics, be treated in a frame of reference in arbitrary motion. But this formal invariance—I shall call it *Kretschmann invariance*—is only possible at the price of the introduction of additional (and unobservable) elements. In contrast, the gauge invariance of Section 16.3 is achieved minimally. The Leibniz group (16.3) and (16.4) is unambiguously determined by the structure observed within each snapshot and by the concept of time. The minimal gauge invariance enhances the predictive power; Kretschmann invariance adds nothing.

16.5. General relativity as a generalization of gauge-invariant dynamics

Gauge-invariant dynamics solves the problem of determining the future from observable initial data. Of course, the type of 'snapshot', or 'simultaneity', can be generalized. For example, the snapshot might show the instantaneous values of a field or the distance relations that hold in a closed three-dimensional Riemannian space. Let us consider the latter case. Instead of the Cartesian 'potentials' r_i, the 'potential' in this case is the metric tensor g_{ij}, i, $j = 1$, 2, 3, of the Riemannian space. The analogue of the gauge group (16.3) is

$$g_{ij}(\tau) \to g_{ij}(\tau) + \Lambda_{(i;j)}(\tau), \tag{16.9}$$

where Λ_i are three arbitrary functions, the round brackets in the suffix denote symmetrization, and the semicolon denotes the covariant derivative with respect to g_{ij}. The Leibniz group for this problem is (16.9) and (16.4), and is not far short of being the general covariance group of four dimensions, i.e., the invariance group of general relativity. However, there is no mixing of (16.9) and (16.4).

The reason for (16.4) is that there is nothing intrinsic in any two snapshots to say how far apart they are 'in time'. But the division of the Leibniz group into (16.9) and (16.4) corresponds to the pre-relativistic notion that absolute simultaneity still has meaning. In the operational spirit of Section 16.4, this amounts to the assumption that the simplest form of the equations of motion can be obtained with a single time parameter, this time parameter being the same across the entire universe. In a post-relativistic approach, such a view cannot be maintained; one must consider the possibility that the separation in 'time' between the snapshots is not only unknown but also position dependent in general. Then (16.9) and (16.4) can no longer be kept separate and it is necessary to go over to the general covariance group in four dimensions.

Limiting the discussion to pure geometrodynamics (i.e. when no matter fields are present), the problem therefore is to find a generalization of the action (16.6). It was shown by Baierlein *et al.* (1962) that the action principle of GR can be put in the form

$$S[g_{ij}, N^k] = \int d\tau \int d^3x \{R[g] G^{ijkl}[g](\dot{g}_{ij} - 2N_{(i;j)})(\dot{g}_{kl} - 2N_{(k;l)})\}^{1/2}. \tag{16.10}$$

Here, R is the scalar curvature of the three-dimensional Riemannian spaces, $G^{ijkl} = \frac{1}{2}g(g^{il}g^{kj} + g^{ik}g^{lj} - 2g^{ij}g^{kl})$ is DeWitt's metric, $g = \det\|g_{ij}\|$, and N_i is the shift; $\dot{g}_{ij} = dg_{ij}/d\tau$; and indices are raised and lowered with g_{ij}. The summation convention is assumed.

The action (16.10) is a generalization of eqn (16.6). The variation with respect to the shift N_i (the 'thin-sandwich' problem described by Wheeler (1964)) is exactly analogous to the horizontal stacking described in Section 16.4. The shift enters eqn (16.10) in the way it does solely on account of (16.9). The really distinctive structure of GR dictated by the mixing of (16.9) and (16.4) into the four-dimensional covariance group is reflected by the appearance of DeWitt's metric G and the fact that R multiplies the remainder of the integrand at each point and there is only a single integral over space. Were invariance only with respect to (16.9) and (16.4) separately required, eqn (16.10) could be replaced by a much less specific expression.

Once the horizontal stacking has been performed, i.e., the thin-sandwich problem has been solved,† the vertical stacking is performed by *defining* a position-dependent lapse:

$$N = \tfrac{1}{2}R^{-1/2}(k_{ij}k^{ij} - k^2)^{1/2}, \qquad (16.11)$$

where $k_{ij} = dg_{ij}/d\tau - 2N_{(i;j)}$. This is a position-dependent generalization of Poincaré's proposition that 'time' is chosen to make the laws of dynamics simple.

In gauge-invariant dynamics, the horizontal and vertical stacking of the snapshots leads to the construction of absolute space and time. In geometrodynamics, the 'heap' of three geometries is stacked analogously into a Ricci flat four-dimensional space. Moreover, in allowing arbitrary position-dependent 'time' separation between simultaneities, GR is a *non plus ultra* as regards predictive power. For this reason, one must agree with Wheeler (1964) that, at least in the case of a closed universe, GR is a faithful realization of Machian ideas.

16.6. Quantum mechanics and general relativity

As mentioned in Section 16.1, the attempts to quantize GR have been made under the assumption that QM has a basic structure with which one cannot tamper and to which GR must be matched. However, if we consider what features of Newtonian kinematics are essential to QM and then realize that gauge-invariant dynamics and GR are constructed in such a way as to eliminate precisely these features, we must question such an approach.

There are two features of Newtonian kinematics that play an essential role in QM: the phase space and the distinguished time variable. These are assumed to exist before any dynamics takes place in them. In the modern treatment of dynamics, for example, the phase space with its symplectic form is assumed given. Many different Hamiltonian evolutions can take place on one and the same phase space. In this view, the *momentum* of a particle exists prior to any dynamical law that it may satisfy. The basic structure of QM is set up prior to the dynamics, just as absolute space and time are assumed to exist prior to Newton's laws.

But this is not the case in gauge-invariant dynamics. It is only after the horizontal and vertical stacking (Section 16.4) that we recover the 'space–time' structure that corresponds to the arena provided by Newtonian absolute space and time. Moreover, the stacking procedure is

---

† Unfortunately, not much is known about the solution of the thin sandwich problem. Uniqueness has been proved under certain restrictions by Belasco and Ohanian (1969). This at least shows that the kind of non-uniqueness inherent in Newtonian dynamics is not encountered in GR.

carried out with one specific Lagrangian and the specific system (taken to represent the entire universe) one is considering. If divorced from these, a particle can be associated with neither a direction nor a magnitude of momentum.

We have seen that the kinematic structure of one-particle Newtonian dynamics may well have a Machian origin. It is a small step from this to the idea that some of the most basic features of QM have a Machian origin too. In particular, there is a suggestive correspondence between the failure of predictive power in Newtonian dynamics associated with the 'invisibility' of the energy and angular momentum of a system and the fact that wave packets in QM are constructed by superposition of eigenfunctions corresponding to different values of the quantum analogues of these very same quantities.

Smolin (see Chapter 10) has attempted to derive the one-particle Schrödinger equation in a many-particle global framework, but otherwise there seem to have been few attempts made in this direction.

16.7. The Hamiltonian constraints of general relativity

When GR was cast into Hamiltonian form as a preliminary to canonical quantization (Dirac 1958; Arnowitt *et al.* 1962), it was found that the dynamics of GR is not controlled by a conventional Hamiltonian but by so-called Hamiltonian constraints (Dirac (1964); see Kuchař (1981) for a modern review and an extensive bibliography). According to Kuchař, the presence of these constraints 'incredibly complicates the implementation of the canonical quantization programme'. In this connection, it is interesting to examine the literature to see what attitude the investigators took to the problem of dealing with these constraints. Overall, the tendency seems to have been to look for similar constraints in systems that have already been successfully quantized and hope that these would give guidance in the case of GR. However, the constraints with which comparisons were made arise from what I have called Kretschmann invariance (Section 16.4), i.e., a formal invariance achieved by adding to the original dynamical elements of the theory without in any way increasing the predictive power of the theory.

This is well demonstrated by the so-called super-Hamiltonian constraint, which was claimed to be analogous to the constraint that arises in parametrized particle dynamics (Dirac 1964; Lanczos 1949). This similarity was taken as a guide to the way in which one should attempt to identify the time variable in the quantization of GR, the point being that in parametrized particle dynamics 'time' is included among the ordinary dynamical variables, and it was concluded that in GR 'time' is hidden among the dynamical variables, which in the Hamiltonian formulation are the components of the spatial metric g_{ij}, i, $j = 1$, 2, 3, and the momenta conjugate to them. However, I want to argue here that the super-Hamiltonian constraint of GR is rather an indication that there is no 'time' at all in GR. To do this, it will be necessary to consider two different constraints: the one that arises in parametrized particle dynamics and derives from a Kretschmann invariance and the one that arises in gauge-invariant dynamics and derives from a minimal invariance.

Let $\mathcal{L}(q_i, \dot{q}_i, t)$, $i = 1, \ldots, 3n$, be the Lagrange function of n Newtonian point particles with kinetic energy T and potential energy U. Parametrize the particle paths by an arbitrary time label τ ($\mathrm{d}q_i/\mathrm{d}\tau = q_i'$) and adjoin the absolute time $t = q_0$ to the q_i's. The new Lagrangian $\bar{\mathcal{L}} = \bar{\mathcal{L}}(q_i, q_i', t, t')$ is homogeneous of degree one in the 'velocities' q_i' and t', so that the action is invariant with respect to (16.4). It is well known that as a result the Hamiltonian corresponding to $\bar{\mathcal{L}}$ vanishes identically (Dirac 1964). To obtain a Hamiltonian description, it is necessary to introduce a constraint. Namely, the Hamiltonian description of the parametrized form of the dynamics is derived from the principle

$$\delta I = 0, \quad I = \sum_{i=0}^{3n} \int [\pi_i q_i' - N(\pi_0 + H)] \, \mathrm{d}\tau, \tag{16.12}$$

where H is the Hamiltonian corresponding to $\mathcal{L}$ (the 'physical' Hamiltonian) and $\pi_i = \partial \bar{\mathcal{L}}/\partial q_i'$. Variation with respect to the Lagrange multiplier N gives the constraint

$$\mathcal{H} \equiv \pi_0 + H = 0. \tag{16.13}$$

Variation with respect to π_0 gives the 'lapse'

$$N = t', \tag{16.14}$$

which remains an arbitrary function not determined by eqn (16.12). This is what is known as the parametrized form of dynamics. In it, the absolute time t and $-H$ appear as an extra pair of canonical co-ordinates. The parameter invariance of eqn (16.12) is a typical Kretschmann invariance, achieved by adding t and $-H$ to the dynamical variables. The predictive power of the theory is unchanged.

Now suppose the system is conservative, i.e. $\mathcal{L}$ does not contain t explicitly. Then t is an ignorable co-ordinate (see Lanczos 1949) and can be eliminated by the Routhian procedure, giving Jacobi's principle for the path in configuration space:

$$\delta J = 0, \quad J = \int (E - U)^{1/2} s' \, \mathrm{d}\tau, \quad \mathrm{d}s^2 = \sum_{i=1}^{3n} m_i \, \mathrm{d}q_i \, \mathrm{d}q_i, \tag{16.15}$$

where E is the constant total energy: $E = T + U$. Thus, in eqn (16.15), E

is regarded as a constant, but U is configuration dependent. In Newtonian dynamics, the *speed* at which the system moves through its configuration space is found from the energy equation by quadrature after the 'orbit' problem (16.15) has been solved.

Note that the principle (16.12) yields solutions of the original problem with *all* values of the energy, but eqn (16.15) yields only those with the value E. In particular, if $E = 0$ we recover our result of Section 16.3. Let us assume this is the case. Now the variational principle (16.15) is still invariant under (16.4), so that the Hamiltonian corresponding to eqn (16.15) again vanishes. The only way we can obtain a Hamiltonian description now is to take the action

$$I_0 = \sum_{i=1}^{3n} \int (\pi_i q_i' - N_0 H)\, d\tau. \qquad (16.12')$$

Variation with respect to N_0 gives

$$H = 0, \qquad (16.13')$$

and consistency enforces

$$N_0 = 1. \qquad (16.14')$$

So far as I know, a super-Hamiltonian constraint like eqn (16.13') has not hitherto been considered in the literature, though it seems to be more appropriate for comparison with the constraint in GR. The difference between eqns (16.13) and (16.13') is made clear by the two-snapshot initial-value problem considered in Section 16.2 and the distinction made in Section 16.4 between Kretschmann invariance and minimal invariance. The variables which occur in eqn (16.13) are *heterogeneous*. Namely, the momenta corresponding to the ordinary dynamical variables are all directly observable (we can assume for the sake of the discussion that the horizontal stacking has already been performed) whereas the momentum corresponding to the time variable $(-H)$ is not observable. For given values of the remaining (observable) dynamical variables, the numerical value of H is undetermined. This is true in both the conservative and the non-conservative case, and it corresponds to the one-parameter uncertainty in the dynamical evolution from observable initial conditions noted in Section 2. In contrast, in eqn (16.13') there is no such time variable. All the variables are observable, and the future is uniquely determined by the observable data. Since GR shares this property with gauge-invariant dynamics, one must consider the possibility that the attempt to find a 'time' variable in GR is misconceived.

As an alternative, one could at least look at the quantization of the simplest problem in gauge-invariant dynamics that has constraints of the same intrinsic structure as GR: the three-body problem of celestial dynamics in the centre-of-mass frame with vanishing total angular momentum (analogue of the super-momentum constraint) and zero total energy (super-Hamiltonian constraint). In particular, if a 'time' is to be found among the ordinary dynamical variables, it will have to be constructed from among the genuine observables of the system, i.e. the three relative distances between the bodies (thus, 'time' could be either the perimeter or the area of the triangle formed by them). This makes it clear that quantization of GR differs in a fundamental way from the quantization of all other systems hitherto considered, a fact that has perhaps been obscured rather than illuminated by formal Kretschmann-type analogies.

Acknowledgements

Especial thanks to Karel Kuchař for many recent discussions. Thanks also for earlier discussions to H. Goenner, D. Liebscher, D. Raine, L. Smolin, and above all Bruno Bertotti, with whom many of these ideas were developed. I also thank A. Ashtekar for discussing some unpublished notes on this subject.

References

Arnowitt, R., Deser, S., and Misner, C. W. (1962). In *Gravitation: an introduction to current research* (ed. L. Witten). John Wiley, New York.
Baierlein, R. F., Sharp, D. H., and Wheeler, J. A. (1962). *Phys. Rev.* **126**, 1864–5.
Barbour, J. B. (1982). *Br. J. Phil. Sci.* **33**, 251–74.
—— and Bertotti, B. (1977). *Nuovo Cimento* B **38**, 1–27.
—— —— (1982). *Proc. R. Soc. Lond.* A **382**, 295–306.
Belasco, E. P. and Ohanian, H. C. (1969). *J. Math. Phys.* **10**, 1503–7.
Dirac, P. A. M. (1958). *Proc. R. Soc. Lond.*, A **246**, 326–43.
—— (1964). *Lectures on quantum mechanics* Belfer Graduate School of Physics, Yeshiva University, New York.
Einstein, A. (1916). *Ann. Phys., Lpz.* **49**, 769–822. (Engl. transl. in *The principle of relativity*). Dover, New York, (1952).
Kretschmann, E. (1917). *Ann. Phys., Lpz.* **53**, 575–614.
Kuchař, K. (1981). In *Quantum gravity 2, a second Oxford symposium* (ed. C. J. Isham, R. Penrose, and D. W. Sciama). Oxford University Press, Oxford.
Lanczos, C. (1949). *The variational principles of mechanics*. Toronto University Press, Toronto.
Leibniz, G. W. (1716) in *The Leibniz–Clarke correspondence* (ed. H. H. Alexander). Manchester University Press, Manchester (1956), p. 69.
Mach, E. (1872). *Die Geschichte und die Wurzel des Satzes von der Erhaltung der Arbeit*. (Engl. transl. *History and root of the principle of conservation of energy*. Open Court, Chicago 1911). (See also Mach E. (1960). *The science of mechanics*. Open Court, La Salle.)
Newton, Sir Isaac (1686). *Philosophiae naturalis principia mathematica*.
Poincaré, J. H. (1905). *Science and hypothesis* Walter Scott, London.
Wheeler, J. A. (1964). In *Relativity, groups and topology* (ed. B. DeWitt and C. DeWitt). Gordon and Breach, New York.

17

Quantum time–space and gravity

David Finkelstein and Ernesto Rodriguez

17.1. Introduction

It is far-fetched to describe quantum gravity by a quantum (q) pseudo-metric tensor g depending on classical (c) coordinates x, a theory we call here cq for its hybrid classical-quantum nature. After all, consider how we determine these variables.

We may determine coordinates of an event by four explosions which hurl good clocks in all possible directions at all possible speeds. The four readings on the four clocks that reach an event are its coordinates.

We may determine the interval between events by setting off a similar explosion at one event; the reading on the clock that reaches the other event is the interval.

It would be arbitrary to say that the four variables resulting from the first procedure are c-numbers, while that from the second would be a q-number. The cq theory does not go so far; it claims that because the shrapnel consists of clocks, it acquires a quantum indeterminacy from the gravitational field. Were the shrapnel charged, it would pick up the quantum nature of the electromagnetic field it traverses as well. To determine c coordinates, some reliable and continuous number-generators insensitive to gravity and any other q field must replace the clocks. This still seems far-fetched, in view of the universal coupling to gravity. Gravity determines propagation and the characteristic surfaces of all other fields.

It seems more likely that gravity and its light cones reflect a fundamental microscopic structure of time space, and that the q nature of gravity derives from that of this microstructure.

In the c theory, the co-ordinates belong to the level of the manifold, the third level in the hierarchy of theories of Table 17.1. The usual (cq) theory of fields, applied to gravity, assumes that the first three levels are c and the last three are q. We have been exploring a physics which is q all the way down, with a q correspondent for each of these levels. We report (Sections 17.2 and 17.3) on the present state of our q gravity. One problem we all have is giving meaning to a psi vector for the universe. We explain (Section 17.4) how and why we do not do this.

Table 17.1. Levels of classical field theory

Level	Structure
1. Set theory	Number of events in the world
2. Topology	Connections between events
3. Manifold theory	Co-ordinates
4. Pseudometric geometry	Pseudometric tensor
5. Field kinematics	Field variables at each event
6. Dynamics	Action principle

17.2. Quantum set theory

We start one level before Table 17.1 with the predicate algebra, level 0. In c theories this is Boolean, in q theories it is the non-Boolean lattice of projections in a Hilbert space. Therefore we have constructed a set theory with such a q predicate algebra. The framework for the c theory from which we start is the class F of all finite sets, regarded as the phase space or sample space of a random finite set FSet. The Cantor bracket operation $s \rightarrow \{s\}$ is a map $F \rightarrow F$. F is generated from a distinguished element, the null set, by bracket and union operations. Despite this simplicity, F is a universal language for c mathematical physics, if we regard continua as approximations to large finite objects. We set out to make a similarly universal language for q mathematical physics. The world is not a set, since it obeys the principle of quantum superposition; perhaps it is a quantum set.

We end up with a Clifford–Hilbert space S regarded as the Hilbert space of a q set QSet. In addition to the usual structure of a Clifford algebra and a Hilbert space, S is provided with one linear operator B ('bracket') as an element of structure. The result of applying B to a vector v in S is written also as

$$Bv := \{v|$$

where the bracket symbol $\{|$ fulfills the essential functions of both the bracket symbols of Cantor ($\{ \}$) and Dirac ($\langle |$) combined. S contains the Clifford unit 1, and is generated from 1 by the operations of B, Clifford product and sum. The correspondence between the concepts of c and q set theory is shown in Table 17.2 and explained below.

The disjoint union of c sets mentioned in Table 17.2 is defined to be their union if they are disjoint, and undefined otherwise. The Boolean union is the intersection less the union. Thus Boole's law $A + A = 0$ becomes Clifford's law $ss = 1$. The vector 1 in S stands for 'nothing' but the vector 0 in S stands for nothing, and would be written as a blank

Table 17.2. Classical and quantum set concepts

Classical	Quantum
Set	Vector in S
Class of sets	Subspace of S
Null set	1
Non-set	0
Disjoint union	Grassman product
Boolean union (xor)	Clifford product
$\{s\}$	$\{s|$
Intersection	Not unique
Union	Not unique
Cardinality	Grassmann grade
Multiplicity	Dimension
N-simplex	$N + 1$-adic
[Minkowski pseudometric]	Clifford norm
[Lorentz group]	Clifford group

space but for typographical ambiguity. Multiplicity is the cardinality of a class of sets regarded as a predicate about FSet. It measures *a priori* probability. It is evidently easy to confuse q sets with q classes and cardinality with multiplicity, despite their great formal differences, since Von Neumann did so in his discussions of quantum 'sets' (which were actually classes).

The q correspondents of the union and intersection of c sets are frame-dependent, and omitted.

In c theory, an abstract N-simplex is a set of $N + 1$ vertices; formal averages of the vertices define the points of the simplex. We interpret these averages as incoherent superpositions of the vertices, and take coherent superpositions in the q theory. An '$N + 1$-adic' means a member of S with Grassmann grade $N + 1$.

17.3. Quantum gravity

The bottom line of Table 17.2 is the pay-off. In the c theory, the pseudometric g is still wildly free when the set of time–space is given; the Minkowski pseudometric and the Lorentz group are not concepts of classical set theory at all but of a higher level theory. The q set is born with a pseudometric. The Clifford norm or signature we identify with the local Minkowski norm, the Clifford group with the local Lorentz group.

Therefore we take the world to be a q simplicial complex, a q set of q simplices. Each member of S maximally describes such a complex. We therefore call the members of S 'plexors'. We posit no new laws in going from the level of q sets to q topology, which embraces q pseudometric geometry and the kinematics of gravity as well.

17.4. Quantum dynamics

There are two prevalent ways to formulate quantum theory, and for ease of reference let us call them briefly the good way and the bad. Our own preference will gradually become clear in the following discussion.

In the good formulation, a psi vector represents 'the entire situation' (Bohr) and more specifically an initial process of preparation (Heisenberg) chosen and carried out by the observer C on the system under study S. A complete experiment is described by an initial psi representing an initial determination, a final psi representing a final determination, and possibly a Hamiltonian representing intermediate processes. No psi evolves from the initial psi to the final psi; both psis are selected by the observer. Quantum theory gives the amplitude for such an experiment to have a positive outcome as a matrix element of an evolution operator between the initial and final vectors.

Indeed, a psi vector gives infinitely more information about the observer than the observed. A photon polarization vector gives one bit of information about the photon: it passed through the polarizer instead of being absorbed. The same psi vector gives infinity bits about the polarizer of the observer: It is turned to such-and-such an angle. As the system is enlarged, the disparity between its information capacity and the information represented by a psi vector for the system increases geometrically. For example, a particle can store information in its position x, but this is vastly less than the information in a Schrödinger wave-function $\psi(x)$.

The bad formulation aims to attribute to every system a unique though possibly unknown state in quantum theory just as in classical theory. Since this is impossible for coupled systems thanks to the principle of superposition, we fall back on isolated systems. Since now the psi vector is an attribute of the system, it must evolve, and in order to agree with the final psi vector chosen by the observer it must collapse, in an acausal (indeterminate) and non-local way. But this non-locality is harmless because the psi vector itself is unobservable and cannot be used to send signals.

In the good formulation, an electron gun emits electrons and ideally is described by a psi vector.

In the bad formulation, an electron gun emits psi waves, which later collapse into other psi waves.

We do not know the origin of the bad formulation. Collapsing wave functions appear in Von Neumann's book on quantum mechanics, but there it is made explicit that a psi vector does not belong to an individual but to an ensemble, and an ensemble may represent the process by which it can be produced. Nevertheless, one feature of the bad formulation is already present in Von Neumann's theory: he says that observation *changes* a psi vector. In the good formulation, observation is *represented by* a psi vector. (Or, more graphically, by a dual vector, a bra.)

The good formulation and the bad are completely equivalent in familiar examples, but the relation between the bad formulation and the experience it describes is so indirect and different from ordinary language that it may cause trouble in new situations. A ψ vector refers to an observer and, like a probability distribution, or an entropy, has a relative character that is sometimes called subjective. In particular, if we write down a psi vector for the universe, the good formulation makes us ask what process of determination it stands for; evidently this process can only be one of self-determination, since there is no observer outside the universe. Can self-determination be maximal?

The equivalent question in the bad formulation is: What would make a general psi vector for the universe collapse into the one we have written down?

An atom has many processes of determination, only some of which are maximal (pure cases, of entropy zero) and are associated with psi vectors. These processes are carried on by us, outside the atom, and it is we who absorb the entropy that must leave the atom. Presumably, a maximal determination of a system takes a much vaster outer system, and maximal self-determination is impossible.

There is no doubt that determinations of the universe are possible. When observer C determines system S, the metasystem $C \times S$ is determining itself to some degree, but only to a degree that leaves C quite undetermined, though correlated to S. Maximal self-determination is forbidden under penalty of the second law of thermodynamics.

Accordingly we do not write plexors for the universe, but for the system under study. Such a plexor describes a history of the system, and we may write the amplitude for a plexor E describing an experiment as a linear functional

$$a = D^* E.$$

D is another plexor expressing the dynamical law. D may contain 'external fields' expressing our intervention during the experiment, and it is tempting, since so much of the universe is outside the system, to hypothesize an ultimate Mach principle: the entire form of D is determined by the influence of the outer system on the system observed. But (as with the original Mach principle) this may be an overstatement; there may be a part of D that is intrinsic, representing a basic natural law.

More generally, we combine D and E into one world plexor W representing both the process carried out on the system under study and the dynamical law. By 'world' we mean the entire system under study, typically excluding the observer. W gives maximal information about the topology and geometry of the world, the degrees of freedom in the world, the dynamics of these degrees of freedom, and the experiment being carried out. A fixed functional *Amp* extracts the amplitude for the process from the world plexor W:

$$a = AmpW.$$

We have practised writing plexors for elementary quantum mechanics problems and are now attempting field theory. The q simplicial analogue of Kaluza–Klein theory is the most attractive model at present. To approximate the world as the product of an external four-dimensional q complex with a compact internal q complex, we have developed a natural product for q complexes. We are presently exploring various dynamical plexors for such a world, suggested by the Einstein action of general relativity and the Regge action for polyhedral relativity.

17.5. Self-reference in field theory

We have all wondered whether some unconscious assumption still blocks the development of physics, playing the role today that the implicit axioms of absolute time and logic played before the advents of relativity and quantum theory. Here is what seems a plausible candidate for such an invisible axiom.

In c mechanics there can be a clean cut between the variables of the observer C and the observed S; C variables commute with S variables. This is possible also in non-relativistic quantum mechanics and field theory, provided that C and S interact through instantaneous potentials, as in the Von Neumann theory of measurement. But in relativistic quantum field theories of the cq kind, C and S are made of the same stuff; the whole system is not a product $C \times S$, but rather S, which comprises C as well. The usual assumption that a relativistic quantum field theory possesses maximal (pure state, zero entropy) determinations implies therefore that the system determines itself maximally.

Relativistic quantum field theories of the cq kind unconsciously assume the possibility of unlimited self-determination.

17.6. Third relativity

The problems of the previous two sections seem to call for what J. A. Wheeler has called a third relativity.

In a *relativity of the first (or c) kind*, different observers have different *names* for the possibilities of the system. I call 'here' what you call 'there'. The relativities of Bruno, Galileo, and Einstein (special and general) are of this kind. Relativity 1 produces bilingual dictionaries called co-ordinate transformations to relate its observers. When the White Knight tells Alice what the name of his poem is called, he is operating on the level of relativity 1.

In a *relativity of the second (or cq) kind*, different observers have different *possibilities* for the system. I determine momentum, you determine position. Each possibility corresponds to a basis vector for the Hilbert space of the system. It is common to call a Hilbert space basis a *frame*. Then the relations between such observers are frame transformations. A co-ordinate transformation is a special case of a frame transformation where the frame vectors are merely permuted or changed in phase but not superposed. In general, relativity 2 relates its observers not causally by dictionaries but acausally by transition amplitudes. When the White Knight tells Alice the name of his poem, he is operating on the level of relativity 2.

In a *relativity of the third (or q) kind*, different observers have different *systems*. For example, they may see each other as part of the system, though not themselves. We call the relation between such observers a system transformation. When the White Knight tells Alice his poem, he is operating on the level of relativity 3.

Co-ordinate and frame transformations are special cases of system transformations where the system does not change. We have not succeeded in formulating the interesting case where the observer and system both change significantly. Some of the questions are answered in the quantum theory of open systems; but the changes in the observer remain problematical.

Yet the need seems genuine. Since it takes nearly the whole universe to carry out the determination process represented by a psi vector for the rest, the only way we can study the whole universe is by shifting the boundary between observer and observed until every part of the universe gets its turn to be observed. Neither the relativity 1 of c theory nor the relativity 2 of cq theory deal with such shifts.

Some of the parallels we conjecture between the three developments are shown in Table 17.3. Line 2 gives what is transformed in the relativity. Line 3 tells how the transformations are described. By $alg(L)$ for a linear space L we mean the algebra of linear operators on L. The algebra on Line 3 includes linear transformations of statistical operators, encountered in the quantum theory of open systems. The paradoxes of relativities 1 and 2 arise from new non-commutativities they introduce in their transformations. We do not find such a new non-commutativity in

Table 17.3. Correspondences among the three relativities

Relativity 1	Relativity 2	Relativity 3
c	cq	q
Co-ordinates	Frame	System
Permutation group	Unitary group	$alg(alg(S))$
c	h	tau
Time	Reality	Law, system
Twin paradox	Two-slit paradox	Two-observer paradox

relativity 3 as yet, but a non-invertibility instead. Line 4 gives the physical constant that 'goes to zero' (or infinity) in the correspondence limit. Here *tau* is the fundamental time or chronon. Since we now find gravity on the first level of the theory, we consider *tau* to be approximately the Planck time; previously we had hoped it would be much larger. The familiar particles are huge and light, like blimps, not small and dense and simple as we had hoped. Quantum gauge theories, and therefore, we presume, suitable Kaluza–Klein theories, give a plausible description of the bag of these blimps. Line 5 shows some old absolutes that each kind of relativity relativizes. Line 6 names an apparent paradox of each relativity, a violation of previously unconscious assumptions. The last four entries in the third column are based on analogies and not to be trusted; God does not often tell the same joke twice.

17.7. Summary

A q theory of time–space leads naturally to a q theory of gravity. It describes time–space as a q simplicial complex, using the language of q set theory. The problem arises of how the universe should be maximally described in a q theory. We claim that it should not. To describe the universe we should move the cut between observer and observed until every part of the universe takes its turn at being observed. This requires a third kind of relativity, where we count all classical theories of relativity as of the first kind, and transformation theory in the sense of Dirac as of the second kind. What becomes relative in relativity 3 is not merely the co-ordinates of the possibilities of the system (relativity 1), nor the possibilities of the system (relativity 2), but the system itself and the dynamical law.

Acknowledgement

This chapter is based on work supported by National Science Foundation Grant PHY 780-9661. The name 'third relativity' is taken from recent writings of J. A. Wheeler.

References

Background details and references will be found in

Finkelstein, D. and Rodriguez, E. (1984). *Int. J. Theor. Phys.* **23**, 1065.
Rodriguez, E. (1984). Ph.D. Thesis. Georgia Institute of Technology.

18

Quantum topo-dynamics in higher dimensions

Y. Aharonov and M. Schwartz

A few years ago we presented a one dimensional quantum lattice model that has bag-like excitations and showed the phenomenon of confinement without being related to gauge theories (Aharonov and Schwartz 1982). The model is described by the following Hamiltonian:

$$H = \frac{g}{4} \sum_i (1 + \sigma^z_{i+1})(1 - \sigma^x_i)(1 - \sigma^z_{i-1})$$

$$+ (1 - \sigma^z_{i+1})(1 - \sigma^x_i)(1 + \sigma^z_{i-1}) + h \sum_i (1 - \sigma^z_i), \quad (18.1)$$

where the σs are Pauli matrices.

The basic property of the above Hamiltonian is that a spin can be flipped only if its two nearest neighbours are of opposite signs. It is easy to show that a region of negative sign cannot be created spontaneously and cannot be torn into two separated regions; neither can two regions be united to form a single region. Each region of negative spins can be viewed therefore as a separate entity, a property that is preserved by the dynamics. In fact we can say that the Hamiltonian preserves the topological properties of the state it acts upon. We were able to find a very useful description by going to the continuum limit obtained by defining $g = G/a^2$ and $H = ha$ where G and H are kept finite and a, the lattice spacing, tends to zero. The equation describing a single entity with end points at R_1 and R_2 is the Schrödinger equation for the amplitude $\psi(R_1, R_2)$ of the entity having these end points.

$$-\frac{G}{4}\left[\frac{\partial^2 \psi}{\partial R_1^2} + \frac{\partial^2 \psi}{\partial R_2^2}\right] + H\,|R_1 - R_2|\,\psi = E\psi, \quad (18.2)$$

together with a boundary condition

$$\frac{\partial}{\partial(R_1 - R_2)}\,\psi|_{R_1 = R_2} = 0, \quad (18.3)$$

or, in the case of n entities,

$$-\frac{G}{4}\sum_{i=1}^{n}\frac{\partial^2\psi}{\partial R_1^{(i)^2}}+\frac{\partial^2\psi}{\partial R_2^{(i)^2}}+H\sum_{i=1}^{n}\left|R_1^{(i)}-R_2^{(i)}\right|\psi=E\psi, \qquad (18.4)$$

with the boundary conditions

$$\frac{\partial}{\partial(R_1^{(i)}-R_2^{(j)})}\psi\big|_{R_1^{(i)}=R_2^{(j)}}=0, \qquad (18.5)$$

where $R_1^{(i)}$ and $R_2^{(i)}$ are the end points of the ith entity. In order to describe real-bag-like structures that may have applications in high energy physics as well as in condensed matter and biology where we encounter structures that preserve their topological properties over a long period of time, we need to go over to higher dimensions.

The problem that arises is whether the ideas presented in the one-dimensional case can be carried over to higher dimensions.

Recently we were able to show that it is possible to construct Hamiltonians that are local and that preserve the topological properties of the states they act upon in more than one dimension. An example is

$$H=\mathcal{H}\{\sigma_i^z\}+g\sum_i T_p^i(1-\sigma_i^x)T_p^i \qquad (18.6)$$

where T_p^i is a topology preserving projector, that enables a spin to flip due to the action of σ^x only when this operation does not change the topological properties of the state. Our important observation concerning T_p^i is that it is a local operator depending on σ_j^z where j is in the near vicinity of i. The actual form of T_p^i and the number of neighbour shells needed depends on the lattice, but in all cases it is small. (For the cubic lattices the number of neighbour shells needed does not exceed the space dimensionality.) So far, we used special forms of $\mathcal{H}$ that ensured the existence of entities that consist of one or two sites of negative spins and obtained some interesting results (Aharonov and Schwartz 1984). It seems to us, however, that a much more interesting issue is the description of extended entities. The one-dimensional example suggests that such a description will be obtained by constructing a continuum limit. On the other hand it is clear from the beginning that such an undertaking may be hampered by difficulties related to the fractal nature of the surfaces enclosing our entities and by the possibility of touching of two points on the same surface.

The description we attempt therefore is by no means a continuum limit of the equations generated by the Hamiltonian (18.6). It may be viewed more as an analogy of the Schrödinger eqn (18.2) that is obtained on one hand by coarse graining and on the other hand by keeping in mind the fractal nature of the surfaces. Boundary conditions similar to eqn (18.3) representing the fact that two points on the outer surface of the entity cannot touch are neglected.

Let us consider a cluster of negative spins consisting of N sites on a d-dimensional lattice and having M boundary points. The amplitude of obtaining a given cluster is a functional of the boundary (analogous to the one dimensional amplitude depending on the end points). We will assume, following the accumulated knowledge in statistical mechanics, that the important clusters contributing to the energy eigenfunctions have a surface-to-volume ratio that is of order 1, and that surface to volume ratio is almost independent on the clusters for those important clusters. In fact we may expect that for large N

$$\frac{M}{N} = \alpha + O\left(\frac{1}{\sqrt{N}}\right). \tag{18.7}$$

The parameter α may depend on the lattice and the topological properties of the clusters under consideration.

In order to obtain now a continuum description, let us consider a region of negative spins enclosed in a surface S. The amplitude of obtaining such an entity is $\psi\{S\}$. Let the original discrete structure be introduced by N points within the surface with a density of $1/a^d$ and M surface points with a density of $1/a^{d-1}$. We assume now in analogy with the one-dimensional case that applying the Hamiltonian to $\psi\{S\}$ yields

$$H\psi\{s\} \approx -g\sum_{i=1}^{m} \psi\{s+\delta s_i\} + \psi\{s-\delta s_i\} - 2\psi\{s\} + hN\{s\}\psi\{s\} \tag{18.8}$$

where the surfaces $s + \delta s_i$ and $s - \delta s_i$ are obtained respectively by enclosing the clusters obtained by adding and removing a unit of size a^d at the surface point i to the original cluster and perpendicular to S (Fig. 18.1). We try an *ansatz* for a solution of the eigenvalue equation

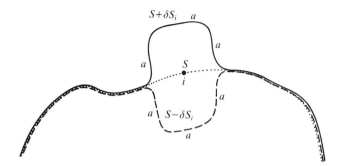

Fig. 18.1. A two-dimensional example. The lines S, $S + \delta S_i$, and $S - \delta S_i$ coincide out of the vicinity of i.

associated with eqn (18.8)

$$\psi\{s\} = \oint_s F(r, v) \cdot ds = \oint_s \nabla\phi(r, v) \cdot ds, \qquad (18.9)$$

where the integration is over the closed surface S and $V = N\{s\}a^d$ is the volume enclosed in the surface. The equation obtained to leading order in a is

$$-ga^2 \oint \nabla(\nabla \cdot F(r, v) \cdot ds - g\alpha a^d v \oint \frac{\partial^2 F}{\partial v^2}(r, v) \cdot ds$$

$$+ \frac{h}{a^d} V \oint F(r, v) \cdot ds = E \oint F(r, v) \cdot ds. \quad (18.10)$$

Using the divergence theorem and the fact that eqn (18.10) holds for *all* surfaces enclosing a volume V we conclude that

$$-ga^2 \nabla^2(\nabla^2\phi) - g\alpha a^d v \frac{\partial^2}{\partial v^2} \nabla^2\phi + \frac{h}{a^d} v \nabla^2\phi = E \nabla^2\phi. \qquad (18.11)$$

Separating variables, $\phi = \Omega(v)R(r)$, we obtain

$$-ga^2\nabla^2(\nabla^2 R) = E_k \nabla^2 R \qquad (18.12)$$

and

$$-g\alpha a^d v \frac{\partial^2}{\partial v^2} \Omega(v) + \frac{h}{a^d} v\Omega(v) = E_I\Omega(v), \qquad (18.13)$$

where the sum of the kinetic energy E_k and the internal energy E_I is the total energy E. Obviously $R(r) = e^{iq \cdot r}$ and $E_k(q) = ga^2 q^2$. Putting eqn (18.13) in the form

$$-\frac{\partial^2}{\partial v^2} \Omega(v) - \frac{E_I}{g\alpha a^d} \frac{1}{v} \Omega(v) = -\frac{h}{g\alpha a^{2d}} \Omega(v), \qquad (18.14)$$

we see that $\Omega(v)$ obeys an s state hydrogen atom Schrödinger equation. The nth eigen energy obeys

$$\frac{1}{2n^2} \left[\frac{E_I^{(n)}}{g\alpha a^d} \right]^2 = \frac{h}{g\alpha a^{2d}}, \qquad (18.15)$$

so

$$E_I^{(n)} = \sqrt{2hg\alpha}\, n \qquad (18.16)$$

and the characteristic volume is the 'Bohr volume'

$$V_0 = \frac{2g\alpha a^d}{E_I^{(1)}} = \sqrt{\frac{g\alpha}{h}}\, a^d. \qquad (18.17)$$

We see that the motion of the entity consists of a centre of mass translation with kinetic energy ga^2q^2 and an internal volume oscillation with a frequency $\omega = (2/\hbar)\sqrt{hg\alpha}$. The parameter $2ga^2/\hbar^2$ is to be identified as m, the mass of the entity. The characteristic volume is given in terms of M and ω by the following expression

$$V_0 = \frac{\hbar}{\omega}\,\alpha Ma^{d-2}. \tag{18.18}$$

In two dimensions we can choose the parameter g and h in such a way that ω and V_0 remain finite when a tends to zero. In more than two dimensions this is not possible; either ω remains finite when a tends to zero and the entity becomes point like, or if the entity remains extended the oscillation frequency tends to zero.

In condensed matter physics any description of clusters or aggregates includes also a typical short range parameter, the radius of the excluded volume.

The same is also true in the description of space itself that can be seriously considered as a continuum only on scales larger than the Planck length.

Our theory suggests now a length scale many orders of magnitude larger than the elementary length in the theory. Such a length scale is associated with a volume oscillation frequency that is related to that longer scale.

Acknowledgement

This chapter is based on work supported in part by the US–Israel Binational Science Foundation.

References

Aharonov, A. and Schwartz, M. (1982). *Phys. Rev. Lett.* **48**, 1137.
Aharonov, A. and Schwartz, M. (1985). *Phys. Lett.* **157** B, 57.

19

Constructing a bit string universe: a progress report

H. Pierre Noyes, Michael J. Manthey, and Christoffer Gefwert

19.1. Introduction

The work presented here, although the centre of serious enquiry for three decades (Bastin 1966; Bastin and Kilmister 1952, 1954*a,b*, 1955, 1957, 1959; Bastin *et al.* 1979; Manthey and Moret 1983; Noyes 1984; Noyes *et al.* 1983; Parker-Rhodes 1981; Stein 1978, 1980, 1981), has not achieved its final form. In particular, we have not yet seen how to construct relativistic quantum mechanics with a clear understanding of *what* results, when achieved, would terminate the program. Hence we have not yet met the requirements that one of us (CG) sees as necessary for such a program (Gefwert 1983, 1984*a,b,c*). Yet this somewhat inchoate theory has suggested a model which allows us to generate a bit string universe constructively from a couple of simple operations and a few well understood symbols—a universe that can model more simply, and in some cases more accurately than conventional physics, 'facts' currently accepted by physicists. Many of those involved in this research believe we have more than that, but they have not made their case compelling either inside their current home† or in the broader communities we inhabit (physics, philosophy, mathematics, computer science, . . . ?). So what this paper is about is to get any imaginative reader to think that our approach might be exciting, and fun—a strategy suggested to one of us (HPN) by Wheeler some time ago.

The fun is not far to seek once one can engage in the initial surrender of disbelief. GUTS, SUPERGRAVITY, and other theories of physics with (it is hoped) fewer parameters than those currently required as empirical input, yet more than the minimal three *we* need (for connection, *not* adjustment, to experiment), are avidly sought within 'conven-

† The Alternative Natural Philosophy Association, founded 1979; for information contact the ANPA Secretary–Treasurer, 15905 Bear Creek Road, Boulder Creek, CA 95006, USA.

tional' frameworks. Mathematics has to face the challenge of whether its theorems are 'computable'—if not why not?—or the relevance of the question, and whether a 'proof by computer' that cannot be checked step by step by humans is a 'proof'. Computer scientists have to ask whether the demonstrable (and practically important) uncertainties encountered in concurrent programming are more than a problem to be designed around, but might set limits to their theories, whether concurrency itself involves problems that are beyond the scope of a 'universal computer' or 'Turing machine' even if they could build one, whether 'randomness' is a 'meaningful' concept when there is no algorithm for computing a random number, etc. Our approach involves all of these questions, so can be fun even where we cannot be certain of our 'answers'.

19.2. Program Universe 2

Our computer algorithm (Program Universe 2)† starts from nothing (in the computer, other than program and available memory) and generates a growing universe characterized by two cardinals: SU, $N \in integers$. For computer operations any element of the universe may be simulated by an ordered string of the symbols 0, 1 containing N such symbols which we can call $U[i]$, $i \in 1, \ldots, SU$. We use two operations to increase these cardinals: (1) PICK, which picks any string from the universe with probability $1/SU$ and a second string (shown to be different by discrimination—see below) with the same prior probability and generates a string by discrimination; if the new string is not already in the universe it is adjoined and SU is increased by one. If the string produced by PICK is already in the universe we invoke (2) TICK which picks a bit for each $U[i]$, randomly chosen between the two symbols 0, 1, adjoins it at the head of the string, and hence increases N by one; the code then returns to PICK. The flow is thus $PICK \rightarrow [novel(adjoin) \quad ELSE \ contained(TICK)] \rightarrow PICK \ldots$.

That this seemingly trivial program can produce any interesting structure is due to the subtleties implied by the discrimination operation which (sometimes) creates new strings. For two *ordered* bit strings of length N, symbolized by $S_i = (\ldots, x_i, \ldots)_N$ where $x_i \in [0, 1]$, $i \in [1, N]$, *discrimination* is defined for bits by $D_N S_i S_j = (\ldots, x_i +_2 x_j, \ldots)_N$, where $+_2$ is binary addition (exclusive or), or for integers by $D_N S_i S_j = (\ldots, (x_i - x_j)^2, \ldots)_N$. This allows us to think of the elements of this universe, which we have sometimes called *Schnurs*, as ordered strings of bits *or* of integers from the set [0, 1], an ambiguity we exploit in defining conserved quantum numbers.

† For Program Universe 1 see Noyes *et al.* (1983); Program Universe 2 has also been coded by M. J. Manthey in Pascal with concurrency, an effort which made this report possible.

To get the program started we assign the first string in the universe the value R (i.e. a random choice between 0 and 1) and the second again the value R, provided only it differs from the first. We now enter the main program at PICK, and continue till doomsday. We say that each tick follows an *event*. Note that by this specification of events, and the integral ordering of the ticks (even if, outside of the computer simulation it turns out to be unknowable) we have abandoned the concept of *simultaneity*, and not just 'distant simultaneity' as is customary in special relativity.

Discrimination (which we for current purposes write as $+$, with $a + a = 0$, and a, b linearly independent (l.i.) iff $a + b \neq 0$) creates sets which close under discrimination called discriminantly closed subsets (DCsS). For example, if a and b are l.i., the set $\{a, b, a + b\}$ closes, since any two when discriminated yield the third. Similarly, if c is l.i. of both a and b, we have the DCsS $\{a, b, c, a + b, \ b + c, \ c + a, \ a + b + c\}$. Provided we call singletons such as $\{a\}$ DCsSs as well, it is clear that from n l.i. strings we can form $(2^n - 1)$DCsS, since this is simply the number of ways we can choose n distinct thing one, two, . . . , up to n at a time.

The first construction of the hierarchy Bastin (1966) started from *discrimination* using ordered bit strings as already defined. Starting from strings with two bits ($N = 2$) we can form $2^2 - 1 = 3$ DCsSs, for example: $\{(10)\}$, $\{(01)\}$, $\{(10), (01), (11)\}$. To preserve this information about discriminate closure we map these three sets by non-singular, linearly independent 2×2 matrices which have only the members of these sets as eigenvectors, and which are linearly independent. The non-singularity is required so that the matrices do not map onto zero. The linear independence is required so that these matrices, rearranged as strings, can form the basis for the next level. Defining the mapping by $(ACDB)(xy) = (Ax + Cy, Dx + By)$ where $A, B, C, D, x, y \in 0, 1$, using standard binary *multiplication*, and writing the corresponding strings as $(ABCD)$, three strings mapping the discriminate closure at level 1 are (1110), (1101), and (1100) respectively. Clearly this rule provides us with a linearly independent set of three basic strings. Consequently these strings form a basis for $2^3 - 1 = 7$ DCsSs. Mapping thse by 4×4 matrices we get 7 strings of 16 bits which form a basis for $2^7 - 1 = 127$ DCsSs. We have now organized the information content of 137 strings into 3 levels of complexity. We can repeat the process once more to obtain $2^{127} - 1 \simeq 1.7 \times 10^{38}$ DCsSs composed of strings with 256 bits, but cannot go further because there are only 256×256 linearly independent matrices available to map them, which is many too few. We have in this way generated the critical numbers $137 \simeq hc/2\pi e^2$ and $1.7 \times 10^{38} \simeq hc/2\pi G m_p^2$ and a hierarchical structure which terminates at four levels of complexity: (2, 3),

$(3, 7)$, $(7, 127)$, $(127, 2^{127} - 1)$. It should be clear that the hierarchy defined by these rules is *unique*, a result achieved in a different way by John Amson (Bastin *et al.* 1979).

In the context of Program Universe, since the running of the program provides us with the strings and also an intervention point (adjoin the novel string produced by discrimination from two randomly chosen strings) where we can organize them conceptually without interfering with the running of the program, we can achieve the construction of a representation of the hierarchy in a simpler way. The procedure is to construct first the basis vectors for the four levels by requiring linear independence both within the levels and between levels. Since adding random bits at the head of the string will not change the linear independence, we can do this at the time the string is created, and make a pointer to that $U[i]$ which is simply i, and which does not change as the string grows.

Once this is understood, the coding is straightforward, and has been carried through by one of us (MJM). Each time a novel string is produced by discrimination, it is a candidate for a basis vector for some level. All we need do is find out whether or not it is l.i. of the current (incomplete) basis array, and fill the levels successively. Calling the basis strings $B_l[m]$ where $l \in 1, 2, 3, 4$ and $m \in 1, \ldots, B[l]$ with $B[1] \ldots B[4] = 2, 3, 7, 127$, we see that the basis array will be complete once we have generated 139 l.i. strings. Since the program fills the levels successively, it is easy to prove that if we discriminate two basis strings from *different* levels we must obtain one of the basis strings in the highest level available during the construction, or level 4 when the construction is complete, i.e. *if $i \neq j$ and both $<l_{last}$ then $B_i + B_j = some\ B_{last}$.*

Once we have 139 l.i. basis strings, which will happen when the bit string length N_{139} is greater than or equal to 139, we can ensure the generation of some representation of the combinatorial hierarchy by going to TICK. Then the only alteration of these N_{139} initial bits that can occur from then on will be the filling up, by discriminate closure, of any of the remaining elements of the hierarchy in this representation as a consequence of the continuing random discriminations. Since we keep on choosing strings at random and discriminating them, discriminate closure ensures that we will eventually generate all $2^{127} + 136$ elements of the hierarchy [BUT NO MORE]. Of course there will eventually come to be many different strings with the same initial bits, N_{139}. We fix this number and call it *LL* for 'label length'; our program automatically generates ensembles of strings labelled by some (unknown) representation of the hierarchy. The coding which provides pointers to these ensembles is again straightforward. From now on we will call the first *LL* bits in a string the *label*, and the remaining $(N - LL)$ bits the *address*. Finally we note that when the label array is complete we know that among the labels L_i at any one level we can find exactly $B(i)$ l.i. strings and no more; it becomes arbitrary which of the many possible choices we make, so the 'basis' becomes a structural fact and does not single out any particular strings. It follows immediately that *if $i \neq j$ and both $<l_{last}$ then $L_i + L_j =$ some L_{last}*.

19.3. Labels and quantum number conservation

An event has been defined when the universe has bit-length N and (when the labels have closed at bit-length LL) at address bit-length $A = N - LL$ as the failure to produce a new string by discrimination. This can occur only when two or more discriminations happen *before* the next tick. Sequentially, the first of the two discriminations we consider resulted from picking S_1 and S_2 at random and generating a string not yet contained in the universe $S_3 = D_N S_1 S_2$. The second discrimination can lead to an event in two different ways. In the first case we pick any two of these three strings again. Since $S_1 + S_2 + S_3 = 0_N$, this second discrimination *necessarily* must yield a string already in the universe. Then the program takes us to TICK, and all the strings in the universe will be augmented by one random bit at the head. Clearly this can happen in three different ways, but we cannot tell without further information which one of the three occurred. The second class of events generated by the program occurs when we pick two strings S_4 and S_5 both of which are different from the first three strings considered, but which on discrimination yield one of the first three strings, that is $D_N S_4 S_5$ is equal to S_1 or S_2 or S_3. Again, any one of these three possibilities will lead to TICK and we cannot know which of the three occurred. We can, however, calculate the *relative* probabilities between the two classes of events; in a subsequent paper we will show how this enables us to calculate *coupling constants*.

Since the levels close, they convey no *dynamic* information; this must come from the addresses, and the labelled address ensembles. By construction, addresses are random bit strings (other than through correlation with the label). But they still have a structure we have not yet exploited, the number of zeros N^0 and the number of ones N^1 in any address string. This allows us to define for each string, or string segment, a parameter bounded by -1 and 1 $v_d = (N^1 - N^0)/(N^1 + N^0)$ which we call *d-velocity*. For simplicity we consider first only two bit labels for which $(10) + (01) + (11) = (00)$, implying three possible configurations for the discriminations that occur in our definition of event: (1) $(10)v_{10} + (01)v_{01} \leftrightarrow (11)v_{11} \leftrightarrow (10)v'_{10} + (01)v'_{01}$; (2) $(10)v_{10} + (11)v_{11} \leftrightarrow (01)v_{01} \leftrightarrow (10)v'_{10} + (11)v'_{11}$; and (3) $(11)v_{11} + (01)v_{01} \leftrightarrow (10)v_{01} \leftrightarrow (11)v'_{11} +$

$(01)v'_{01}$. To these labels we have added address strings symbolized by v whose significance we will develop below.

We can now see, in this limited environment, how quantum number conservation comes about. We call $a = (10)$ a *particle* label, $\bar{a} = (01)$ an *antiparticle* label and $q = (11)$ a *quantum* label. Then the three configurations can be thought of as three primitive scattering processes: (1) $a + \bar{a} \leftrightarrow a + \bar{a}$, (2) $q + \bar{a} \leftrightarrow q + \bar{a}$ and (3) $a + q \leftrightarrow a + q$. We see that if we represent the label as $(b_1 b_2)$ then $b_1 - b_2$ can be used to define the particle quantum number as $+1$, 0, -1 and that the number of particles minus the number of antiparticles is *conserved*. Further, if we link up tick-separated events to make more complicated diagrams, this fact will persist. Thus we have established our first discrete conservation law—so far only for level 1 labels.

If we could treat paired descriptors like (10), (01) along a string as independent quantum numbers, the interpretation given in the last section would generalize to any label with even label length; we could map onto a linear, finite Hilbert space without much effort. The hierarchical connection between levels does not allow us to do this because, for example, a label with string length 4 allows 16 symbols, four of which are linearly independent, while the hierarchy requires us to use only (any l.i. choice of) three of these and hence only seven non-null strings. Nevertheless, at level 2, it is possible to define conserved quantum numbers for two in–two out processes. Since the demonstration is by exhibition, and not by an elegant proof, we omit it here. Using an appropriate connection between the particle, antiparticle dichotomy and velocity direction, level 2 label event structure allows two conservation laws. One implementation allows these to be the difference between the number of particles and antiparticles; for a string (b_1, b_2, b_3, b_4) this is $d = b_1 - b_2 + b_3 - b_4$. The second quantum number in this pair takes twice the component of spin along the velocity direction to be $2s_z = b_1 + b_2 - b_3 - b_4$. Then, as can be seen from the first two columns of Table 19.1 we have two fermions and two antifermions, two helicity 1 quanta, and an $s_z = 0$ interaction with no 'antiparticle'. The natural interpretation is the electron–positron–gamma–coulomb system. Alternatively, we could identify the second of these quantum numbers with twice the 'z component' of isospin. Then we define $2U_z = -2b_1 + b_2 + 2b_3 - b_4$, and find, consistent with the usual definition $V_z = I_z + U_z$, that we have seven of the eight quantum numbers of the basic SU_3 octet. Our scheme even becomes an octet when we add the label (0000) which is degenerate as to these quantum numbers with (1111) (and which will eventually be generated in this position by the continuing discriminations); of course in our context this label is meaningful only when it is part of a longer string.

Mapping level 2 to level 3 produces four bits which can take on all 16

Table 19.1. Two interpretations for level 2 quantum numbers

	$(b_1b_2b_3b_4)$	d	$2s_z$ or $2I_z$	$2U_z$	$2V_z = 2(I_z + U_z)$
STRING:	1110	+1	+1	+1	+2
	0010	+1	−1	+2	+1
	1100	0	+2	−1	+1
	1111	0	0	0	0
	0000	0	0	0	0
	0011	0	−2	+1	−1
	1101	−1	+1	−2	−1
	0001	−1	−1	−1	−2

$d = b_1 - b_2 + b_3 - b_4$; $2s_z = b_1 + b_2 - b_3 - b_4$
$2I_z = b_1 + b_2 - b_3 - b_4$; $2U_z = -2b_1 + b_2 + 2b_3 - b_4$; $2V_z = -b_1 + 2b_2 + b_3 - 2b_4$

independent possibilities and additional parts of the string which have only three independent basis vectors, like level 2, giving $8 \times 16 = 128$ possibilities from which we must subtract the null string to leave 127. It is therefore natural to take the octet as the colour octet and the 16 states as what we can get from two different spin $\frac{1}{2}$ fermions with their antiparticles—the up and the down quarks? If this works out in detail, level 1, 2, and 3 then give the first generation of the standard model (two component electron neutrinos; electrons, positrons and vector quanta with a long-range scalar interaction; coloured up and down quarks with their antiparticles).

Turning to level 4, we note that we can obtain a level 4 basis string by putting together three basis strings from different lower levels providing us immediately with $2 \times 3 \times 7 = 42$ basis vectors, and by repeating this structure three times in longer strings end up with 126 of the 127 basis vectors for level 4. Since each of these has, internally, the same quantum number structure, this looks suspiciously like the three generations which have been found experimentally (electron, up, down; muon, charm, strange; tau, truth, beauty). The 127th basis vector is needed for cross-generational interactions. Once we understand it in terms of our dynamics, we will have our own constructive candidate for 'grand unification'. The one string which couples everything in the same way, and which occurs with a prior probability of $1/(2^{127} + 136)$, then represents the Newtonian gravitational interaction, and the starting point for our theory of 'supergravity'. But to make that development possible we must first connect our scheme to space–time.

19.4. Reconstruction of relativistic quantum mechanics

To go from quantum number conserving label-specified scattering events to a quantum scattering theory we assume that the bits added to a particular labelled ensemble between two scattering events (separated by $b = N^1 + N^0$ ticks) conserving that label define a random walk ensemble (Stein 1978, 1980, 1981) with $p = \langle N^1/b \rangle$ the probability of a step in the + direction. Then the step in the opposite direction has probability $q = 1 - p = \langle N^0/b \rangle$ and (for fixed step length and time interval) the most probable position moves with velocity $v = p - q$, which coincides with our definition of 'd-velocity'. We further assume that we can attribute a mass m_L (or mass distribution) to each label. Then we can *define* two digital quantities scaled by m_L by $E^L_d = m_L c^2 \times (N^1 + N^0)/2(N^1 N^0)^{1/2}$ and $p^L_d = m_L c \times (N^1 - N^0)/2(N^1 N^0)^{1/2}$. Then $E^2 - p^2 c^2 = m^2_L c^4$ independent of our digital definition of v_d for any label L. However, $p_d c/E_d = v_d = c(N^1 - N^0)/(N^1 + N^0)$ consistent with our original definition.

So far our random walks, and energy-momenta are in $1 + 1$ Minkowski space. However, thanks to our postulated (later to be articulated) connection between masses and labels, we have a *metric* space. Further, we have only four basic types of event (involving the four levels of the hierarchy) so can construct a $3 + 1$ space which is rotationally invariant (no way to pick a reference direction). We think this is obvious, and find detailed justification unilluminating.

So far we have, dimensionally speaking, only $[M]$ relative to an arbitrary finite mass and $[L/T]$ limited by c; the theory is still scale invariant. To introduce a unit of length we take the step length in our random walk to be $l = hc/E$, which in the unique zero velocity frame (for any otherwise undistinguished bit string segment $\langle N^1 \rangle = \langle N^0 \rangle$ and hence $\langle v = 0 \rangle$ by construction) reduces to h/mc. Clearly our Lorentz invariance requires h to be a universal constant. We now can (once we have made the appropriate connecting links) 'measure' the distance between events starting from some reference event. But, in contrast to the unique zero velocity frame given by our theory, there is no way to single out one reference event, or origin in space. Consequently our theory has Poincaré invariance, not just Lorentz invariance, and at least at large distance must have momentum conservation. Thus our theory has the usual free particle relativistic kinematics—the starting point for an S-matrix theory. Since we are developing elsewhere a relativistic finite particle number quantum mechanical scattering theory (Lindesay 1981; Lindesay and Markevich 1983; Noyes 1982; Noyes and Lindesay 1983; Noyes and Pastrana 1983), we will not pursue that construction further here. What remains is to establish the usual quantum mechanical interference phenomena in space.

The laboratory paradigm we start with is two 'counters' with volumes $\Delta x \Delta y \Delta z$ whose geometrical dimensions are measured by standard macroscopic techniques and a time resolution Δt measured by standard clocks. When two counters separated by a macroscopic space and time interval larger than the volumes and time resolutions of the counters have fired, some random walk connecting those two volumes has occurred. The connection to the bit string universe is the understanding that what we have called an *event*, and connected constructively to relativistic quantum scattering theory, initiates the chain of happenings that end in the firing of a counter or equivalent natural 'event' (Noyes 1984; Noyes *et al.* 1983). But we do not know within those macroscopic volumes where this random walk started and ended.

To meet this problem, we construct an ensemble of 'objects' (i.e. labelled ensembles with a specified *d-velocity*) all characterized by the same vector velocity v and the same label (or mass) chosen in such a way that, after k steps, each of length $l = (h/mc)[1 - (v/c)^2]^{1/2} = hc/E$, the peak of the random walk distribution will have moved a distance l in the direction of v. Our basic 'quantization condition' is $E = hc/l$, which defines a second length by $p = h/\lambda$. We take as our unit of time the time to take one step, $\delta t = l/c$. Once 'time' is understood in this digital sense, the velocity of the peak of each subensemble in this coherent ensemble has a velocity c/k. We call this *coherent* ensemble of ensembles a *free quantum particle* of mass m, velocity v, and momentum $p = mv/[1 - (v/c)^2]^{1/2}$. There is a second 'velocity' associated with this ensemble of ensembles, namely that with which 'something' moves at each step always in the direction v. We call this v_{ph}; clearly $v_{ph} = kc$, and $vv_{ph} = c^2$. Since this velocity exceeds the limiting velocity it cannot support any direct physical interpretation, and in particular any which would allow the supraluminal transmission of information; of course it can provide for the supraluminal correlations experimentally demonstrated in EPR experiments. Associated with each of the two velocities and the label (or mass) there are two characteristic lengths $\lambda_{ph} = l = hc/E$; $\lambda = kl = h/p$.

Now we consider two basis states for a spin $\frac{1}{2}$ fermion which we write as $(10)p$ and $(01)p$ where p stands for an address ensemble triple. We have seen that such a 'particle' can scatter from another and lead to a final state in a different direction. But there are only two possible states in the new direction. To preserve the (asymptotic) rotational invariance of our theory therefore requires that the new state be expressable as a *coherent* sum of the two states referring to the new direction. Then Lorentz invariance leads directly to the usual spin $\frac{1}{2}$ formalism using two-component spinors and Wigner rotations. The operational consequences can be followed through and lead to the usual density matrix formulation with all the 'interference' phenomena reduced to probability statements (Noyes *et al.* 1983). The key to this is that our *asymptotic* definition of momentum, plus coherence between spin label states and the momentum *direction*—which is *also* an asymptotically measurable quantity—allows us to *understand* coherent superposition, the amplitude squared rule, and therefore the wave–particle dualism within the framework of the bit string universe.

19.5. Understanding the mass scale and mass unit

To understand the mass scale we make use of the original hierarchy identification $\hbar c/Gm_p^2 = 2^{127} + 136 \approx 1.7 \times 10^{38}[1 \pm 0(1/137)]$, which differs from experiment only to order $1/137$. Hence we have the choice of taking the accepted proton mass as our mass unit and correcting the gravitational constant, or vice versa. Until our dynamics is further developed, we cannot do either. For the electromagnetic coupling we have as our first result $\hbar c/e^2 = 137 \pm 0(1/137)$. Since we must calculate in the physical 'gauge' we can reasonably state that this is the value for electrostatic interactions. Then we can calculate finite 'renormalization' corrections due to spin dependence as order $1/137$ corrections to the fine structure constant. For this the finite particle number relativistic quantum scattering theory should suffice, not only to get definite results but even to kill this theory if it is wrong.

Meanwhile, we are justified in using this theoretical number without correction to calculate the rest energy of the electron from its electrostatic energy $m_e c^2 = \langle q^2 \rangle \langle 1/r \rangle$. Since this calculation, initially achieved by Parker-Rhodes (1981), has been published several times (Bastin *et al.* 1979; Parker-Rhodes 1981; Noyes 1984; Noyes *et al.* 1983), we are brief here. The minimal meaningful distance in a zero velocity system with spherical symmetry is the Compton radius $h/2m_p c$; r must start from this value, and scales a random variable y greater than or equal to one. Similarly, since charge is conserved, $\langle q^2 \rangle = (hc/2\pi \times 137)\langle x(1-x) \rangle$. Hence $m_p/m_e = 137\pi/\langle x(1-x) \rangle \langle 1/y \rangle$. Since we have now established our space as necessarily three-dimensional, the discrete steps in y must each be weighted by $(1/y)$ with three degrees of freedom. Hence $\langle 1/y \rangle = [\int_1^\infty (1/y)^4 \, dy/y^2]/[\int_1^\infty (1/y)^3 \, dy/y^2] = 4/5$. Since the charge must both separate and come together with a probability proportional to $x(1-x)$ at each vertex, the other weighting factor we require is $x^2(1-x)^2$. For one degree of freedom this would give $\langle x(1-x) \rangle = [\int_0^1 x^3(1-x)^3 \, dx]/[\int_0^1 x^2(1-x)^2 \, dx] = 3/14$. Once the charge has separated into two lumps each with charge squared proportional to x^2 or $(1-x)^2$ respectively, we can then write a recursion relation (Bastin *et al.* 1979; Noyes 1984; Noyes *et al.* 1983; Parker Rhodes 1981), $K_n = [\int_0^1 [x^3(1-x)^3 + K_{n-1}x^2(1-x)^4] \, dx]/[\int_0^1 x^2(1-x)^2 \, dx]$ and hence $K_n = 3/14 + (2/7)K_{n-1} = (3/14)\sum_{i=0}^{n-1} (2/7)^i$. Therefore, invoking again the three degrees of freedom, we must take $\langle x(1-x) \rangle = K_3$ and we obtain the Parker-Rhodes result $m_p/m_e = 137\pi/[(3/14)[1 + (2/7) + (2/7)^2](4/5)] = 1836.151497 \ldots$. Since the electron and proton are stable for at least 10^{31}

years we identify this ratio with m_p/m_e in agreement with the experimental value 1836.1515 ± 0.0005; thus we claim to calculate the basic mass ratio scale for the theory. Whether this mass ratio calculation remains unchanged when we go on to level 4 and we must show how to calculate the masses of *unstable* baryons and bosons from our dynamical theory is under investigation.

We believe that so far as QED, and more generally quantum field theory, goes our finite particle number scattering equations will either succeed or fail in a few years. So far as we can see at this stage they contain the same physics for finite processes as the conventional approach, and have the advantage of being automatically unitary. Gravitation differs from QED and related theories modelled on it in that the field itself is a source of the field. In the weak field limit the spin 2 theory of gravitation has the right characteristics, as has been known for some time (Weinberg 1972). But Noyes understands from Chris Isham† that the passage from this microtheory to the macroscopic Einstein theory is plagued with infinities. It should be clear by now that if our scheme works this will not be our problem. It is primarily for that reason that we are calling our work to your attention in this volume.

19.6. Summary

We start from the symbols 0, 1, binary addition, *sequence* represented by the integers, and a random operator R which gives us either 0 or 1 with equal prior probability. From these we construct the discrimination operation for ordered bit strings and the strings themselves employing Program Universe 2. We expect that when this program is fully evaluated it will provide an algorithmic definition of *events* sequentially ordered by the integers but accessible for purposes of interpretation only by statistical arguments. We use the *combinatorial hierarchy* to organize the information content of the early stages of the construction into four *levels* characterized by the cumulative cardinals 3, 10, 137 and $2^{127} + 136$. When the information carrying capacity of this construction is exhausted, we use these elements as *labels* to organize the growing universe of strings into *labelled ensembles* of *addresses*.

Accepting this growing universe of labelled bit strings as adequately constructed, we use the concept of *d-velocity* to *order* the connections between tick-separated events. We show that the labels support an interpretation as the conceptual carriers of conserved quantum numbers between events. By assigning a mass to each label (to be calculated at a later stage) we show that we can define *d-energy* and *d-momentum* in

---

† Private communication.

such a way that our definition of *d-velocity* can be identified formally with a limiting velocity and the masses as invariants. We construct a $3 + 1$ dimensional momentum–energy space which, for large bit string segments, is approximately continuous and Lorentz invariant. Using this basis we construct a momentum space scattering theory with conserved quantum numbers and 3-momentum conservation.

Granted 3-momentum conservation, either as derivable or as an additional postulate, we claim to have connected our bit string universe to laboratory practice by our counter paradigm and to have conventional relativistic particle kinematics available to us. The random walk paradigm then provides us with the basic Einstein–de Broglie quantization condition. We extend this to a wave theory by noting that at any level we will have somewhere in the label string the independent pair (10), (01) which can be used as a dichotomous variable referring to a conserved quantum number. But the algebraic sign of this quantum number is then correlated with the algebraic sign of the *d-velocity* in the address string. We claim that our construction of a three $3 + 1$ asymptotic space with rotational and Lorentz invariance, plus our scattering theory, then lead to the usual directional properties of spin. Having uncovered a directional (algebraically signed) internal quantity we then have all that is needed for the understanding of quantum interference phenomena, even though our basic theory is discrete. We therefore claim to have arrived at an *understanding* of the wave–particle dualism in our context.

If it be granted that we have successfully established an understanding of 'free particle' de Broglie waves and correctly identified the source of the quantization condition in terms of random walks, we claim that we have a firm basis for constructing a relativistic quantum scattering theory for any finite number of 'particles'. The 'coupling constants' which give the probabilities of scattering events relative to the 'free particle' basis then can be assigned, because they are based on the random elements in Program Universe 2, as the prior probability assignments of these events relative to the 'free particle' basis. This, we claim, justifies Bastin's original identification of the hierarchy cardinals as the 'scale constants' of physics, Parker-Rhodes' calculation of the proton–electron mass ratio, and the hierarchy connection between the gravitational constant and the unit of mass required by the theory.

Once this task is accomplished, we will be able to go on to establishing our own route to 'grand unification' and 'supergravity'. Here we are, up to a point, playing a currently fashionable game. The main difference is that most contemporary approaches take the space–time continuum and quantum field theory as the framework within which to pose the problem. Any intuited or postulated or 'derived' structure for the interaction Lagrangian which is not in conflict with the conventional connection to experiment is fair. In contrast, since we have a specific connection between our quantum numbers, the wave–particle dualism, constructed 'space–time' and a rigid *exoskeleton* set by the combinatorial hierarchy, our choices are much more limited. We therefore can hope to discover in a finite time whether our theory is *necessarily* in conflict with experiment, or not.

Our final point is that by making *velocity* basic, rather than space and time, we believe we have the correct fundamental starting point for unifying macroscopic quasi-continuous measurement with a digital model. Further, our ticking universe allows us to fuse the special relativistic concept of *event* with the unique and indivisible events of quantum mechanics, *and* with the events which complementarily limit our understanding of cosmology. Whatever else survives from this attempt to construct a digital model for the universe, we are convinced that this is the correct place to connect relativity with quantum mechanics in a fundamental way. We close by remarking that the cosmological implications of the model are not in obvious conflict with experience.

Acknowledgement

This chapter is based on work supported by the USA Department of Energy, contract DE-AC03-76SF00515.

References

Bastin, E. W. and Kilmister, C. W. (1952). *Proc. R. Soc.* A**212**, 559.
—— —— (1954*a*). *Proc. Camb. Phil. Soc.* **50**, 254.
—— —— (1954*b*). *Proc. Camb. Phil. Soc.* **50**, 278.
—— —— (1955). *Proc. Camb. Phil. Soc.* **51**, 454.
—— —— (1957). *Proc. Camb. Phil. Soc.* **53**, 462.
—— —— (1959). *Proc. Camb. Phil. Soc.* **55**, 66.
Bastin, T. (1966). *Studia Philosoph. Gandensia* **4**, 77.
——, Noyes, H. P., Amson, J., and Kilmister, C. W. (1979). *Int. J. Theor. Phys.* **18**, 455.
Gefwert, C. (1983). *A participator: the metaphysical subject* SLAC-PUB-3277.
—— (1984*a*). *The proposition-as-rules idea* SLAC-PUB-3303.
—— (1984*b*). *On the logical form of primitive recursive functions* SLAC-PUB-3334.
—— (1984*c*). *On the logical form of mathematical language* SLAC-PUB-3344.
Lindesay, J. V. (1981). Ph.D. thesis, Stanford. (Available as SLAC report no. 243.)
Lindesay, J. V. and Markevich, A. (1984). In *Few body problems in physics*, Vol. II (ed. B. Zeitnitz). North-Holland, Amsterdam.
Manthey, M. J. and Moret, B. M. E. (1983). *Communications of the ACM* **26**, 137.

Noyes, H. P. (1982). *Phys. Rev.* C**26,** 1858.

—— (1984). In *The wave–particle dualism* (ed. S. Diner *et al.*). Reidel, Dordrecht.

—— and Lindesay, J. V. (1983). *Austral. J. Phys.* **36,** 601.

—— and Pastrana, G. (1984). In *Few body problems in physics,* Vol. II (ed. B. Zeitnitz). North-Holland, Amsterdam.

—— Gefwert, C., and Manthey, M. J. (1983). In *Proceedings 7th International Congress on logic, methodology and philosophy of science* Salzburg (In press). (For a more detailed account see *Towards a constructive physics* SLAC-PUB-3116.)

Parker-Rhodes, A. F. (1981). *The theory of indistinguishables* Synthese Library, vol. 150. Reidel, Dordrecht.

Stein, I. (1978). Seminar at Stanford.

—— (1980). Paper at the 2nd annual meeting of the Alternative Natural Philosophy Association. Cambridge.

—— (1981). Paper at the 37rd annual meeting of the Alternative Natural Philosophy Association. Cambridge.

Weinberg, S. W. (1972). *Gravitation and cosmology* Wiley, New York.

20

Hawking's wave function for the universe

Don N. Page

A physical model of a system, such as the universe, consists of three parts: (1) physical variables, (2) dynamical laws, and (3) boundary conditions.

The physical variables are those needed to describe the state of the system. If the system is considered to be classical, the variables are those needed to specify its point in phase space (e.g. position and momentum for a single point particle) at each instant of time. If the system is quantum mechanical, the variables can usually be taken to be the (complex) values of the wave function (up to an overall phase) for all points of the configuration space (e.g. positions) of the corresponding classical system.

The dynamical laws are local relations between the physical variables and are generally expressed as differential equations. For a classical system, they may be expressed as Hamilton's equations derived from a Hamiltonian or as the Euler–Lagrange equations derived from an action. For a quantum system, they may be expressed as Schrödinger's equation, using an operator form of the Hamiltonian, or they may be given as the result of a path integral between nearby instants of time, using the classical action. Typically, the dynamical laws give the time evolution of the system and determine the physical variables at all times if they are specified at one time.

The boundary conditions specify the values of the physical variables that remain independent after the dynamical laws are imposed. In other words, they select the unique state of the system from the many otherwise allowed. If the dynamical laws determine the time evolution, then the boundary conditions can be all the independent physical variables at one specific time. (If this is chosen to be the earliest time of interest, the boundary conditions are initial conditions, but for an invertible time evolution they could just as well be chosen to be final conditions or conditions at any other time.)

Theoretical physics has largely concentrated on discovering the physical variables and dynamical laws of the universe. Many pieces have been found and have been shown to agree well with observations (e.g. electromagnetic fields obeying Maxwell's equations, gravitational fields obeying the Einstein equation, and electron fields obeying the Dirac equation), giving encouragement to the hope that all of the pieces can be found and put together in some coherent way such as a 'unified field theory' (supergravity?).

In contrast, the boundary conditions of the universe are known much less precisely. In fact, many physicists believe they can never be known exactly, but must instead be highly arbitrary. The main reason for this attitude seems to be that subsystems or parts of the universe can be studied with essentially the same physical variables and dynamical laws as those the whole universe is presumed to have (except that a possible wave function for the whole must be replaced by a density matrix for each part, and the parts are not isolated but have interactions between them), but the boundary conditions vary greatly between the parts and are obviously not the same as for the whole. Thus the reductionist methodology of dividing the universe into separate parts for analysis has been very successful for discovering physical variables and dynamical laws, but it has not given nearly so much hope that the boundary conditions can also be discovered. Some physicists have no hope at all in this and so feel that a unified field theory would be the ultimate achievement physics can make.

Yet we do know that there must be enormous restrictions on the boundary conditions of the universe. Cosmological observations show that the large-scale universe is highly isotropic, homogeneous, and very nearly balanced between expanding forever and eventually recollapsing. Even more remarkable, were it not so familiar, is the fact that virtually all subsystems that have been examined show distinct arrows of time, all pointing in the same direction. This extreme time asymmetry of the observed universe, summarized by the second law of thermodynamics, cannot be explained by the known physical variables and local dynamical laws, which are time symmetric (in the CPT sense), but must be a consequence of the boundary conditions of the universe. Thus the observed properties of the universe show that its boundary conditions are not completely arbitrary but instead are very special, though our finite observations of limited precision cannot pin down the boundary conditions precisely.

Stephen Hawking (Hawking 1982; Hartle and Hawking 1983; Hawking 1983, 1984) has made a specific proposal for the boundary conditions of the universe that would in principle determine its state uniquely. The proposal is that the wave function $\psi(g_{ij}, \phi)$ for any compact spatial hypersurface with three-geometry or metric $g_{ij}(x^k)$ and matter field configuration $\phi(x^k)$ on it is given by a path integral over all compact Euclidean (positive-definite metric) four-geometries, each with action I_E, whose only boundary is the hypersurface in question:

$$\psi(g_{ij}, \phi) = \int d[g_{\mu\nu}] \, d[\phi] \exp(-I_E[g_{\mu\nu}, \phi]). \qquad (20.1)$$

This proposal replaces the usual prescription of integrating over four-geometries between an initial hypersurface and the one in question and thus eliminates the arbitrary choice of boundary conditions on the initial hypersurface. Hawking describes his proposal by saying 'The boundary conditions of the universe are that it has no boundary'.

One problem which arises is that the Euclidean action I_E can be made arbitrarily negative by real conformal perturbations, which makes the path integral (20.1) divergent if taken literally over all real Euclidean metrics, as Gibbons *et al.* (1978) showed. These authors suggest integrating instead over imaginary conformal perturbations, so, strictly speaking, Hawking's wave function is not obtained precisely from a Euclidean path integral. The revised prescription allows $\psi(g_{ij}, \phi)$ to oscillate, whereas the strictly Euclidean path integral is, formally at least, manifestly positive. The oscillations turn out to be important for a classical Lorentzian space–time interpretation of the wave function (Hawking 1983, 1984), so one might say the observed indefinite metric signature of our universe is a consequence of the indefinite Einstein–Hilbert action of the gravitational field.

Hartle and Hawking (1983) have shown that the wave function with a suitable choice for the measure obeys the Wheeler–DeWitt equation

$$\left[-G_{ijkl} \frac{\delta^2}{\delta g_{ij} \, \delta g_{kl}} + g^{1/2} \left(-{}^3R + 2\Lambda + 16\pi G T_{nn}\left(\frac{\delta}{\delta \phi}, \phi \right) \right) \right] \psi(g_{ij}, \phi) = 0, \qquad (20.2)$$

where T_{nn} is the energy density or component of the stress–energy operator doubly normal to the three-dimensional hypersurface with metric g_{ij} and determinant g, and

$$G_{ijkl} = \tfrac{1}{2} g^{-1/2} (g_{ik} g_{jl} + g_{il} g_{jk} - g_{ij} g_{kl}) \qquad (20.3)$$

is the metric on the superspace of all such hypersurface geometries. At each point on the hypersurface, G_{ijkl} has signature $(-+++++)$, the single minus sign occurring for conformal perturbations which have the 'wrong' sign in the action. Hawking (1984) has argued that one can impose a gauge condition on the choice of hypersurfaces so that there is but one independent negative coefficient of the second functional derivatives over the entire hypersurface, assuming T_{nn} obeys a positive energy condition so that the $\delta^2/\delta\phi^2$ terms in it all have negative coefficients. The resulting Lorentzian signature makes the Wheeler–DeWitt eqn (20.2) a hyperbolic differential equation, with a variable mass-squared given by the terms $g^{1/2}(-^3R + 2\Lambda + 16\pi G T_{nn}(0, \phi))$ without functional derivatives. When this mass-squared is positive, it often leads to a wave function oscillating in the time-like variable $g^{1/2}$, which can be interpreted to be a superposition of classical Lorentzian four-dimensional space–times (Hawking 1983, 1984).

Hawking's proposal for using only compact four-geometries with no other boundaries in the path integral (20.1) now amounts to boundary conditions on the Wheeler–DeWitt eqn (20.2). It is convenient to give them as initial conditions at $g^{1/2} = 0$, which forms a (singular) Cauchy surface for eqn (20.2). The configurations having $g_{ij} = g^{1/3}\tilde{g}_{ij}$ with $\tilde{g}_{ij}$ and matter fields ϕ regular as $g^{1/2} \to 0$ can bound compact Euclidean four-geometries whose volume and action I_E go to zero with $g^{1/2}$, so one would expect the wave function (20.1) for these configurations to depend purely on the measure when $g^{1/2} = 0$ and be unity there for a suitable choice of measure. Such configurations with $\tilde{g}_{ij}$ and ϕ regular at $g^{1/2} = 0$ do not form an entire Cauchy surface for eqn (20.2), but some minisuperspace models suggest it may be sufficient to require that ψ be regular for small $g^{1/2}$ and approach unity as $g^{1/2}$ is decreased to zero at fixed regular $\tilde{g}_{ij}$ and ϕ.

Because no one knows how to solve the Wheeler–DeWitt eqn (20.2) on the full infinite-dimensional extended superspace of three-dimensional metric and matter-field configurations, it is useful to look at the equation on minisuperspace configurations, which are restricted configurations described by only a finite number of parameters, with all of the other degrees of freedom of the full superspace frozen out. Here the minisuperspace model of a $k = +1$ Friedmann–Robertson–Walker (FRW) round three-sphere geometry minimally coupled to a homogeneous massive scalar field, discussed by Hawking (1983, 1984), will be described and analysed further. The two parameters describing these configurations, and thus co-ordinatizing the corresponding minisuperspace, are taken to be a and ϕ, where a is the radius of the three-sphere and $2^{-1/2}\sigma^{-1}\phi$ is the value of the scalar field, with $\sigma^2 = 2G/3\pi$. The mass of the scalar field is taken to be $\sigma^{-1}m$.

When the cosmological constant is set equal to zero, the Wheeler–DeWitt equation has the form (Hawking 1983, 1984)

$$\frac{1}{2}\left(\frac{1}{a^p}\frac{\partial}{\partial a}a^p\frac{\partial}{\partial a} - a^2 - \frac{1}{a^2}\frac{\partial^2}{\partial\phi^2} + a^4m^2\phi^2\right)\psi(a, \phi) = 0, \qquad (20.4)$$

where the unspecified exponent p represents some of the uncertainty of the factor ordering. Since the value of p makes little qualitative difference in the solution, I shall set $p = 1$ so that if one defines $\alpha = \ln a$, the equation can be rewritten in the particularly simple form

$$\left(\frac{\partial^2}{\partial \alpha^2} - \frac{\partial^2}{\partial \phi^2} + e^{6\alpha} m^2 \phi^2 - e^{4\alpha}\right)\psi = 0,\tag{20.5}$$

the Klein–Gordon equation of a particle of variable mass-squared

$$\mu^2 = e^{4\alpha}(e^{2\alpha} m^2 \phi^2 - 1) = a^4(a^2 m^2 \phi^2 - 1)\tag{20.6}$$

in the two-dimensional Lorentz metric

$$ds^2 = -d\alpha^2 + d\phi^2.\tag{20.7}$$

As discussed above, one would expect that Hawking's prescription for the wave function gives the boundary condition $\psi(a = 0, \ \phi \ \text{finite}) = 1$ with the appropriate measure. This corresponds to a boundary condition at I^- of the metric (20.7). Here we can argue that $\psi = 1$ also along $\mathscr{I}^-$, the past null infinity of eqn (20.7) (Hawking and Ellis 1973), giving sufficient Cauchy data to solve eqn (20.5) everywhere in the (α, ϕ) plane. The argument is that one can fill in the three-sphere with any weakly curved four-geometry and slowly varying scalar field with gravitational action of the order of $-a^2$ and scalar field action of the order of $a^4 m^2 \phi^2$. If both of these have magnitudes that are small compared with unity, and if such interpolating configurations dominate the path integral (20.1), one would expect the result to be determined primarily by the measure and to be unity for the appropriate measure. Thus we should be able to set $\psi \cong 1$ for $\alpha \ll -1$ and $\alpha \ll -\frac{1}{2}\ln|m\phi|$. This includes not only I^-, which has $\alpha = -\infty$ and $\alpha + |\phi| = -\infty$, but also $\mathscr{I}^-$, which has $\alpha = -\infty$ with $\alpha + |\phi|$ finite.

Now we may proceed with integrating eqn (20.5) forward in the time-like co-ordinate α. For $\alpha < -\ln|m\phi|$ or $a^2 m^2 \phi^2 < 1$, the mass-squared in eqn (20.6) is negative, leading to exponential growth of the wave function with α or a. For $\alpha \ll -\ln|m\phi|$, we may neglect the third term in eqn (20.5) and obtain the approximate solution

$$\psi(a, \phi) \approx I_0(\tfrac{1}{2}a^2), \qquad a^2 \ll \frac{1}{m^2 \phi^2},\tag{20.8}$$

where

$$I_0(z) = \sum_{n=0}^{\infty} \frac{(z/2)^{2n}}{(n!)^2}\tag{20.9}$$

is the zeroth-order modified Bessel function of the first kind, which gives the asymptotic behaviour

$$\psi(a, \phi) \sim (\pi a^2)^{-1/2} e^{a^2/2}, \qquad 1 \ll a^2 \ll \frac{1}{m^2 \phi^2}.\tag{20.10}$$

However, for $a^2 \gg 1$ one must also take into account the ϕ dependence. The exponentially growing part of the wave function for $ma \gg 1$ spills over into $a^2 m^2 \phi^2 > 1$ where $\mu^2 > 0$, and for $\phi^2 \ll 1$ has the asymptotic form

$$
\psi \sim \text{const.}(ma)^{-1/4 - m^2/8} \exp\left\{ \left[\tfrac{1}{2}a^2 - \tfrac{1}{2}ma \right.\right.
$$

$$
+ \frac{1}{16}(3 + 4m^2 + m^4)(ma)^{-1} + O(ma)^{-2} \Big]
$$

$$
+ \left[-\frac{1}{2}ma^3 + \frac{3}{4}a^2 - \frac{3}{16}(m^{-1} + 2m)a \right.
$$

$$
\left. - \frac{3}{32}(2m^{-2} - 1 + m^2) + O(m^{-3}a^{-1}) \right]\phi^2
$$

$$
+ \left[-\frac{9}{32}ma^3 + \frac{45}{128}a^2 + \frac{9}{256}(3m^{-1} - m)a + O(m^{-2}) \right]\phi^4
$$

$$
+ \left[-\frac{27}{256}ma^3 + \frac{9}{512}a^2 + O(m^{-1}a) \right]\phi^6
$$

$$
+ \left[-\frac{243}{16\,384}ma^3 + O(a^2) \right]\phi^8 + O(\phi^{10}) \Big\} . \tag{20.11}
$$

Although the wavefunction grows exponentially with a^2 at $\phi^2 = 0$, it becomes very sharply peaked around $\phi = 0$, with the dependence on ϕ at large ma^3 being approximately Gaussian with r.m.s. width $(ma^3)^{-1/2}$. It drops below unity roughly along the curves $\phi = \pm(ma)^{-1/2}$, which are outside the curves $\phi = \pm(ma)^{-1}$ where $\mu^2 = 0$, but which are approaching the α- or a-axis asymptotically.

There is also an oscillatory part of the wave function in the region $a^2 m^2 (\phi^2 + 1) \gg 1$. For $\phi^2 \gg m^{-2}$, there is negligible growth in the wave function with α in the region $\alpha < -\ln|m\phi|$ where $\mu^2 < 0$, so one may neglect the last term in eqn (20.5). Then if $\phi^2 \gg 1$ as well, the solution is approximately

$$
\psi(a, \phi) \cong J_0(\tfrac{1}{3}a^3 m\phi), \qquad \phi^2 \gg m^{-2} + 1, \tag{20.12}
$$

where $J_0(z) = I_0(iz)$ is the ordinary Bessel function of the first kind, which gives the asymptotic form for $a^3 m |\phi| \gg 1$,

$$
\psi(a, \phi) \sim (\tfrac{1}{6}\pi a^3 m |\phi|)^{-1/2} \cos(\tfrac{1}{3}a^3 m |\phi| - \tfrac{1}{4}\pi). \tag{20.13}
$$

In much of the rest of the oscillatory region, one can use an eikonal or WKB approximation for the wave function. One needs to match the exponentially growing solution (20.8) for $\mu^2 < 0$ to an oscillating solution for $\mu^2 > 0$ where the boundary ($\mu^2 = 0$ or $a^2 m^2 \phi^2 = 1$) is space-like in the metric (20.7), which means $\phi^2 > 1$. One can write the oscillating solution as

$$\psi = Re(Ce^{iS}), \tag{20.14}$$

where C is a complex variable and S is a real solution of

$$-(\nabla S)^2 \equiv \left(\frac{\partial S}{\partial \alpha}\right)^2 - \left(\frac{\partial S}{\partial \phi}\right)^2 = \mu^2 \equiv e^{6\alpha}m^2\phi^2 - e^{4\alpha}. \tag{20.15}$$

Then the minisuperspace Wheeler–DeWitt eqn (20.5) takes the form

$$Re[(2\nabla S \cdot \nabla C + C\Box S - i\Box C)ie^{iS}] = 0. \tag{20.16}$$

If S varies sufficiently rapidly compared with ∇S, there is a solution with C slowly varying so $\Box C$ may be neglected in eqn (20.16). One may express this solution as

$$C \cong \exp \int \tfrac{1}{2}\mu^{-2}\Box S \, \mathrm{d}S, \tag{20.17}$$

where the integral is taken along the trajectories normal to the lines of constant S, which have tangent vectors

$$\frac{\mathrm{d}}{\mathrm{d}S} = (\nabla S)^{-2} \, \nabla S = \mu^{-2}\left(\frac{\partial S}{\partial \alpha}\frac{\partial}{\partial \alpha} - \frac{\partial S}{\partial \phi}\frac{\partial}{\partial \phi}\right). \tag{20.18}$$

These trajectories give solutions of the classical equations of motion

$$\dot{\alpha}^2 = m^2\phi^2 + \dot{\phi}^2 - e^{-2\alpha}, \tag{20.19}$$

$$\ddot{\phi} + 3\dot{\alpha}\dot{\phi} + m^2\phi = 0, \tag{20.20}$$

where the overdot means $\mathrm{d}/\mathrm{d}t$ in the $k = +1$ Lorentzian FRW metric

$$\mathrm{d}s^2 = \sigma^2(-\mathrm{d}t^2 + e^{2\alpha}\,\mathrm{d}\Omega_3^2), \tag{20.21}$$

with $\mathrm{d}\Omega_3^2$ being the metric on a unit three-sphere and α and ϕ being functions of t only. Equation (20.15) is then the Hamilton–Jacobi equation for eqns (20.19) and (20.20), with

$$\frac{\mathrm{d}}{\mathrm{d}t} = e^{-3\alpha}\mu^2\frac{\mathrm{d}}{\mathrm{d}S} = e^{-3\alpha}\left(\frac{\partial S}{\partial \alpha}\frac{\partial}{\partial \alpha} - \frac{\partial S}{\partial \phi}\frac{\partial}{\partial \phi}\right). \tag{20.22}$$

Matching the growing solution for $\mu^2 < 0$ to the oscillating solution for $\mu^2 > 0$ gives $S = -\tfrac{1}{4}\pi$ along the curves $e^{2\alpha}m^2\phi^2 = 1$ where $\mu^2 = 0$. If one starts at $t = 0$ at a point on the upper curve with $\phi = \phi_0 \gg 1$ and $\dot{\phi} = 0$ there, the classical trajectory is described for $\phi \gg 1$ by

$$e^{\alpha} \approx m^{-1}\phi_0^{-1}\cosh[m\phi_0 t - \tfrac{1}{6}m^2t^2 - \tfrac{1}{3}\ln(1 - \tfrac{1}{3}m\phi_0^{-1}t)], \tag{20.23}$$

$$\phi \approx \phi_0 - \tfrac{1}{3}\phi_0^{-1}[\ln\cosh(m\phi_0 t) + \tanh^2(m\phi_0 t)]. \tag{20.24}$$

For
$$m^{-1}\phi_0^{-1} \ll t \le t_1 = 3m^{-1}\phi_0 - m^{-1}\phi_0^{-1}(1 - \ln 2), \qquad (20.25)$$

where the three-dimensional curvature term $e^{-2\alpha}$ is negligible in eqn (20.19), and where $\phi \gg 1$, eqns (20.23) and (20.24) take the form

$$\phi \approx \tfrac{1}{3}m(t_1 - t), \qquad (20.26)$$
$$a = e^\alpha \approx \tfrac{1}{2}m^{-1}\phi_0^{-2/3}\phi^{-1/3}\exp[\tfrac{3}{2}(\phi_0^2 - \phi^2)]. \qquad (20.27)$$

Thus the scalar field ϕ decreases fairly linearly with time in this region, whereas the radius e^α of the three-sphere grows exponentially by an enormous factor for $\phi_0 \gg 1$ as in the chaotic inflation model (Linde 1983). This huge expansion makes the spatial curvature $e^{-2\alpha}$ negligible for a very long time, so eqns (20.19) and (20.20) with $e^{-2\alpha}$ dropped become those of expanding $k = 0$ FRW model until a much later time. If one defines by

$$\phi \equiv \frac{2\cos\theta}{3\tau}, \qquad \dot\phi \equiv -\frac{2m\sin\theta}{3\tau}, \qquad (20.28)$$

a phase angle θ for the scalar field and a quantity τ which is inversely proportional to the square root of the energy density and hence is proportional to the Hubble time

$$\frac{dt}{d\alpha} \equiv \dot\alpha^{-1} = \frac{3\tau}{2m}, \qquad (20.29)$$

these $k = 0$ equations give

$$\frac{d\theta}{d\alpha} = \tfrac{3}{2}(\tau - \sin 2\theta), \qquad (20.30)$$

$$\frac{d\tau}{d\theta} = \frac{2\tau\sin^2\theta}{\tau - \sin 2\theta}. \qquad (20.31)$$

These equations are a good approximation for the $k = 1$ FRW equations when $e^{-2\alpha} \ll m^2\tau^{-2}$.

The appropriate solution to eqn (20.31) at small θ is

$$\tau = \sin 2\theta(1 + \sin^2\theta) + O(\theta^7). \qquad (20.32)$$

Using this in eqns (20.28)–(20.30) and matching to eqns (20.26) and (20.27) gives

$$t \approx -m^{-1}\theta^{-1} + t_1 + O(\theta), \qquad (20.33)$$
$$\alpha \approx -\tfrac{1}{6}\theta^{-2} + \tfrac{1}{3}\ln\theta + \tfrac{3}{2}\phi_0^2 - \tfrac{1}{3}\ln(\tfrac{8}{3}m^3\phi_0^2) + \tfrac{5}{18} + O(\theta^2). \qquad (20.34)$$

One can now use eqns (20.32)–(20.34) as starting values and integrate eqns (20.29)–(20.31) numerically to arbitrarily large θ, τ, t, and α. ϕ drops below unity for $\theta \gtrsim t_1$ or $t \gtrsim t_1$ and then oscillates with extrema

$$\phi_n = (-1)^n c_n / n \tag{20.35}$$

at $\theta_n = n\pi$, where the c_ns are all numbers of order unity obtainable from $\tau_n = \tau(\theta_n)$ by

$$c_n = \frac{2\theta_n}{3\pi\tau_n}. \tag{20.36}$$

For example, a numerical integration gives

$$c_1 \approx 0.13634966. \tag{20.37}$$

For large θ, eqn (20.31) gives

$$\tau \approx \theta - \tfrac{1}{2}\sin 2\theta + \text{const.} + O(\theta^{-1}), \tag{20.38}$$

so for large n

$$c_n = \frac{2}{3\pi} + O(n^{-1}) \approx 0.21220659 + O(n^{-1}). \tag{20.39}$$

For large θ, eqns (20.29) and (20.30) also imply

$$t = m^{-1}[\theta + \text{const.} + O(\theta^{-1})], \tag{20.40}$$

$$\alpha = \tfrac{2}{3}\ln\theta + \text{const.} + O(\theta^{-2}). \tag{20.41}$$

This corresponds approximately to a matter-dominated $k = 0$ FRW model with scale factor $e^\alpha \propto t^{2/3}$ (Hawking 1983, 1984).

Since $m^2\phi^2 + \dot{\phi}^2$ drops as t^{-2}, it eventually becomes comparable to the more slowly dropping $e^{-2\alpha}$ in eqns (20.19), and eqns (20.29)–(20.31) must be replaced by the $k = 1$ FRW equations. Because $\dot{\alpha}^2$ is then $\ll m^2$, eqn (20.20) implies

$$m^2\phi^2 + \dot{\phi}^2 \approx e^{\alpha_{max} - 3\alpha} \tag{20.42}$$

with some constant α_{max}, so eqn (20.19) gives the $k = +1$ matter-dominated FRW behaviour with the radius e^α reaching a maximum at $\alpha = \alpha_{max}$ and then recontracting. From eqn (20.34) one can see that

$$\alpha_{max} \approx \tfrac{9}{2}\phi_0^2 - \ln(m\phi_0^2) + O(1), \tag{20.43}$$

so for $\phi_0 \gg 1$ the maximum size $e^{\alpha_{max}}$ is enormous, roughly the cube of the size given by eqn (20.27) when ϕ first crosses zero.

Once the classical trajectories have been found, one can integrate eqns (20.22) and (20.17) to find S and C for the WKB approximation (20.14). For $\phi \gg 1$, one gets

$$S \approx \tfrac{1}{3}m^{-2}\phi^{-2}(e^{2\alpha}m^2\phi^2 - 1)^{3/2} - \tfrac{1}{4}\pi, \tag{20.44}$$

$$C \approx Ae^{-\alpha}(e^{2\alpha}m^2\phi_0^2 - 1)^{-1/4}, \tag{20.45}$$

where A may be determined by a WKB match of the oscillating solution for $\mu^2 > 0$ to the exponentially growing solution (20.8) for $\mu^2 > 0$ at the point $\phi_0(\alpha, \phi)$ where the appropriate classical trajectory through (α, ϕ) hits one of the curves $\mu^2 = 0$. From eqn (20.27) one can see that

$$\phi_0^2 \approx \phi^2 + \tfrac{2}{3}\alpha + \tfrac{1}{9} \ln[m^6\phi^2(\phi^2 + \tfrac{2}{3}\alpha)^2] + O(1), \qquad (20.46)$$

in terms of which one finds

$$A \approx [6^{1/2} + 2 \exp(\tfrac{1}{3}m^{-2}\phi_0^{-2}) - 2]\pi^{-1/2}. \qquad (20.47)$$

For $\phi \lesssim 1$ but $me^\alpha \gg 1$ so that there is at least one classical Lorentzian trajectory from $\phi_0^2 \gg 1$ which satisfies the $k = 0$ approximation $e^{-2\alpha} \ll m^2\phi^2 + \dot\phi^2$ at the point of interest, S has the form

$$S \approx \tfrac{1}{3}me^{3\alpha}f(\phi), \qquad (20.48)$$

where

$$\left(\frac{df}{d\phi}\right)^2 = 9(f^2 - \phi^2) \qquad (20.49)$$

with

$$f(\phi) = |\phi + \tfrac{1}{18}\phi^{-1} + O(\phi^{-3})| \qquad (20.50)$$

for large ϕ^2. If one solves for S by integrating eqn (20.22) along a classical trajectory starting at some positive $\phi_0 \gg 1$, the trajectory never drops below $\phi \approx -c_1$ given by eqn (20.37). This is reflected in the fact that if one integrates eqn (20.49) down from large positive ϕ, $f^2 - \phi^2$ goes to zero at $\phi = -c_1$ so that the integration cannot be continued to more negative values of ϕ. (Of course, one can choose a trajectory which starts at negative ϕ_0 to get a different WKB solution which works for $\phi < -c_1$, in fact for all $\phi < c_1$.)

$\phi \approx -c_1$ is where nearby classical trajectories, which start at nearby large values of ϕ_0, first cross. It thus represents a caustic, where $\Box S$ diverges and the WKB approximation (20.14) breaks down. If one modifies the WKB approximation to allow it to be continued along the trajectories, one gets further caustics at each extremum at ϕ. These are nearly horizontal lines in the (α, ϕ) plane which asymptotically approach $\phi = \pm\phi_n$ given by eqn (20.35) for $\alpha \gg \ln(m^{-1}n)$. For each point between the nth and $(n + 1)$th caustic for each sign of ϕ, that is, for $\alpha \gtrsim \ln|m^{-1}n|$ and $(n + 1)^{-1}c_{n+1} \lesssim |\phi| \lesssim n^{-1}c_n$, there are $2n + 1$ trajectories of the form discussed above with $\dot\alpha > 0$, each of which can contribute a WKB-like solution. Hence for $|\phi| \ll 1$, the wave function is a superposition of roughly $[4/(3\pi)]|\phi|^{-1}$ oscillating solutions for $e^{2\alpha}m^2\phi^2 > 1$, or roughly $(4m/3\pi)e^\alpha$ solutions for $e^{2\alpha}m^2\phi^2 < 1$, in addition to the exponential piece (20.11). It is hard to estimate accurately the total contribution of all of these components, but it does seem likely that for large α even the oscillatory part of the wave function will be concentrated near $\phi = 0$. It is also apparent that the wave function is not strongly damped for large e^α, so there is a significant probability for this minisuperspace model universe to be very large compared with the Planck size.

Once one has an approximation for the wave function ψ of a minisuperspace model of the universe, one can use it to calculate the conditional expectation value of any observable O as

$$E(O \mid C) = \langle \psi | P_C O P_C | \psi \rangle / \langle \psi | P_C | \psi \rangle, \qquad (20.51)$$

where P_C is the projection operator onto the condition C. Since observations made within the universe can only test these conditional expectation values (or conditional probabilities if O is also a projection operator), the absolute expectation values are untestable and the wave function need not be normalized or even normalizable. (The FRW wave function analysed above certainly is not.)

Eventually one would like to have a model sufficiently realistic that one can calculate (given suitable conditions necessary for our existing and making observations) the resulting conditional probabilities for the large-scale universe to be observed to be highly isotropic, homogeneous, nearly balanced between expanding forever and recollapsing, and having a strong arrow of time. The FRW model described above is of course not sufficiently realistic to give definitive answers to any of these questions, and it is too simple to answer at all any but the third question. However, it is encouraging to note that with a natural choice of measure for the wave function, the model does predict that the conditional probability distribution of the criticality density parameter

$$\Omega = (m^2 \phi^2 + \dot{\phi}^2) / \dot{\alpha}^2 \qquad (20.52)$$

at fixed energy density is entirely concentrated at $\Omega = 1$ (Hawking and Page 1985). That is, observations made at any fixed energy density will give unit probability for the universe to be infinitely large and hence spatially flat and precisely balanced between expanding forever and recollapsing. It will be interesting to see whether more realistic models for Hawking's wave function can confirm this prediction and explain other observations we make of the cosmos.

Acknowledgements

The ideas of this paper originated largely with the lectures by Stephen Hawking at the 1983 NATO Summer School on Relativity, Groups, and Topology in Les Houches, France (Hawking 1983), and with further discussions with him at the University of Cambridge. Comments by various people such as Hawking and Ian Moss at the Oxford Quantum Gravity Discussion Conference have been influential. Appreciation is expressed for the hospitality of the Aspen Center for Physics, where this work was completed and written up. Financial support was also provided by NSF Grants PHY-8117464 and PHY-8316811 and by an Alfred P. Sloan Fellowship.

References

Gibbons, G. W., Hawking, S. W., and Perry, M. J. (1978). *Nucl. Phys.* B **138,** 141–50.

Hartle, J. B. and Hawking, S. W. (1983). *Phys. Rev.* D **28,** 2960–75.

Hawking, S. W. (1982). In *Astrophysical cosmology: proceedings of the study week on cosmology and fundamental physics* (ed. H. A. Brück, G. V. Coyne, and M. S. Longair). Pontificiae Academiae Scientiarum Scripta Varia, Vatican.

—— (1983). In *Relativity, groups and topology II* (ed. B. S. DeWitt and R. Stora). North-Holland Physics Publishing, Amsterdam.

—— (1984). *Nucl. Phys.* B**239,** 257–76.

—— and Ellis, G. F. R. (1973). *The large scale structure of space–time*. Cambridge University Press, Cambridge.

—— and Page, D. N. (1985). *Nucl. Phys.* B (to be published).

Linde, A. D. (1983). *Phys. Lett.* **129B,** 177–81.

21

Canonical quantization of black holes

P. Hajicek

One of the characteristic problems of quantum gravity is that the causal structure of space–time is not given *a priori*, but is a function of the metric—one of the dynamical variables of the theory. This leads to well-known principal difficulties in all methods of quantization (except for Euclidean quantum gravity). Let us notice that not only general relativity suffers from this; supergravity as well as the Kaluza–Klein type of theory will encounter the same difficulty. Of course, one will not see much trouble as long as one stays near some fixed classical solution and approximates the true light cones by the light cones of the solution. However, this is dangerous even in the classical theory. (Christodolou and Schmidt 1979).

The problem will manifest itself in the strongest way if we consider processes which include a collapse to a black hole, or a quantum analogy of these. From the classical theory we know that the causal structure depends strongly on whether the collapse takes place or not. Thus, one expects that the quantum theory of the full nonlinear gravidynamics will meet additional difficulties, if one does not use the Euclidean method (Hawking 1979).

In this paper, we want to study the so-called constraint-free canonical quantization method (Ashtekar *et al.* 1974). We will choose a gauge and exclude all dependent and gauge variables by means of the constraints and gauge conditions so that only the true dynamical variables and their unconstrained dynamics will undergo a quantization.

The choice of gauge can be described as follows: one selects four components of the dynamical field to play the role of the time t and the space co-ordinates x^k. This part of the field will, therefore, not be quantized (turned to operators). Hence, it seems that the so-called equal time commutation relations can be well-defined by means of this structure: the $t = $ const. hypersurfaces, and the co-ordinates x^k on them will not become fuzzy. However, the difficulty with the dynamical space–time will now emerge in a different form. The range of the co-ordinates x^k cannot, in general be fixed, because the dynamical development can change them. We have no guarantee that the hypersurfaces $t = $ const. are everywhere space-like. We do not know whether each kinematically allowed space–time admits a foliation by Cauchy hypersurfaces and, if so, whether $t = $ const. are such hypersurfaces. Let us call these and related questions 'the foliation problem'.

I am going to describe a solution to the foliation problem for a simplified model of gravidynamics—the so-called Berger–Chitre– Moncrief–Nutku (BCMN) model (Berger *et al.* 1974). Let us first briefly introduce the model. It results from the Einstein–Maxwell system to which an uncharged scalar field is minimally coupled. All dynamical degrees of freedom are frozen except for the spherically symmetric ones. We have the following variables: a metric g_{ab} on a two-dimensional manifold M' (t-r-surface of the original four-dimensional space–time), a real scalar field ϕ on M' (r-co-ordinate), and the real scalar field ψ on M' (the original scalar). The action has the form (Thomi *et al.* 1984)

$$I = \frac{1}{2} \int d^2x \, |g|^{1/2} \left[\frac{1}{G} + g^{ab} \, \partial_a \phi \, \partial_b \phi + \frac{1}{2} R \phi^2 \right.$$
$$\left. + G^2 \frac{Q^2 + P^2}{\phi^2} - \phi^2 g^{ab} \, \partial_a \psi \, \partial_b \psi + m^2 \phi^2 \psi^2 \right]. \quad (21.1)$$

Here, G is Newton's constant, Q and P are the electric and the magnetic charges of the possible black hole in the middle, g is the determinant of g_{ab}, R is the curvature scalar corresponding to g_{ab} and m is the mass of ψ.

We are going to employ the original BCMN gauge. The foliation condition reads

$$n^a \, \partial_a \phi = 0; \quad (21.2)$$

n^a is the unit normal vector to the $t = $ const. hypersurface. The gauge condition is

$$x = \sqrt{G} \, \phi. \quad (21.3)$$

This is possible because ϕ is an auxiliary field and the function $\phi(t, x)$ can be specified arbitrarily (there is no dynamics for it in eqn (21.1)). Each such specification fixes the co-ordinate x along the $t = $ const. hypersurfaces. The differential eqn (21.2) must be augmented with a boundary condition; we choose

$$\lim_{x \to \infty} g_{00} = -1. \quad (21.4)$$

Thus, only those $t = $ const. hypersurfaces are elements of our foliation which reach the infinity i^0 (we assume that the space–time is asymptotically Minkowskian).

For example, in the Kruskal space–time, the conditions (21.2), (21.3) and (21.4) select the Schwarzschild co-ordinates t and r from all possible co-ordinate systems. The foliated region consists of the two external Schwarzschild space–times. We observe that the foliated part of the space–time M' is not the whole of M'. We also can see another typical property of the BCMN foliation: the $t = $ const. hypersurfaces intersect each other at the double apparent horizon (simultaneously future and past). We are going to cut all the surfaces $t = $ const. at such a point and consider only that part of M' which is a neighbourhood of one particular $\mathcal{I}$; we denote this part by M. In our example, M has the following nice properties:

(1) M is itself asymptotically Minkowskian;
(2) M is globally hyperbolic;
(3) the BCMN foliation of M is everywhere space-like.

The property (1) is necessary if one is to construct a well-defined scattering theory in M. The properties (2) and (3) show that the dynamics in M is well-defined and the foliation is regular.

However, the Schwarzschild space–time is a very special field system: it is a solution of the full system of classical field equations and even as such it is very special (e.g. $\psi = 0$). The foliation problem is whether there is a gauge condition such that the properties (1), (2), and (3) are valid for *any* kinematically possible field trajectory.

In Hajicek (1984b,c), I have shown that the BCMN gauge is such a gauge. The starting assumption is the validity of the constraints and the gauge propagating equations (equations for Lagrange multipliers; here the lapse and shift functions). These equations read as follows. The Hamiltonian constraint is:

$$\frac{\partial}{\partial x}\left(\frac{x}{\gamma}\right) = -x^2[(n^a\,\partial_a\psi)^2 + (m^a\,\partial_a\psi)^2] + 1 - m^2x^2\psi^2 - G^2\frac{Q^2+P^2}{x^2},$$

(21.5)

the momentum constraint is

$$\frac{\partial\gamma}{\partial t} = 2\alpha\gamma^{3/2}x(n^a\,\partial_a\psi)(m^a\,\partial_a\psi),$$

(21.6)

and the gauge propagating equations are

$$\frac{1}{\alpha^2\gamma}\frac{\partial}{\partial x}(x\alpha^2) = x^2[(n^a\,\partial_a\psi)^2 + (m^a\,\partial_a\psi)^2] + 1 - m^2x^2\psi^2 - G^2\frac{Q^2+P^2}{x^2},$$

$$\beta = 0.$$

(21.7)

Here

$$\gamma = g_{11}, \qquad \alpha = \sqrt{-g_{00} + \frac{g_{01}^2}{g_{11}}}, \qquad \beta = g_{01},$$

$$n^a = \left(\frac{1}{\alpha}, -\frac{\beta}{\alpha\gamma}\right),$$

$$m^a = \left(0, \frac{1}{\sqrt{\gamma}}\right), \tag{21.8}$$

so that α and β are respectively the lapse and shift functions and m^a is the unit tangential vector to the $t = $ const. hypersurfaces (Hajicek 1984a).

It was shown in Hajicek (1984c) that the three eqns (21.5), (21.6), and (21.7) are equivalent to the following *tensorial* equation

$$\frac{\delta I}{\delta g_{ab}} = 0, \tag{21.9}$$

and that this will be the case for any choice of foliation and gauge.

The foliation satisfying eqn (21.2) will be non-space-like at the points where the gradient $\partial_a\phi$ is non-space-like and only there. As $\partial_a\phi$ is always space-like near infinity, the foliation can become time-like only, if it crosses the boundary of the region in which the gradient $\partial_a\phi$ is timelike; at this boundary, $\partial_a\phi$ will be null. Hence, it is sufficient to show that the foliation avoids the points of M' in which $\partial_a\phi$ is null. However, these points are apparent horizons. Indeed, let p be such a point. If $\partial_a\phi$ is non-zero at p, then the surface $\phi = $ const. is tangential to either an ingoing or an outgoing null hypersurface at p. Thus, the divergence of the null rays of this hypersurface is zero at p. If $\partial_a\phi$ is zero at p, then both ingoing and outgoing null hypersurfaces through p have zero divergence at p. In any case, p is an apparent horizon, and it is, therefore of interest to investigate the properties of the apparent horizons which follow from the eqns (21.5), (21.6) and (21.7).

Equation (21.9), which is equivalent to them, implies the following:

Theorem 1. Let $\mathscr{H}$ be the outgoing (ingoing) null hypersurface through a future (past) apparent horizon p. Then, the divergence of the null geodetic generators of $\mathscr{H}$ is non-positive (non-negative) to the future (past) of p.

For the proof, see Hajicek (1984c). The validity of this theorem is, of course, well-known for general solutions of Einstein equations with a large class of sources (see, e.g. Hawking and Ellis 1973). The only new thing here is that we have used just the constraints and the gauge propagating equations in order to prove it. From Theorem 1, the usual causal properties of apparent horizons follow immediately:

(a) future (past) apparent horizons are not visible from $\mathscr{I}^+(\mathscr{I}^-)$;

(b) the world tube of an apparent horizon is non-time-like (see, e.g. Hawking and Ellis 1973).

So far, we have derived properties which are independent of the particular choice of gauge. For the BCMN gauge, the following theorem was proved in Hajicek (1984c):

Theorem 2. The BCMN foliation avoids apparent horizons.

Thus, the local part of the foliation problem is settled. What about the global hyperbolicity of M? If we assume that the data are regular everywhere along $t = $ const., then the answer to this question depends solely on the internal boundary of these hypersurfaces. It was shown in Thomi *et al.* (1984) that the nature of this boundary can be prescribed and two possibilities have been considered: a regular centre and an internal infinity (extremal black hole). In any case, nothing can go through the boundary, so the space–time is globally hyperbolic.

Finally, in Hajicek (1984b) I have shown the following

Theorem 3. If M' is asymptotically Minkowskian and M is the part of a neighbourhood of $\mathscr{I}$ in M' which is foliated by the BCMN foliation condition, then M is asymptotically Minkowskian.

Hence, we can conclude that there is no problem with foliation in the BCMN model and the way is open to apply the canonical quantization method to it. At the end, let us discuss the possible extension of these results to other models and theories, as well as the consequences for the theory of the Hawking effect.

An important point is that Theorem 1 is independent of the foliation condition. As it can be used in the proof of the next two theorems, it could turn out that there is a whole class of foliations with properties similar to those of the BCMN gauge. This could be useful for a generalization to the four-dimensional general relativity, because the BCMN gauge is based on a very special feature of spherically symmetric space–times.

The picture which Theorems 1, 2, and 3 give concerning the kinematically possible field trajectories is quite different from the current ideas about the expectation value of the metric in a space–time with collapse. According to these ideas, a black hole will appear and will then gradually evaporate by the so-called Hawking radiation. The future apparent horizon which will form will have a time-like world tube and will be visible from $\mathscr{I}^+$. The crucial question in this respect is which causal properties of the kinematically possible trajectories can survive the quantization and can, in this way, appear as properties of the expectation value of the metric. Naively, it could seem that all such properties must survive, because the expectation value of the metric can be calculated as a path integral average over the kinematically possible trajectories. However, this is not true in general. For example, the classical energy density of, say a Klein–Gordon field, is everywhere non-negative for any kinematically possible trajectory but its expectation value can be negative at some points.

In Hajicek (1984*b*), this question was discussed at some length. In particular, a trick was used due to Ashtekar and Horowitz (1982), which is based on the following fact: if a given general property of all kinematically possible trajectories can be considered as a property of the configuration space of the system, then it will survive the quantization. If, for example, the absence of apparent horizons in *M* can be completely characterized as the absence of critical points of the field ϕ along the Cauchy hypersurfaces, as was shown in Hajicek (1984*c*), then this property will survive. This argument, however, has some weak points, which were also discussed in Hajicek (1984*b*).

In any case, the nature of the Hawking effect *in space–times with collapse* remains an unsolved problem. For example, all derivations so far published of the effect use the simplifying assumption that all influence of the collapsing matter on the external world goes solely through the gravitational field which it produces (Hawking 1975; Boulware 1976). If this is true, a large loss of information about the collapsed matter must result independently of whether a horizon forms or not, just because gravity is universal. However, in reality, there *are* differences between particles of the fields of which the collapsing matter consists and the other fields. The most simple example of such a difference is that the former sort of created particle produces gravity *linearly*, because the background classical solution contains a large field of the kind, whereas the latter sort only produces *quadratic* effects. In this or some other way, some information about the collapsing matter could be saved. More research is necessary to answer these questions, and I hope that the BCMN model will be an efficient tool in this respect.

Acknowledgement

This chapter is based on work supported in part by Schweizerischer Nationalfonds.

References

Ashtekar, A. and Geroch, R. (1974). *Rep. Prog. Phys.* **37,** 1211.
—— and Horowitz, G. T. (1982). *Phys. Rev.* D **26,** 3342.
Berger, B. K. *et al.* (1972). *Phys. Rev.* D **5,** 2467.
Boulware, D. G. (1976). *Phys. Rev.* D **13,** 2169.
Christodolou, D. and Schmidt, B. G. (1979). *Convergent and asymptotic iteration methods in general relativity*. Preprint MPI-PAE/Astro 177.

Hajicek, P. (1984*a*). *Spherically symmetric systems of fields and black holes. II. Apparent horizon in canonical formalism.* Bern preprint BUTP-84/2.
—— (1984*b*). *Spherically symmetric systems of fields and black holes. IV. No room for black hole evaporation in the reduced configuration space?* Preprint MPA 124.
—— (1984*c*). *Spherically symmetric systems of fields and black holes. V. Pre-dynamical properties of causal structure.* Bern preprint BUTP-84/28.
Hawking, S. W. (1975). *Commun. Maths. Phys.* **43,** 199.
—— (1979). In *General relativity: an Einstein centenary survey* (ed. S. W. Hawking and W. Israel). Cambridge University Press, Cambridge.
—— and Ellis, G. F. R. (1973). *The large scale structure of space–time.* Cambridge University Press, Cambridge.
Thomi, P., Isaak, B., and Hajicek, P. (1984). *Spherically symmetric systems of fields and black holes. I. Definition and properties of apparent horizon.* Bern preprint BUTP-84/1.

22

Correlations and causality in quantum field theory

Robert M. Wald

One of the most intriguing effects which occurs in quantum theory is the Einstein–Podolsky–Rosen (EPR) phenomenon. One prepares a system so that correlations exist in the quantum state over space-like related regions of space–time. In the simplest example, one prepares systems A and B to be in the joint state $\Psi = (1/\sqrt{2})(|a_1\rangle\,|b_1\rangle + |a_2\rangle\,|b_2\rangle)$ (where $|a_1\rangle$ and $|a_2\rangle$ and $|b_1\rangle$ and $|b_2\rangle$ are orthonormal) and then moves these systems far apart. (One natural way to do this for, say, a two-photon system is by the decay of an excited atom, as described in Chapter 1.) Given that the total system is in state Ψ, there is no way of knowing in advance whether A is in state $|a_1\rangle$ or $|a_2\rangle$—the probability of each is $\frac{1}{2}$—nor is there any way of knowing B's state prior to a measurement. Nevertheless, the results of measurements on A and B are completely correlated. Once an observer of subsystem A measures it to be in state $|a_1\rangle$, he knows for certain that the observer of subsystem B will find it to be in state $|b_1\rangle$, even though the observer of B is in a space-like related region. If one attempts to describe the 'underlying reality' of what is happening in this process—a dangerous and unjustified thing to do in quantum theory—the invocation of causality violation appears to be unavoidable. However, no physical violation of causality actually occurs since the two observers cannot use the EPR phenomenon to communicate information to each other.

In this note, I wish to point out that the above phenomenon occurs in an essential way in quantum field theory. The vacuum fluctuations of a quantum field display EPR-like correlations in space-like separated regions and these correlations play a key role in radiation phenomena. They account for how particles, which are 'spread out over all space', can be emitted from sources of compact space–time support without violating causality.

Rindler quantization provides a very useful and insightful means of analysing this phenomenon, and we shall employ it in the investigation below. Although we shall be concerned here only with effects occurring in Minkowski space–time, the similar correlations occurring near black holes play a key role in the particle creation process which takes place there. For this and other reasons, I believe it is quite possible that a deeper understanding of the correlations present in vacuum fluctuations may be of importance for bridging the gap that presently exists between quantum theory and gravitation.

I shall begin by briefly reviewing some of the properties of the vacuum state of a free quantum field and the description of this state given by Rindler quantization. A simple model of a quantum mechanical system which radiates quanta of the field then will be introduced, and the issues of correlations and causality will be analysed. The results and discussion presented below are based on research done in collaboration with W. G. Unruh, and further technical details can be found in Unruh and Wald (1984).

The usual intuitive notion of the ordinary (Minkowski) vacuum state, $|0_M\rangle$, of a quantum field is that of space–time being 'completely empty'; in the usual view only when particles are present does any non-trivial dynamics of the field occur. This view is quite misleading. A very high degree of dynamical activity occurs when the field is in the state $|0_M\rangle$. However, this dynamical activity—known as vacuum fluctuations—is such that it cannot induce transitions upward in energy of inertial quantum mechanical systems which are coupled to the field. Thus, inertial quantum mechanical systems which are initially in their ground state remain there if the field is in the state $|0_M\rangle$ and thus it may appear that 'nothing is occurring'. However, such inertial systems which are initially in excited states or—as we shall see in more detail below— accelerating quantum mechanical systems which are initially in their ground state (with respect to the notion of energy conjugate to the time of the accelerating observer) dynamically interact in a non-trivial way with the vacuum fluctuations.

The occurrence of vacuum fluctuations for a free quantum field can be understood in terms of the well-known behaviour of a harmonic oscillator in ordinary quantum mechanics. Consider, for simplicity, a Klein–Gordon field, ϕ, which classically satisfies the equation,

$$(\partial_a \partial^a - m^2)\phi = 0 \tag{22.1}$$

We Fourier transform ϕ with respect to spatial variables, but not with respect to time:

$$\tilde{\phi}(t, k) = \frac{1}{(2\pi)^{3/2}} \int e^{-ik \cdot x} \phi(t, x) \, d^3x \tag{22.2}$$

Then, classically, each 'mode', $\tilde{\phi}$, satisfies the same equation as does the position of an ordinary harmonic oscillator, namely

$$\frac{d^2\tilde{\phi}}{dt^2} + (k^2 + m^2)\tilde{\phi} = 0 \tag{22.3}$$

The quantum theory of ϕ is obtained by treating each amplitude $\tilde{\phi}(\boldsymbol{k})$ as an independent harmonic oscillator. This leads to the following expression for the field operator, ϕ, in the Heisenberg representation

$$\phi = \sum_i (F_i a_i + \bar{F}_i a_i^\dagger). \tag{22.4}$$

Here the $\{F_i\}$ are a basis of positive frequency solutions of eqn (22.1) which are orthonormal in the Klein–Gordon inner product,

$$(F_i, F_j) \equiv i \int_\Sigma [\bar{F}_i \, \partial_a F_j - F_j \, \partial_a \bar{F}_i] n^a \, d\Sigma = \delta_{ij} \tag{22.5}$$

The operator coefficients a_i and $a_i^\dagger$ are the called annihilation and creation operators (although in the case of an ordinary harmonic they are usually called lowering and raising operators) and satisfy

$$[a_i, a_j] = 0 = [a_i^\dagger, a_j^\dagger] \tag{22.6}$$

$$[a_i, a_j^\dagger] = \delta_{ij} \tag{22.7}$$

Note that in writing eqn (22.4) we have converted the formal integral over $\boldsymbol{k}$ of the Heisenberg operator expression for the amplitude $\tilde{\phi}$ of the plane wave modes into a (better mathematically defined) discrete sum over an arbitrary basis of normalized positive frequency modes.

As in the case of a simple harmonic oscillator, the ground state, Ψ_0, satisfies,

$$a_i \Psi_0 = 0 \tag{22.8}$$

We interpret Ψ_0 as representing the vacuum state $|0\rangle$, and interpret the state obtained by acting on $|0\rangle$ with n creation operators as representing an n-particle state. A key point to bear in mind is that $|0\rangle$ is *not* an eigenstate of the field operator ϕ, just as the ground state of an ordinary harmonic oscillator is not an eigenstate of the position operator X. Thus, we have the 'vacuum fluctuations' mentioned previously.

More generally, the above ideas can be carried over to define the theory of a free Klein–Gordon field in a stationary, globally hyperbolic, space–time (Ashtekar and Magnon 1975; Kay 1978). Again, we have

$$\phi = \sum_i (f_i b_i + \bar{f}_i b_i^\dagger) \tag{22.9}$$

where $\{f_i\}$ comprise an orthonormal basis of positive frequency solutions and b_i and $b_i^\dagger$ satisfy eqns (22.6) and (22.7). Again, the state Ψ_0 satisfying $b_i \Psi_0 = 0$ for all i is interpreted as the vacuum state.

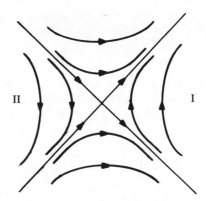

Fig. 22.1. A space–time diagram of Minkowski space–time, showing the orbits of a boost Killing field. These orbits are time-like in the regions I and II bounded by the planes (shown in the figure) where the Killing field becomes null. Since regions I and II are globally hyperbolic, we may view each of them as a globally hyperbolic, static space–time, where the notion of 'time translation symmetry' is now defined via Lorentz boosts.

An interesting application of this more general quantization prescription is obtained by returning to Minkowski space–time. Minkowski space–time possesses a ten-parameter group of isometries—the Poincaré group. Many of these isometries have time-like orbits in regions of Minkowski space–time. However, aside from the usual time translations, only one class—the Lorentz boosts—satisfies the additional property that the regions in which the orbits are time-like are globally hyperbolic, as illustrated in Fig. 22.1. Thus, we may view region I (or region II) of Minkowski space–time as a static, globally hyperbolic space–time in its own right, where the notion of 'time translations' is now defined by the one-parameter family of Lorentz boosts. Since the orbits of the Lorentz boosts are uniformly accelerating world lines, this coincides with the natural notion of 'time translation symmetry' that a uniformly accelerating observer would attribute to Minkowski space–time.

Thus, we now have two procedures for defining the quantum theory of a Klein–Gordon field in regions I and II of Minkowski space–time. First, we have the prescription (22.4) which uses the ordinary notion of time translation symmetry and defines ϕ over all of Minkowski space–time. We shall refer to this prescription as 'Minkowski quantization'. Second, we have the prescription (22.9) which uses the Lorentz boosts symmetry and applies (separately) to regions I and II only. This second prescription is known as 'Rindler quantization' (Fulling 1973).

Minkowski and Rindler quantizations should describe the same physical theory. Therefore, our two formulae for the field operator ϕ—namely eqn (22.4) for all of Minkowski space–time and eqn (22.9) in regions I and II separately—should agree in the regions where they both apply. Equating the two expressions for ϕ, we obtain a linear relation between the Minkowski annihilation operator a_i and the Rindler annihilation and creation operators b_{Ij}, $b_{Ij}^\dagger$, and b_{IIj}, $b_{IIj}^\dagger$ for regions I and II,

$$a_i = \sum_i (\alpha_{ij}^I b_{Ij} + \beta_{ij}^I b_{Ij}^\dagger + \alpha_{ij}^{II} b_{IIj} + \beta_{ij}^{II} b_{IIj}^\dagger). \tag{22.10}$$

The key point is that the notions of positive frequency with respect to inertial and accelerating time are inequivalent. More precisely, suppose F is a solution of eqn (22.1) on all of Minkowski space–time and which is purely positive frequency with respect to the Minkowski time translation parameter t. If we restrict F to region I and Fourier transform it with respect to the relevant time translation parameter there—namely, the Lorentz boost parameter τ—then F will, in general, have a non-zero negative frequency part. As a direct mathematical consequence, the coefficients β_{ij}^I and β_{ij}^{II} in eqn (22.10) will not vanish identically. But this implies that the Minkowski vacuum $|0_M\rangle$ and Rindler vacuum $|0_R\rangle$ are physically different states:

$$|0_M\rangle \neq |0_R\rangle, \tag{22.11}$$

since the state which satisfies $a_i \Psi = 0$ for all i is not the same as the state which satisfies $b_{Ii} \Psi = b_{IIi} \Psi = 0$ for all i.

Using eqn (22.10) and the formulae for α_{ij}^I, β_{ij}^I, α_{ij}^{II}, and β_{ij}^{II}, one can explicitly determine the Rindler state which corresponds to the Minkowski vacuum. Details of the calculation can be found in Unruh and Wald (1984) and the earlier references cited therein. The final result is,

$$|0_M\rangle = \prod_j \left\{ (1 - e^{-2\pi\omega_j/a}) \sum_{n_j} e^{-\pi n_j \omega_j/a} |n_j, I\rangle \otimes |n_j, II\rangle \right\}, \tag{22.12}$$

where the product is taken over a complete set of Rindler modes labeled by j which have (nearly) definite frequency ω_j, and a is the acceleration of the world line along which the Lorentz boost parameter τ agrees with proper time. The state $|n_j, I\rangle$ represents n_j particles in the mode j in region I, and $|n_j, II\rangle$ represents n_j particles in the corresponding mode in region II.

If one is interested only in phenomena occurring in region I, one can construct from the pure state (22.12) of the joint system (consisting of the field in region I and II) a density matrix describing region I alone. One thereby obtains the famous result (Unruh 1976) that $|0_M\rangle$ corresponds precisely to a thermal density matrix of Rindler particles at temperature

$$kT = \hbar a/2\pi. \tag{22.13}$$

However, our main interest is in the correlations between phenomena in region I and II. Equation (22.12) displays in a clear way the existence of EPR-type correlations in $|0_M\rangle$. Regions I and II are space-like related. In each of these regions, the number, n_j, of Rindler particles in mode j does not have a definite value. Nevertheless, if an observer in region I finds n_j Rindler particles in mode j, then with a probability equal to one, n_j such Rindler particles will be found in region II.

The role of these correlations in physical phenomena can be seen more clearly when one introduces a quantum mechanical system which is coupled to the quantum field ϕ. Such systems have frequently been introduced previously to serve as particle detectors and thereby to give a physical interpretation of Rindler quanta as the particles which a uniformly accelerating observer would 'see'. However, our interest here is to introduce a system coupled to the field to serve as a model of a radiator of (ordinary, Minkowski) particles of the field ϕ. Nevertheless, for this purpose we shall consider the simple uniformly accelerating 'particle detector' model of Unruh and Wald (1984). As shown there, such a particle detector acts as a radiator of Minkowski particles. Indeed, a main point of that reference was to show how the accelerating observer's view of this system as a detector of Rindler quanta is consistent with the inertial observer's view of the system as an emitter of Minkowski quanta. Despite the fact that our interest in this system here arises from its 'emitter' properties, I will refer to the system as a 'detector' in the discussion below.

Since the uniformly accelerating particle detector system is 'time translation invariant' with respect to Lorentz boost time τ rather than inertial time t, it is easiest to specify the dynamics of this system and its interaction with the quantum field in terms of a Hamiltonian canonically conjugate to τ in region I. As described in more detail in Unruh and Wald (1984), we take our detector system to have two levels (with basis states $|\downarrow\rangle$ and $|\uparrow\rangle$), with total Hamiltonian

$$H = H_0(\phi) + \Omega A^\dagger A + \varepsilon(\tau) \int \phi(x, \tau)[\psi(x)A + \bar{\psi}(x)A^\dagger]\sqrt{-g}\, d^3x, \quad (22.14)$$

where $H_0(\phi)$ is the Hamiltonian of the free Klein–Gordon field, and A and $A^\dagger$ are given by,

$$A|\downarrow\rangle = A^\dagger|\uparrow\rangle = 0 \qquad (22.15)$$

$$A^\dagger|\downarrow\rangle = |\uparrow\rangle, \qquad A|\uparrow\rangle = |\downarrow\rangle \qquad (22.16)$$

The relation of this model to the 'particle in a box' detector of Unruh (1976) is given in Unruh and Wald (1984).

If initially our quantum mechanical system is in its ground state $|\downarrow\rangle$ and the Klein–Gordon field is in state Ψ (so that the joint detector-field system initially is in state $|\chi_{in}\rangle = |\downarrow; \Psi\rangle$ then to first order in the coupling constant ε, the final state in the interaction picture will be (Unruh and Wald 1984),

$$|\chi_{out}\rangle = |\downarrow; \Psi\rangle - ib_I(\gamma)|\uparrow; \Psi\rangle \qquad (22.17)$$

where $b_I(\gamma)$ is the region I Rindler annihilation operator for a particular Rindler mode γ (whose norm is proportional to ε) determined by the 'spatial sensitivity function', $\psi(x)$, of the detector (see eqn (22.14)) and the length of time during which the detector is turned on.

Equation (22.10) can be inverted to express $b_I(\gamma)$ in terms of Minkowski annihilation and creation operators. We obtain

$$b_I(\gamma) = a(\Gamma_1) + a^\dagger(\Gamma_2) \qquad (22.18)$$

where Γ_1 and Γ_2 are certain Minkowski modes (with norm proportional to ε), whose definition is given in Unruh and Wald (1984). Consequently, if the initial state of the Klein–Gordon field is the Minkowski vacuum, to first order in ε the final state is,

$$|\chi_{out}\rangle = |\downarrow, 0_M\rangle - i\|\Gamma_2\| |\uparrow, 1\rangle \qquad (22.19)$$

where $|1\rangle$ denotes the state with one particle in the mode $\Gamma_2/\|\Gamma_2\|$. Equation (22.19) shows that our system acts as a radiator of a Minkowski particle in state $\Gamma_2/\|\Gamma_2\|$ when it makes a transition from the state $|\downarrow\rangle$ to the state $|\uparrow\rangle$.

Now, the quantum detector is confined entirely to region I and hence should not be able to exert any causal influence in region II. Nevertheless, when it makes a transition, it emits a (Minkowski) particle in state $\Gamma_2/\|\Gamma_2\|$. But this particle is 'spread out over all space', and, in particular, its expected stress-energy in region II is positive. (Indeed, the amplitude of Γ_2 is greater in region II than in region I; see Unruh and Wald 1984). Thus, the excitation of our detector system in region I appears to have 'caused' the quantum field energy to increase in region II. It should be noted that although we have considered here only a particularly simple model of a radiating system, the same phenomenon must occur in all radiation processes, as *every* Minkowski particle mode is inherently 'spread out over all space' since a positive frequency solution of eqn (22.1) cannot vanish in any open region of space–time.

As discussed in more detail in Unruh and Wald (1984), neither the presence of the Minkowski particle partly in region II nor the increase in expected energy in region II—both of which arise in association with the excitation of the detector in region I—can be used to violate causality, i.e. to communicate information between regions I and II. While in region II, a observer would have no way of knowing whether he was observing effects of a particle emitted by the detector in region I or merely observing vacuum fluctuations. Indeed, the coupled field-detector system is manifestly causal when expressed in the Heisenberg representation.

What is occurring is precisely the EPR phenomenon. The excitation of the detector in region I is correlated with effects occurring in region II. These correlations between the detector and field can be traced directly to the correlations occurring over space-like separated regions in the initial state $|0_M\rangle$ of the quantum field. Interaction of the field with the detector simply transfers some of these already existing vacuum correlations to the detector in a causal manner.

The correlations present in $|0_M\rangle$ were already made manifest by eqn (22.12). Their existence also can be seen from the well-known fact that the vacuum expectation value of the anticommutator of field operators is non-vanishing at space-like related events. In particular, in the massless case, we have

$$\langle 0_M| \phi(x)\phi(y) + \phi(y)\phi(x) |0_M\rangle = \frac{1}{4\pi^2\sigma}, \qquad (22.20)$$

where σ is the square of the geodesic distance between x and y. Hence, since $\langle 0_M| \phi(x) |0_M\rangle = 0$ for all x, we have,

$$\langle 0_M| \phi(x)\phi(y) + \phi(y)\phi(x) |0_M\rangle \neq 2\langle 0_M| \phi(x) |0_M\rangle \langle 0_M| \phi(y) |0_M\rangle, \qquad (22.21)$$

which expresses the presence of correlations in the values of the field at space-like related events. This implies that knowledge of the field at y will alter the probability distribution for the field at x. On the other hand, the commutator of field operators *does* vanish when x and y are space-like related:

$$\phi(x)\phi(y) - \phi(y)\phi(x) = 0. \qquad (22.22)$$

This implies that when the altered probability distributions for the field at x are weighted by the probability of the outcome of the measurement at y and summed, one obtains the original probability distribution. Hence, the decision to measure the field at y does not result in a change in the probability distribution for the field at x, and one cannot make use of the existence of the field correlations at x and y to communicate information between these events.

The above considerations suggest a new viewpoint on emission processes in quantum field theory. The usual intuitive picture is of a field quantum emerging in a direct manner from its source. However, at least for the model considered above no such direct emission appears to be occurring. Indeed, as concluded in Unruh and Wald (1984), it would seem more accurate to view the detector as interacting with the vacuum fluctuations of the quantum field in such a way that correlated fluctuations in space-like related regions are liberated to become real (Minkowski) particles.

The above considerations also contain some possible implications for cosmology. In our discussion, we restricted attention to the correlations occurring in Minkowski space–time when the field in the state $|0_M\rangle$. However, for any state of a quantum field in curved space–time having a finite expected value of its stress-energy tensor, the expectation value of the anticommutator function will be of the Hadamard form (see e.g. Fulling *et al*. 1978), i.e. for x sufficiently near y we have

$$\langle \phi(x)\phi(y) + \phi(y)\phi(x) \rangle = \frac{A}{\sigma} + B \ln \sigma + C, \qquad (22.23)$$

where A, B, and C are smooth functions and σ, again, is the square of the geodesic distance between x and y. Thus, correlations in the field over space-like separated regions similar to those for $|0_M\rangle$ (see eqn (22.20) above) occur for any physically reasonable state in any space–time. In particular, the initial state of the universe must have possessed such correlations.

It is customary to assume in cosmology that any two physical phenomena occurring outside each other's horizon must be entirely unrelated. However, we have now seen that although no causality violation occurs, significant correlations in such physical phenomena can (and, in some cases, must) occur, i.e. two phenomena taking place outside each other's horizon need not be statistically independent. The possible cosmological implications of such correlations appear worthy of further investigation.

Acknowledgements

As previously mentioned, the above discussion is based upon research done in collaboration with W. G. Unruh and reported in Unruh and Wald (1984). Partial support by NSF grant PHY 80-26043 to the University of Chicago is acknowledged.

References

Ashtekar, A. and Magnon, A. (1975). *Proc. R. Soc. Lond.* A**346**, 375.
Fulling, S. A. (1973). *Phys. Rev.* D**7**, 2850.
——, Sweeny, M., and Wald, R. M. (1978). *Commun. Math. Phys.* **63**, 257.
Kay, B. S. (1978). *Commun. Math. Phys.* **62**, 55.
Unruh, W. G. (1976). *Phys. Rev.* D**14**, 870.
—— and Wald, R. M. (1984). *Phys. Rev.* D**29**, 1047.

23

Self-duality and spinorial techniques in the canonical approach to quantum gravity

Abhay Ashtekar

23.1. Introduction

One of the most fascinating recent developments in the mathematical literature related to quantum gravity is the discovery of $\mathcal{H}$-spaces and nonlinear gravitons by Newman (1975) and Penrose (1976). (For a review, see, e.g. Ko *et al*. 1981.) The fact that Einstein's equation becomes so simple in the self-dual (or anti self-dual) case that one can have a full grasp on the structure of its general solutions is in itself striking. What is even more astonishing is that the classical scattering matrix in this case is trivial: In spite of nonlinearities, the 'characteristic data' on the future null infinity of an $\mathcal{H}$-space is related to that at past null infinity in the same way as in the *linear* theory! (Ludvigsen *et al*. 1981.) This powerful result tempts one to the conclusion that there is something physically deep about self-duality. A specific proposal already exists: Penrose has suggested that certain self-dual solutions should be thought of as representing one-particle states of ('dressed') nonlinear gravitons.

In spite of the aesthetic appeal of these ideas, they have had little impact on the traditional approaches to quantum gravity. The main reason, I feel, is that these approaches are based on notions and techniques which cannot accommodate self-duality in a natural way. For definiteness, let us consider the canonical and the path integral methods. The basic entity in these approaches is the 3-geometry which serves as the configuration variable; they both use the configuration or the q-representation. The notion of self-duality, on the other hand, cannot be formulated just in terms of a 3-geometry, and hence the operator representing the self-dual part of the Weyl curvature is cumbersome in the q-representation.

However, except in certain special cases of quantum cosmologies, the

q-representation itself is faced with a number of difficulties. Let us consider a few examples. 3-geometries, $\{q_{ab}\}$, have three degrees of freedom per space-point and the wave functions $\psi(\{q\})$ are subject to one equation per point, the Wheeler–DeWitt equation. A basic idea in the canonical approach based on the q-representation was to interpret $\{q_{ab}\}$ as containing the two true dynamical degrees of freedom as well as time and to interpret the Wheeler–DeWitt equation as the 'evolution equation' on the wave function. However, already in the linear approximation, Kuchăr (1970) found that the desired splitting into true degrees and time is impossible and that another representation in which certain combinations of the 3-metric and extrinsic curvature are 'diagonal' is more convenient.† Another problem arises due to the fact that the space of positive-definite metrics is not an affine space but a convex set (in the space of all second rank, symmetric tensor fields): positive definiteness of q_{ab} requires one to modify the Canonical Commutation Relations and confronts one with the delicate issues concerning the correct choice of the self-adjoint extension of the canonically conjugate momentum operator P^{ab} (see e.g. Isham 1984). Finally, the constraints of general relativity pose a problem which, at least potentially, is much more serious. In the classical theory, the constraints are first class; the space of constraint functions is closed under Poisson brackets. However, in general, this does not imply that the commutator algebra of operator constraints must be closed in the quantum theory. This point is perhaps best seen in the framework of geometric quantization (Woodhouse 1980). Given a classical phase space, this provides an unambiguous prescription to construct a Hilbert space—called the space of pre-quantum states—and to associate with each classical observable a (pre-quantum) operator on this space, such that all the Poisson brackets go over to commutators. However, this is not yet the mathematical structure underlying quantum mechanics: roughly speaking, pre-quantum states are functions on phase space rather than the configuration space. Therefore, to go to quantum mechanics, one has to introduce additional conditions—called the polarization conditions—on the pre-quantum states which cut the number of variables on which they depend by half, thereby choosing a specific (e.g. configuration or momentum or Bargmann (1962) or . . .) representation of quantum states. Let us now consider a classical system with first-class constraints. The constraints can be carried over to the pre-quantum level unambiguously: One simply demands that the permissible wave functions must be annihilated by the pre-quantum operators defined by the constraint functions. Since, the

---

† Here the wave function depends on the trace-free, transverse part of the linearized 3-metric and the trace of the linearized extrinsic curvature. However, a direct generalization of this representation to full quantum gravity runs into severe problems.

commutator algebra of pre-quantum operators is the same as the Poisson bracket Lie algebra, there is no consistency problem at this stage. However, a problem can arise in the passage from the pre-quantum to the quantum theory: The polarization conditions need not be compatible with the already imposed operator constraints. The incompatibility, if present, can reduce the number of admissible quantum states and even trivialize the quantum theory altogether by forcing the 'physical' quantum Hilbert space to be zero-dimensional! Thus, in general, the question of whether or not constraints can be successfully incorporated at the quantum level hinges on the choice of the polarization conditions, i.e. of the representation. What is the situation with the q-representation in quantum gravity? As far as I know, the question of whether the vector operator constraints and the Wheeler–DeWitt equations are mutually compatible is still open in the full theory.† Resolution of this issue is clearly necessary to ensure that one has 'enough' solutions to the operator constraints, i.e. 'enough' quantum states. To get an indication, one might ask a substantially weaker question: are there any solutions at all known? Unfortunately, as the matter now stands, in full quantum gravity the answer to even this question is in the negative. The issue is hard to analyse because of the presence of an infinite number of degrees of freedom. One may therefore try to explore constrained systems with only a finite number of degrees of freedom. It turns out that one can construct examples in which the constraints are modelled after those in general relativity for which the q-representation is problematic for the above reasons but which admit other, completely satisfactory representations (Ashtekar and Stillerman 1985; Stillermann 1985).

These problems suggest that it may be fruitful to look for other representations. However, because of the presence of non-trivial constraints, meaningful representations are hard to construct and one needs an external guide.‡ Remarks made at the beginning of this section suggest a natural avenue: One may look for a representation in which the self-dual part (or the anti-self-dual part) of the Weyl curvature is 'diagonal'. This programme was embarked upon last year. The purpose of this article is to summarize its present status.

The programme is being pursued along two different lines: the first line of attack is based on three-dimensional space-like surfaces while the second uses four-dimensional fields on space–time as a whole. Although closely related, the two sets of ideas are not just translations of one

---

† In quantum cosmology, the drastic freezing of the number of degrees of freedom simplifies the problem enormously and removes all the potential problems that exist in the full theory.

‡ The obvious alternative, the momentum representation, is not viable because of the presence of the scalar curvature term in the scalar constraint.

another. Broadly, Dirac's 'operator constraint method' (followed by the Wheeler group) turns out to be more convenient in the three-dimensional method while the 'reduced phase-space method' (pursued by the Bergmann group) fits naturally in the four-dimensional picture. In my talk, I focused on the four-dimensional method and mentioned the three-dimensional approach only briefly. However, from the discussions that followed, it was evident that most workers in canonical quantization feel uncomfortable with this method and a lot of time can be spent just on the preliminaries, since even the basic ideas appear to be somewhat unnatural from the $3 + 1$ viewpoint. Furthermore, thanks to the stimulus that I received from a number of colleagues during and after the conference, it is now the three-dimensional approach that is better developed. Since I cannot satisfactorily treat both approaches within the space here, I will restrict myself to the three-dimensional case in this chapter. This is a better strategy also because some of the key issues in the four-dimensional method are being investigated by the Pittsburgh group and their current status may well change substantially in the near future.

23.2. Three-dimensional approach

This section is divided into three parts. The first fixes notation, introduces the basic notions and recalls certain results obtained by Sen (1982). The second summarizes a new phase-space description of general relativity based on SU(2) spinors. The third part introduces new variables and contains the central ideas for quantization.

23.2.1. Preliminaries

Although most of the ideas in this three-dimensional approach go through even when the space sections are compact, in order eventually to make contact with the $\mathcal{H}$-spaces and nonlinear gravitons we shall focus on the asymptotically flat case.† Let us therefore begin by fixing a 3-manifold Σ, the complement of a compact set which is diffeomorphic to the complement of a closed ball in $\mathbb{R}^3$. Thus, topological complications, if any, are confined to a compact region in Σ. Tensor fields on Σ will carry lower case latin (abstract) indices; V^a for example, denotes a contravariant vector field. Next, we introduce abstract spinor fields. One may do this simply by considering cross-sections of suitable vector

---

† The asymptotically flat case is also conceptually simpler than the spatially compact one. For example, one can introduce interesting observables such as energy-momentum and angular momentum and speak meaningfully about the ground state. Also, in the Hamiltonian formulation, gauge and dynamics can be disentangled. (see e.g. Ashtekar and Horowitz 1982 (sec. 2), 1984).

bundles over Σ. Alternatively, one may consider generalized tensors (in the sense of Ashtekar *et al.* 1982) with upper case latin indices, e.g., $\lambda^{A\cdots B}{}_{M\cdots N}$, which, for the moment are to be thought of simply as 'internal' indices. These fields will be called abstract spinor fields; they will become the usual spinor fields once they are soldered to the tangent spaces of points of Σ via the Infeld–Van der Waerden symbols. We assume that the system of abstract spinors is such that a basis $e^A{}_\mathbf{A}$ and $e^\mathbf{A}{}_A$ exists, where the bold-faced letters, e.g. $\mathbf{A}$, denote numerical indices which take values 1 and 2. Thus, any abstract spinor field λ^A can be expressed as a linear combination, $\lambda^A = \lambda^\mathbf{A} e^A{}_\mathbf{A} \equiv \lambda^1 e^A{}_1 + \lambda^2 e^A{}_2$, of the basis spinor fields $e^A{}_1$ and $e^A{}_2$. ($e^\mathbf{A}{}_A$ is the dual basis, satisfying $e^A{}_\mathbf{A} e^\mathbf{A}{}_B = \delta^A{}_B$ and $e^A{}_\mathbf{A} e^\mathbf{B}{}_A = \delta^\mathbf{B}{}_\mathbf{A}$.) Any two second rank skew spinor fields are proportional. We choose one such (no where vanishing) field and denote it by ε_{AB}. Let ε^{AB} be the unique spinor field such that $\varepsilon^{AB} \varepsilon_{AC} = \delta^B_C$. We allow ourselves to raise and lower the spinor indices with ε_{AB} and ε^{AB}. Thus, $\lambda^A = \varepsilon^{AB}\lambda_B$ and $\lambda_A = \lambda^B \varepsilon_{BA}$. Finally, with each abstract spinor field, e.g. λ_A, we associate another spinor field $\lambda_A^\dagger$ called the hermitian conjugate of λ_A, such that the following properties hold: (i) $[(a + ib)\lambda_A]^\dagger = (a - ib)\lambda_A^\dagger$ for all real numbers a and b; (ii) $(\lambda_A^\dagger)^\dagger = -\lambda_A$; and (iii) $\lambda^{\dagger A}\lambda_A \geqq 0$, equality holding if and only if $\lambda_A = 0$.

Next, we allow ourselves to consider fields with tensor as well as spinor indices, e.g. $T^{a\cdots b}{}_{c\cdots d}{}^{M\cdots N}{}_{P\cdots Q}$. If one regards the spinor indices as 'internal', these are precisely the type of fields one encounters in the Yang–Mills theory. We can define (torsion-free) connections ∇ (which act on both spinor and vector fields) in terms of their general properties: linearity, Leibnitz rule, absence of torsion and the requirements that, for any function f on Σ, $\nabla_a f$ be the exterior derivative of f, *and that ∇ should annihilate ε_{AB}*, i.e. $\nabla_m \varepsilon_{AB} = 0$. (For details, see e.g. Penrose and Rindler (1984), section 4.2). Any two such connections, ∇ and $\bar{\nabla}$, are related by tensor–spinor fields, Γ^c_{ab} and A^M_{ab},

$$(\nabla_a - \bar{\nabla}_a)V_{bA} = \Gamma^c_{ab}V_{cA} + A^M_{aA}V_{bM} \tag{23.1}$$

such that $\Gamma^c_{[ab]} = 0$ and $A_{a[BM]} = 0$. Finally, one can compute the curvature of any connection ∇: there exists a field R^N_{abM} such that

$$\nabla_{[a}\nabla_{b]}\alpha_M = \tfrac{1}{2}R_{abM}{}^N \alpha_N \tag{23.2}$$

for all spinor fields α_M. By construction, we have $R_{abMN} = R_{[ab](MN)}$.

The reason why we have gone into all these details is to exhibit explicitly the structure that can be introduced without any reference to a metric. We now introduce the soldering forms which yield a metric. Since the spin space is two-dimensional, symmetric second rank spinors, $V^{AB} \equiv V^{(AB)}$, form a three-dimensional vector space. Consider isomorphisms σ_m^{AB} (and σ^m_{AB}) between the space of symmetric second rank spinors and the complexified tangent space at each point of Σ, which map the hermitian spinors, $V^{AB} = V^{\dagger AB}$, to real vectors. Then it is easy to verify that

$$\sigma^M_{AB}\sigma^{CD}_m = \delta^C_{(A}\delta^D_{B)} \quad \text{and} \quad \sigma^m_{AB}\sigma^{AB}_n = \sigma^m_n \tag{23.3}$$

and that q_{mn} defined by†

$$q_{mn} = \sigma^{AB}_m\sigma^{CD}_n\varepsilon_{Ac}\varepsilon_{BD} \tag{23.4}$$

is a positive definite metric on Σ. Given a σ^{AB}_m, therefore, one acquires the ability to solder the abstractly defined spinors to the manifold and also to raise and lower the tensor indices with the metric given by eqn (23.4). Note that although σ^{AB}_m yields a unique metric q_{mn}, it cannot be fully recovered from the metric

$$\bar{\sigma}^{AB}_m = S^A_M S^B_N \sigma^{MN}_m \tag{23.5}$$

yields the same q_{mn} as σ^{AB}_m if S belongs to SU(2), i.e. if $\varepsilon_{AB} = S^M_A S^N_B \varepsilon_{MN}$. Finally, we note that given a σ^{AB}_m, there exists a unique torsion-free connection D on spinor and tensor fields which annihilates σ^{AB}_m; i.e. satisfies $D_n\sigma^{AB}_m = 0$. Because of eqn (23.4), this connection is automatically metric-compatible. (Note that if one is given only the metric-connection on tensor fields, one cannot extend it to spinors uniquely unless we are also given a preferred σ^{AB}_m.)

We shall conclude this sub-section by recalling some beautiful results obtained by Sen (1982) which will play an important role in the later discussion. Fix a pair (q_{mn}, K^{mn}) on Σ, consisting of a positive definite metric q_{mn} and a symmetric second rank tensor field K^{mn} (to be thought of as the extrinsic curvature of Σ). Pick a σ^{AB}_m which yields q_{mn} via (23.4) and let D denote the connection which annihilates this σ^{AB}_m. Introduce two new connections $^{\pm}\mathscr{D}$:

$$^{\pm}\mathscr{D}_m\alpha_B := D_m\alpha_B \pm \frac{i}{\sqrt{2}}K_{mB}{}^C\alpha_c \tag{23.6}$$

(Note that, on tensors, $^{\pm}\mathscr{D}$ have the same action as D: $^{\pm}\mathscr{D}_m V^n = D_m V^n$ for all V^n.) Let us compute the curvature tensors $^{\pm}\mathscr{R}_{abM}{}^N$ of these connections, defined by (23.2). A straightforward calculation yields

$$W^{\pm}_{mn} := \varepsilon^{ab}{}_m(^{\pm}\mathscr{R}_{abMN})\sigma^{MN}_n = \pm\frac{i}{\sqrt{2}}\varepsilon^{ab}{}_m D_a K_{bn}$$

$$\mp\sqrt{2}\,(R_{mn} - K^p_m K_{pn} + KK_{mn}) \pm (R - K_{ab}K^{ab} + K^2)q_{mn}, \tag{23.7}$$

---

† Note that, unlike in the conventions normally used by the relativists, here ε_{AB} is fixed independently of the choice of σ^{AB}_m and the metric q_{mn}. (The usual convention is more convenient when one deals with only a fixed conformal class of metrics.)

where R_{ab} is the Ricci tensor of q_{ab}; R, its trace; and K, the trace of K^{ab}. Consequently the requirement that $W^{\pm}_{mn}$ be symmetric yields:

$$D_a(K^{ab} - Kq^{ab}) = 0 \qquad (23.8)$$

while that $W^{\pm}_{mn}$ be trace-free yields:

$$R - K^{ab}K_{ab} + K^2 = 0 \qquad (23.9)$$

Note that eqns (23.8) and (23.9) are precisely the constraint equations of general relativity! Thus, the constraints are coded in simple algebraic properties of the curvature tensors of $^{\pm}\mathcal{D}$. But this is not all. Let us suppose that the constraints are satisfied, i.e. that $W^{\pm}_{mn}$ are symmetric and trace-free. Then, we can use (q, K) as the initial data and obtain a Ricci-free four-metric g_{mn} on $\Sigma \times R$. What is the relation between the Weyl tensor of g_{ab} (evaluated on Σ) and $W^{\pm}_{mn}$? We have:

$$W^{\pm}_{mn} = -\sqrt{2}\,(E_{mn} \pm iB_{mn}). \qquad (23.10)$$

where E_{mn} and B_{mn} are the electric and the magnetic parts, on Σ, of the Weyl tensor of g_{mn}. Thus, the curvature tensors W^{+}_{mn} and W^{-}_{mn} of $^{+}\mathcal{D}$ and $^{-}\mathcal{D}$ capture *precisely* the self-dual and the anti-self-dual parts of the four-dimensional Weyl tensor.[†]

In what follows, for most part, we will need only $^{+}\mathcal{D}$. We shall therefore often drop the suffix $+$ and write this connection as $\mathcal{D}$. It will be referred to as the *Sen connection*.

23.2.2. *An enlarged phase space for GR*

Consider the space of Infeld–Van der Waerden soldering forms σ^m_{AB} (introduced just above eqn (23.3)) which are asymptotically flat in a suitable sense on Σ. This space will serve, initially, as the configuration space of the enlarged phase space description. Let us denote it by $\mathscr{C}$. Clearly, there is a natural projection mapping (given by eqn (23.4)) from $\mathscr{C}$ to the usual configuration space $\mathscr{C}_0$ consisting of (asymptotically flat) positive definite metrics q_{mn}. The inverse image of a q_{mn} will be referred to as a fibre in $\mathscr{C}$. Since SU(2) acts simply and transitively on fibres, each fibre is diffeomorphic to the manifold SU(2). A tangent vector to $\mathscr{C}$ at any point σ^0 can be represented by a field $t^m{}_{AB} \equiv t^m_{(AB)}$ on Σ. Using σ^0, one can convert t^m_{AB} into a second rank tensor field $t_{mn} := t^p_{AB}q_{pm}\sigma^{AB}_n$. Note that, in general, t_{mn} is neither symmetric nor skew. It is straightforward to verify that the infinitesimal change, h_{mn}, in q_{mn} induced by t^m_{AB} is related

---

[†] This result was obtained independently by Sen using SU(2) spinors and by Ashtekar and Horowitz using SL(2, C) spinors. Note, however, that the introduction of spinors is essential. One may imagine working only with tensors and using torsion proportional to $\varepsilon_{mnp}K^{pq}$. However, one cannot both code field equations in the algebraic properties of the resulting curvature tensor and get self-duality. Also, the framework involves certain *ad hoc* elements.

to t_{mn} by $h_{mn} = -2t_{(mn)}$. The skew part of t_{mn} will be denoted by $-\frac{1}{2}F_{mn}$. Thus,

$$t_{mn} = -\frac{1}{2}(h_{mn} + F_{mn}). \tag{23.11}$$

A vector t^m_{AB} for which h_{mn} vanishes is tangential to the fibres; the infinitesimal motion generated by it in $\mathscr{C}$ changes σ^0 without affecting the metric q_{mn}. Since t_{mn} has nine components, the configuration space is enlarged from six functions per space point to nine.

The cotangent vectors are fields M^{AB}_m on Σ (Hermitian in the spinor indices) which are densities of weight one with a suitable fall-off at spatial infinity. The phase space consists of all pairs $(\sigma^m_{AB}, M^{AB}_m)$. It will be denoted by Γ. The natural symplectic structure Ω on Γ is given by†

$$\Omega|_{\sigma^0}\left((\delta\sigma, \delta M), (\delta\sigma, \delta M)\right) := \int_\Sigma (\delta\sigma)^m_{AB}(\delta M)^{AB}_m - (\delta\sigma)^m_{AB}(\delta M)^{AB}_m, \tag{23.12}$$

so that the Poisson bracket between any two functions $F(\sigma, M)$ and $G(\sigma, M)$ is given by

$$\{F, G\} = \int \left(\frac{\delta F}{\delta M^{AB}_m}\frac{\delta G}{\delta\sigma^m_{AB}} - \frac{\delta G}{\delta M^{AB}_m}\frac{\delta F}{\delta\sigma^m_{AB}}\right). \tag{23.13}$$

Since each point of Γ corresponds to 18 functions per space-point and since there are only four physical degrees of freedom in the phase space, one expects seven constraints. This expectation is indeed correct. The constraints are:

$$M_{[mn]} = 0 \Leftrightarrow M^{A(B}_m\sigma^{C)m}_A = 0$$
$$D_M P^{mn} = 0 \tag{23.14}$$
$$(\det q)R - P^{ab}P_{ab} + \frac{1}{2}P^2 = 0.$$

where $M_{mn} = M^{AB}_m\sigma_{nAB}$; $P^{mn} = M^{(mn)}$; D is the derivative operator of σ^m_{AB}; R, the scalar curvature, and $(\det q)$ the determinant of the three-metric q_{mn}. The last two constraints are the familiar ones. The first constraint is new. To see its meaning, let us smear the left side with a symmetric test field N_{BC} to obtain the function

$$F(\sigma, M) := \int_\Sigma N_{BC}M^{AB}_m\sigma_A{}^{Cm} \tag{23.15}$$

---

† Here, we are using the usual somewhat sloppy short-cut. Strictly, since $\delta\sigma$ falls-off only as $1/r$, and, δM only as $1/r^2$, in general, the integral is logarithmically divergent and a more careful treatment is necessary. However, since this is the standard strategy required in all non-Abelian theories, we have skipped the details here.

on the phase space and consider the Hamiltonian vector field X_F:

$$X_F = \int_\Sigma N_B^C \sigma_{AC}^m \frac{\delta}{\delta \sigma_{AB}^m} - N_B^C M_m^{AB} \frac{\delta}{\delta M_m^{AC}}. \qquad (23.16)$$

Since $N_{AB} = N_{(AB)}$, N_A^B belongs to the Lie algebra of SU(2). Thus, the finite canonical transformations generated by the constraint function simply rotates the (free) spinor indices by a SU(2) transformation. Following Dirac's theory of constrained systems, these motions will be regarded as gauge transformations. Thus, as one might have anticipated, given a metric q_{mn}, the freedom in the choice of a compatible soldering form σ_{AB}^m corresponds to gauge.† It is straightforward to verify that the system (23.14) of constraints is first-class.

Using Sen's (1982) ideas, the constraints can be re-expressed in a form which will turn out to be particularly convenient in the next sub-section. Define π_m^{AB} by $M_m^{AB} = (\det q)^{-1/2} (\pi_m^{AB} - \pi_n^{MN}\sigma_{MN}^n \sigma_m^{AB})$. Then π_m^{AB} has zero weight. For each point (σ, M) of Γ, introduce the Sen connection $\mathscr{D}$:

$$\mathscr{D}_m \alpha_B := D_m \alpha_B + \frac{i}{\sqrt{2}} \pi_{mB}{}^N \alpha_N \qquad (23.17)$$

(Actually, this is a slight generalization of eqn (23.6) since $K_{mn} := K_m^{AB}\sigma_{nAB}$ is not required to be symmetric.) Using the curvature tensor $R_{mnA}{}^B$ of this $\mathscr{D}$, the constraints can be re-written as:

$$M_m^{A(B}\sigma_A^{C)m} = 0, \qquad (23.18a)$$

$$\mathscr{R}_{mn}{}^{AB}\sigma_{AB}^m = 0, \qquad (23.18b)$$

$$\mathscr{R}_{mn}{}^{AB}\sigma_{PA}^m \sigma_B^{nP} = 0 \qquad (23.18c)$$

The surface in Γ where eqns (23.18) hold will be referred to as the *constraint sub-manifold* and denoted by $\bar{\Gamma}$.

23.2.3. Choice of polarization/representation

We now introduce the key elements needed for quantization.

The first step is to introduce a new chart on Γ. Fix a flat connection ${}^0\mathscr{D}$.

---

† Since $t_{mn} = -\frac{1}{2}(h_{(mn)} + F_{[mn]})$, the tangent space at any point of $\mathscr{C}$ can be canonically split in to a vertical part (where $h_{mn} = 0$) and a horizontal part (where $F_{mn} = 0$). One may imagine using this split to construct horizontal cross-sections of $\mathscr{C}$, and using one of them—each would be naturally isomorphic to $\mathscr{C}_0$—as the configuration space in place of $\mathscr{C}$. This would be a simple procedure to get rid of the three extra degrees of freedom, keeping the ability to deal with spinors. It turns out, however, that the horizontal flats are not integrable! Thus, this apparently simple gauge fixing procedure is not viable. (Narain 1984.)

Given any point (σ, M) of Γ, we can now introduce a field $\varphi_{mB}{}^C$ on Σ as follows:

$$(\mathcal{D}_m - {}^0\mathcal{D}_m)\alpha_B =: \varphi_{mB}{}^C \alpha_C \qquad (23.19)$$

for all α_C, where $\mathcal{D}$ is the Sen connection of (σ, M). It is easy to check that φ_{mAB} and $\varphi^\dagger_{mAB}$ yield a chart on Γ.† (Note that, since ε_{AB} and ε^{AB} are constants (or, c-numbers) on Γ, spinor indices can be raised and lowered freely. $\varphi^\dagger_{mAB}$ gives the difference between $^-\mathcal{D}$ and $^0\mathcal{D}$.) Change in the choice of $^0\mathcal{D}$ simply changes φ_{mAB} and $\varphi^\dagger_{mAB}$ by constants. Let us now consider vector fields $(\delta\sigma, \delta M)$ on Γ along which φ_{mAB} do not change. The Lie derivative of $\mathcal{R}_{mnAB}$ along such a field clearly vanishes, whence, infinitesimal motions along such fields (which are, in addition, tangential to the constraint surface) leave the self-dual part of the Weyl curvature untouched, changing only the anti self-dual part. The vector fields of Γ along which φ_{mAB} do not change will be therefore called self-dual. Thus, at each point of Γ, the tangent space has a self-dual sub-space and an anti-self-dual sub-space, which together span the full space. We now show that self-dual sub-spaces are in fact isotropic with respect to the symplectic structure, i.e. that φ_{mAB} form a 'complete set of commuting (with respect to the Poisson bracket) variables'. This implies that it is meaningful to speak of a representation in which φ_{mAB}—and hence the self-dual part of Weyl curvature—are all diagonal. This representation will be called self-dual.

We now give a sketch of the proof of the isotropy of the symplectic structure. Fix a point (σ, M) of Γ and let $(\overline{\delta\sigma}, \overline{\delta M})$ lie in the self-dual flat. Then,

$$\left[(\overline{\delta\sigma})^m_{AB} \frac{\delta}{\delta\sigma^m_{AB}} + (\overline{\delta M})^{AB}_m \frac{\delta}{\delta M^{AB}_m}\right] \circ \int_\Sigma f^b_{MN} \varphi^{MN}_b = 0 \qquad (23.20)$$

for all test densities f^b_{MN} of weight one. This translates to the condition that $(\overline{\delta\sigma})_{mn} \equiv -\frac{1}{2}(h_{mn} + F_{mn})$ and $(\overline{\delta M})^{AB}_m = \delta[(\det q)^{1/2}(\pi^{AB}_m - \pi\sigma^{AB}_m)]$ are related by

$$\tfrac{1}{2}\varepsilon^{bd}_c(D_b h_{dm} + \tfrac{1}{2}D_m F_{bd}) = -\mathrm{i}(\delta\pi)_{mBD}\sigma^{BD}_c \qquad (23.21)$$

Using this fact, and the expression (23.12) of the symplectic structure, it is straightforward to show that

$$\Omega((\overline{\delta\sigma}, \overline{\delta M}), (\delta\sigma, \delta M)) = 0 \qquad (23.22)$$

if $(\delta\sigma, \delta M)$ and $(\overline{\delta\sigma}, \overline{\delta M})$ both belong to the self-dual flat.

Several remarks are in order. (i) On the phase space of the n-dimensional harmonic oscillator one can introduce complex co-ordinates

---

† The difference between φ^{AB}_m and $\varphi^{\dagger AB}_m$ yields M^{AB}_m. Their sum yields the connection D. Because of asymptotic flatness, there exists a unique σ^m_{AB} compatible with this D.

$Z_{\mathbf{k}} = \sqrt{m\omega}\, X_{\mathbf{k}} + (i/\sqrt{m\omega})\, P_{\mathbf{k}}$, $\mathbf{k} = 1, 2, \ldots, n$, which are well-adapted for the Bargmann (1962) representation. The co-ordinates φ_{mAB} on Γ are in some ways analogous to the Z; both constitute a system of complex complete set of commuting variables. Since the phase space in both cases is real, one cannot integrate the vector fields $\delta/\delta\bar{Z}$ or $\delta/\delta\varphi^\dagger$ to obtain Lagrangian sub-manifolds of Γ. In the language of geometric quantum mechanics, the polarization in question is complex. (ii) In the case of the harmonic oscillator, constants m and ω enable one to construct quantities linear in X and P with the same physical dimension necessary in the definition of Z. In the gravitational case, such constants are not available, and, since σ and M have different physical dimension, one can not simply consider '$\sigma + iM$'. To obtain φ, therefore, we had to take linear combination of 'space and time derivatives' of σ. (iii) Why did we choose σ^m_{AB} as the configuration variable rather than, say, σ^{AB}_m? It turns out that many choices—including σ^{AB}_m—yield the same symplectic structure at points of $\bar{\Gamma}$. (Note that we are referring to the full symplectic structure, not just its pull back to $\bar{\Gamma}$.) However, they differ off $\bar{\Gamma}$. The present choice of $\mathscr{C}$ (and Γ) is the only one that I know of for which φ_{mAB} (or, $\varphi^\dagger_{mAB}$) form a completing set of commuting variables *everywhere* on Γ. (iv) Note that, in calculations involving functional derivatives of observables on Γ, one has to carefully keep track of contravariant and covariant nature of tensor indices. For instance, $\varphi^m_{AB} := q^{mn}\varphi_{nAB}$ need not form a complete set of commuting variables. Spinor indices, on the other hand, can be raised or lowered freely because ε_{AB} and ε^{AB} do not depend upon σ or M. Thus, φ_{mAB}, φ^{AB}_m, and $\varphi_{mA}{}^B$ all form a complete set of commuting variables.

Although co-ordinates φ_{mAB} and $\varphi^\dagger_{mAB}$ on Γ are useful in bringing out explicitly the self-dual (or the anti-self-dual) nature of the representation that is being constructed, they are inconvenient in another respect: φ and $\varphi^\dagger$ are not canonically conjugate. It is therefore easier to work with variables $\varphi_m{}^{AB}$ and σ^m_{AB} on Γ: the Hamiltonian vector field generated by the function

$$F(\sigma, M) := \int_\Sigma \mathrm{d}v_q f^{AB}_m \sigma^m_{AB}$$

is given by

$$X_F = \frac{i}{\sqrt{2}} \int f^{AB}_m \left(\frac{\delta}{\delta\varphi^{AB}_m} - \frac{\delta}{\delta\varphi^{\dagger AB}_m} \right), \tag{23.23}$$

whence, if $\varphi_m{}^{AB}$ are taken to be the new 'configuration variables', σ^m_{AB} (or, depending on conventions, $(\det q)^{1/2}.\sigma)$ can be thought of as the new 'momenta'. Let us now return to the constraints (23.18). Since the meaning of eqn (23.18a) has already been discussed, let us focus on the (usual) constraints (23.18b) and (23.18c). The first of these two equations is equivalent to requiring

$$\int_\Sigma dV_q N^b (\mathcal{R}_{mb}{}^{AB}) \sigma^m_{AB} = 0, \tag{23.24}$$

for all smooth vector fields N^b of compact support, while the second is satisfied if and only if

$$\int_\Sigma dV_q N (\mathcal{R}_{mb}{}^{AB}) \sigma^m_{AP} \sigma^b_{BQ} \varepsilon^{PQ} = 0, \tag{23.25}$$

for all smooth functions N^b of compact support. Note that the two constraints are now remarkably similar. Yet, since the curvature tensor $\mathcal{R}_{mb}{}^{AB}$ is independent of σ,

$$\mathcal{R}_{mb}{}^{AB} = {}^0\mathcal{D}_{[m}\varphi_{b]}^{AB} + \varphi^{BP}{}_{[m}\varphi_{b]}{}^A{}_P, \tag{23.26}$$

the vector constraint is again 'linear in the momentum' and the scalar constraint is again 'quadratic in momenta'. If we compare (23.25) to the form of the scalar constraint in the q-representation,

$$\int dV_q N [(q_{ab}q_{cd} - \tfrac{1}{2} q_{ac} q_{bd})(\det q)^{-1} P^{ac} P^{bd} - R] = 0, \tag{23.27}$$

we notice that a considerable simplification has occurred: the 'potential' term, involving the scalar curvature R of the three-metric has no analogue in eqn (23.25). I feel that the fact that eqn (23.25) is homogeneous in momenta may be of considerable help in overcoming many of the difficulties encountered in the q-representation.

We can now summarize the status of the quantum theory.† Wave functions are to be functions of φ alone: $\psi \equiv \psi(\varphi)$. Thus, in the $(\varphi, \varphi^\dagger)$-chart, $\delta\psi/\delta\varphi^\dagger = 0$. The quantum constraints can be written as:

$$\int_\Sigma \mathcal{D}_a N^{AC} \frac{\delta}{\delta\varphi^{AC}_m} \circ \psi = 0, \tag{23.28a}$$

$$\int_\Sigma N^b \mathcal{R}_{mb}{}^{AB}(\varphi) \frac{\delta}{\delta\varphi^{AB}_m} \circ \psi = 0, \tag{23.28b}$$

and

$$\int_{\Sigma.} {}^{``} N\varepsilon^{PQ} \mathcal{R}_{mn}{}^{AB}(\varphi) \frac{\delta}{\delta\varphi^{AP}_m} \frac{\delta}{\delta\varphi^{BQ}_n}{}^{"} \circ \psi = 0 \tag{23.28c}$$

---

† Unfortunately, the ideas that follow are still somewhat tentative: variations are possible at certain stages. The discussion is intended more as an outline of the general direction of thinking than as a presentation of the best possible strategy.

for all test fields N_B^C (such that $N_{CB} = N_{(CB)}$), N^b and N. The first two constraints, being linear in momentum, admit a geometric interpretation: eqn (23.28a) essentially says that $\psi(\varphi)$ should not have a free spinor index while eqn (23.28b) says that the wave function should be invariant under the action of the diffeomorphisms generated by smooth vector fields of compact support on Σ. These two operator constraints are mutually consistent; their commutator algebra just mirrors the corresponding Poisson bracket algebra. The last equation is in inverted commas for two reasons: I still do not know what the correct factor ordering is, and, in any case, a regularization procedure is necessary because second functional derivatives at the same point are involved. A discussion of the consistency of this equation will be given in the next section.

To conclude this sub-section, let us consider the Hamiltonian. Fix any asymptotically constant spinor field α^A. Then, the classical Hamiltonian,

$$H(\varphi, \sigma) := -\frac{1}{8\sqrt{2}} \left[\int_\Sigma dV_q N \mathcal{R}_{ab}{}^{MN} \sigma_{MP}^a \sigma_N^{bP} + i \int_\Sigma dV_q N^b \mathcal{R}_{ab}{}^{MN} \sigma_{MN}^a \right]$$

$$+ \frac{1}{4\pi} \oint_{\partial\Sigma} \varphi_{mAB} \alpha^{\dagger A} \alpha^B \, ds^m, \tag{23.29}$$

with $N = \lambda^{\dagger A} \lambda_A$ and $N^m = -\sqrt{2}\, i \lambda^{\dagger A} \lambda^B \sigma_{AB}^m$ generates, on the constraint sub-manifold $\bar{\Gamma}$, infinitesimal canonical transformations corresponding to the evolution along the lapse-shift pair (N, N^b) (Ashtekar and Horowitz 1984). In the quantum theory, on the space of physical states satisfying eqn (23.28), the Hamiltonian operator reduces to the surface term:

$$\mathbf{H} = \frac{1}{4\pi} \oint_{\partial\Sigma} \varphi_{mAB} \lambda^{\dagger A} \lambda^B \, ds^m \tag{23.30}$$

which preserves the self-dual polarization, i.e. is a well-defined operator in the self-dual representation.

23.3. Discussion

Let us summarize the results reported in Section 23.2. We constructed an enlarged phase-space Γ by taking the soldering forms σ_{AB}^m as the configuration variables in place of the metrics q_{mn}. This enables one to introduce certain spin connections, $\mathcal{D}$, first discussed by Sen (1981, 1982). These, in turn, provide a chart $(\varphi_{mAB}, \varphi_{mAB}^\dagger)$ on Γ. On the constraint sub-manifold $\bar{\Gamma}$ of Γ, one can unambiguously compute the Weyl tensor. We found that φ_{mAB} serves as a potential for the self-dual part of the Weyl curvature while $\varphi_{mAB}^\dagger$ plays the same role for the anti-self-dual part. We then showed that φ_{mAB} (or $\varphi_{mAB}^\dagger$) form a complete set of commuting variables on all of Γ. For definiteness, the self-dual representation—in which the φ_{mAB} play the role of the 'configuration variables'—was chosen. The corresponding 'momenta' are the soldering forms σ^m_{AB}. The constraints were then translated to this representation. The resulting equations are eqn (23.24) and eqn (23.25). The vector constraint is again linear in the new momentum and the scalar constraint is again quadratic. We then considered wave functions ψ which depend only on the new configuration variables and translated the classical constraints to operator eqns (23.28). While the constraints linear in momentum are straightforward to translate, further work is still needed to resolve the factor ordering ambiguity and to regularize the constraint which is quadratic in momenta. Nonetheless, as the matters stand, the status of the new representation appears to be at least as good as that of the q-representation. For instance, the Friedmann–Sorkin (1980) result on half-integral spin from gravity could be re-derived in the φ-representation. The weak field limit can be taken and one finds everything as expected. In particular, the close relation to 'linearized $\mathcal{H}$-spaces' is apparent and one can see that one has the right number of degrees of freedom.†

The hope is of course that one can do better. A major—perhaps the major—source of difficulties in the q-representation is the presence of the 'potential', i.e. the scalar curvature term in the scalar constraint. For instance substantial progress could be made in the so-called strong coupling limit (Pilati 1982; Isham 1984) in which this term can be neglected. In the self-dual representation, as we saw, there is simply no counterpart of this term. One may therefore hope that simplifications would occur. To resolve various issues definitively, one must face the regularization and the factor-ordering problems squarely. Nonetheless, it is encouraging that some indications of simplifications already exist. In particular, given any factor-ordering in (23.29c) in which at least one of the $\delta/\delta\varphi$ always appears on the right, one can write down at least some solutions to all operator constraints. Choose any asymptotically constant field f^m_{AB}. Then,

$$\psi(\varphi) := \oint_{\partial\Sigma} f^m_{AB}\varphi^{AB}_m \, \mathrm{d}^2 S \qquad (23.31)$$

solves all of eqns (23.28). (Recall that N and N^b are of compact support.) Comparing with eqn (23.29), one sees that (23.30) is just the 'mass aspect'. One can write down other solutions associated with 'higher multipoles'. In the q-representation, analogous surface integrals involving

† In the exact theory, using eqn (23.21), one can show that the φ_{mAB} variables have the correct number of degrees of freedom.

only q do not solve the Wheeler–DeWitt equations precisely because of the presence of the scalar curvature term in those equations: in fact, as was pointed out in Section 23.1 not a single solution to these equations is known. Finally, we note that, unlike q_{mn}, φ_{mAB} is not subject to any positivity requirement.

Apart from the problems associated with (23.28c), the most important open problem is that of finding a suitable inner-product. The efforts by Kuchăr (1981) at finding suitable structures to construct a Hilbert space from the space of solutions to the Wheeler–DeWitt equation as well as the general group theoretical methods introduced by Isham (1984) encountered difficulties beyond the strong coupling limit. Therefore, it is of considerable interest to apply these ideas to the self-dual representation. (New types of difficulties may arise, however, because while the coefficient of momenta in eqn (23.27), $(\det q)^{-1} (q_{ab}q_{cd} - \frac{1}{2}q_{ac}q_{bd})$, depends only on the q and not their derivatives, the coefficient, $\varepsilon^{PQ} \mathcal{R}_{mn}{}^{AB}$ in eqn (23.25) depends on φ as well as their first derivatives. Nonetheless, as is often the case, new insight can be obtained by looking at old problems from a new angle.) Another possible avenue to obtaining an inner product is to use the structure made available by the 'Heaven on Earth construction' (see e.g. LeBrun 1980) which appears to be closely related to the use of the φ-variables.

Acknowledgements

It is a pleasure to thank Gary Horowitz, Chris Isham, Karel Kuchăr, Anne Magnon, Lionel Mason, Ted Newman, Roger Penrose, and Amitabha Sen for their comments, suggestions and encouragment. Above all, I am grateful to Kumar Narain for his interest, patience and constructive criticisms.

This chapter is based on research supported in part by an Alfred P. Sloan Research Fellowship and by the NSF grant PHY83 10041 to Syracuse University.

Note added in proof

A satisfactory factor ordering in the formal expressions of quantum constraints is now available. The quantum constraints are closed under the commutator bracket and have the same Lie algebra structure as the Poisson bracket Lie algebra of classical constraints. The Hamiltonian operator is compatible with all constraints. Now, the remaining problem is that of finding explicit representations of this algebra.

References

Ashtekar, A. and Horowitz, G. T. (1982). *Phys. Rev.* D**26,** 3342.
——— (1984). *J. Math. Phys.* **25,** 1473.

—— —— and Magnon-Ashtekar, A. (1982). *Gen. Rel. Grav.* **14,** 411.

—— and Stillerman, M. (1985). Geometric quantization and constrained systems. Syracuse University pre-print.

Bargmann, V. (1962). *Proc. Natl. Acad. Sci.* **48,** 199.

Friedman, J. L. and Sorkin, R. (1980). *Phys. Rev. Lett.* **44,** 1100.

Isham, C. J. (1984). In *Relativity, groups and topology 2* (ed. B. S. DeWitt and R. Stora). North-Holland, Amsterdam.

Ko, M., Ludvigsen, M., Newman, E. T., and Tod, K. P. (1981). *Phys. Rep.* **71,** 51.

Kuchăr, K. (1970). *J. Math. Phys.* **11,** 3322.

—— (1981). In *Quantum gravity 2* (ed. C. J. Isham, R. Penrose, and D. W. Sciama) Clarendon Press, Oxford.

LeBrun, C. R. (1980). D.Phil Thesis, University of Oxford.

Ludvigsen, M., Newman, E. T., and Tod, K. P. (1981). *J. Math. Phys.* **22,** 818.

Narain, K. S. (1984). Private communication.

Newman, E. T. (1975). In *General relativity and gravitation* (ed. G. Sharir and J. Rosen). Wiley, New York.

Penrose, R. (1976). *Gen. Rel. Grav.* **7,** 31.

—— and Rindler, W. (1984). *Spinors and space–time.* Cambridge University Press, Cambridge.

Pilati, M. (1982). In *Quantum structure of space–time* (ed. M. Duff and C. J. Isham). Cambridge University Press, Cambridge.

Sen, A. (1981). *J. Math. Phys.* **22,** 1718.

—— (1982). *Phys. Lett.* **119B,** 89.

Stillerman, M. (1985). Ph.D. thesis, Syracuse University.

Woodhouse, N. M. J. (1980). *Geometric quantization* Clarendon Press, Oxford.

24

Quantum fields, curvilinear co-ordinates, and curved space–time†

N. Sanchez and B. F. Whiting

24.1. Introduction

At the previous Oxford Conferences we saw attempts at a direct construction of a quantum theory of gravity which generally consisted of adapting standard techniques from the theory of quantum fields in Minkowski space. In this conference too, several contributions have added to this effort. But as we have seen—in *this* Discussion Conference—it was also intended that we should pause to examine whether the building blocks which have traditionally been used actually have the properties we should require of them for a complete theory of quantum gravity and we have been asked to address those questions which appear for the first time in the consideration of such a theory. Along with attempts to define the precise boundary conditions for quantum gravity and discussions of possible shortcomings and logical ramifications of quantum mechanics, it seems to us that there may still be investigations for problems in flat space–time which can throw light on both classical and quantum properties in curved space–time. Thus, by developing quantum field theory in curvilinear co-ordinates (capable of describing accelerated observers) in flat space–time, yet in a way which can be directly generalized to curved space–time, we hope to indicate some physical and mathematical considerations which may be relevant in the general case.

In the context of treating quantum fluctuations of gravitational and matter fields in curved space–time, a discussion of quantum field theory for curvilinear co-ordinates would seem very appropriate, not least because, locally, acceleration cannot be distinguished from gravitation, but also because the few results which have already been obtained in curved space indicate effects which can also be demonstrated for flat space; and further because flat space provides a much richer (often solvable) testing ground than we expect to be able to make use of for a generally curved space–time. Thus, we consider the formulation of

† At the request of the editors, detailed mathematical calculations have been deliberately excluded from this account, and will be given elsewhere (Sanchez and Whiting 1985).

satisfactory quantum field theories in certain general classes of curvilinear co-ordinate systems which are related by arbitrary functions to the co-ordinates of Minkowski space. Although we primarily consider flat space, examination of many of the available curved space examples supports our suggestion that an essential aspect to the discussion of thermal and quantum properties for accelerated frames in both flat and curved space–times is the description of the 'mapping' relating some 'sub-manifold' to its global analytic extension.

24.2. Formal description

It will be sufficient for our purposes to discuss scalar fields, which may generally be taken to be massive. Frequently, we shall refer to the mapping from particular curvilinear co-ordinates (of complete extent, i.e. over $(0, \pi)$, $(0, 2\pi)$, $(0, \infty)$ or $(-\infty, +\infty)$ as the case may be) to a co-ordinate system (for us, usually Minkowski co-ordinates) giving the global extension of the manifold only partly covered by the curvilinear co-ordinate system, simply as the 'mapping'. We shall refer to the patch covered by the curvilinear co-ordinates as a 'sub-manifold', even if it happens to coincide with the global extension. We shall look at the significance of the 'mapping' in a discussion of scalar fields on the 'sub-manifold' (i.e. it is not questions of covariance which concern us here). Although we are primarily interested in the four-dimensional case, it will be useful first to review results for two space–time dimensions.

We consider, as an example, strictly monotonic but otherwise arbitrary mappings, $f_\pm$, from (x', t') curvilinear co-ordinates to (x, t) Minkowski co-ordinates

$$x + \varepsilon t = l_\varepsilon f_\varepsilon [\alpha_\varepsilon (x' + {}_\varepsilon t')], \qquad \varepsilon = \pm 1,$$

where x', t' range over $(-\infty, +\infty)$ but where $f_\pm(\pm\infty)$ may be finite or infinite, finite values indicating the existence of event horizons for accelerated frames described by $x' = $ constant. At an event horizon $f'(= \partial f/\partial x')$ will necessarily be zero, but in other asymptotic boundary regions $f'_\pm(\pm\infty)$ may be zero, finite, or infinite, and choosing mappings with these different boundary conditions will affect the discussion of the quantum theory. Note that each mapping introduces two (not necessarily equal) length scales l_ε, α_ε^{-1}; whereas the l_ε correspond to constant re-scalings of the Minkowski null co-ordinates, the α_ϵ will be seen to enter in a more intricate manner.

24.3. Classical thermal properties

The results at the classical level can be summarized as follows. Since thermal properties are known to be associated with periodicity in imaginary time (Gibbons and Perry 1978), mappings which are asymptotically exponential naturally give rise to Green functions which, asymptotically, are indicative of asymptotic thermal equilibrium. The unique mapping for global thermal equilibrium is the exponential mapping with $\alpha_+ = \alpha_-$. If the asymptotic mapping in some boundary region is $l_0 \exp[\alpha_0(x' + \varepsilon t')]$, then the temperature associated with that region is $\alpha_0/2\pi$. Non-exponential mappings have non-thermal properties, but one can, in any case, gives a classification of mappings depending on whether α_0 is zero, finite or infinite (Sánchez 1981).

24.4. (Scalar) quantum theory

For a scalar quantum theory, we need to introduce bases of positive frequency wave functions $\phi_k(x, t)$ and $\psi_\lambda(x', t')$ each satisfying the massive Klein–Gordon equation

$$\Box\Phi = m^2\Phi$$

and the naturally defined inner product (on the 'sub-manifold'), denoted $\langle\phi, \phi\rangle$. We point out that the conditions imposed on the mappings are sufficient to ensure that the ψ have suitable asymptotic properties and are complete in the appropriate sense. We also need to define the quantities

$$B_\lambda(k) = \langle\psi_\lambda, \phi_k^*\rangle \quad \text{and} \quad A_\lambda(k) = \langle\psi_\lambda, \phi_k\rangle$$

in terms of the natural inner product where B and A correspond to the β, α Bogoliubov coefficients which indicate a non-trivial change of basis for $\beta \neq 0$. Finally, we will need the amplitudes (which are independent of the mass of the field)

$$N(\lambda, \lambda') = \int_{-\infty}^{\infty} dk B_\lambda^*(k) B_{\lambda'}(k) \quad \text{and} \quad R(\lambda, \lambda') = \int_{-\infty}^{\infty} dk A_\lambda(k) B_\lambda(k).$$

A key point for us here is that these can be conveniently expressed in terms of the inverse mappings

$$x' + \varepsilon t' = F_\varepsilon(x + \varepsilon t),$$

the singularities of which, we have seen, give the boundaries of the 'sub-manifold'.

In the quantum theory we have 'ground states' (we choose not to use the term 'vacuum' in general), $|0\rangle$ and $|0'\rangle$, associated with the ϕ_k, ψ_λ bases respectively. The energy momentum tensor for a quantum scalar field is a formally divergent operator which can be regularized in a variety of ways and rendered finite by a renormalization of the theory:

$$T_{\text{REN}}^{\mu\nu}(x') = \lim_{\varepsilon \to 0} \mathscr{D}^{\mu\nu}(x', \varepsilon)\{\Phi[x', \rho(\varepsilon)], \Phi[x', \sigma(\varepsilon)]\}_{\text{SUB}}$$

('SUB' here indicates that the renormalization would in practice be effected at each order, by some prescribed "subtraction" scheme), where

ε is the regularizing parameter, $\mathscr{D}^{\mu\nu}(x',0)$ is the classical operator for the energy momentum tensor, $\Phi(x',0)$ is the quantum field operator and ρ, σ allow the dependence on ε to be different in the two positions of Φ in $\mathscr{D}^{\mu\nu}$.

The important result in our discussion is that quantities such as

$$\langle 0|\,\Phi^2\,|0\rangle - \langle 0'|\,\Phi^2\,|0'\rangle = 2Re \int d\lambda \, d\lambda' \{N(\lambda, \lambda')\psi_\lambda^* \psi_{\lambda'}$$
$$+ R(\lambda, \lambda')\psi_\lambda \psi_{\lambda'}\} = \mathscr{G}(x')$$

and

$$\langle 0|\,T^{\mu\nu}\,|0\rangle - \langle 0'|\,T^{\mu\nu}\,|0'\rangle = 2Re \int d\lambda \, d\lambda' \{N(\lambda, \lambda')\mathscr{D}^{\mu\nu}(\psi_\lambda^*, \psi_{\lambda'})$$
$$+ R(\lambda, \lambda')\mathscr{D}^{\mu\nu}(\psi_\lambda, \psi_{\lambda'})\} = \mathscr{T}^{\mu\nu}(x')$$

(here $\mathscr{D}^{\mu\nu}$ is the classical operator acting on fields) are not only finite (of course), but depend in a direct and explicit way on the inverse mappings through N and R. Again the exponential mapping is exceptional; it has

$$N(\lambda, \lambda') = \frac{\delta(\lambda - \lambda')}{e^{2\pi\lambda/\alpha} - 1} \quad \text{and} \quad R(\lambda, \lambda') = 0$$

and for $m = 0$, $\mathscr{T}^{\mu\nu}$ is indicative of a boson gas in global thermal equilibrium. For general mappings of the type we consider, the most singular part of N is always of this form, even for the limiting cases of $\alpha_0 \to 0, \infty$.

24.4. Four dimensions

In four dimensions, entirely new classes of mappings (or equivalently, of 'sub-manifolds' of Minkowski space) will become available. We feel we can indicate some general characteristics of asymptotic behaviour essential for the construction of good quantum field theories in the usual sense, and can point to some general properties (Sánchez and Whiting 1985) which do not depend at all critically on the fact that we have been discussing flat space.

A simple example of a mapping in four dimensions consist of

$$y = y', \qquad z = z',$$

coupled with the previously discussed two-dimensional transformations. $N(\lambda, \lambda')$ and $R(\lambda, \lambda')$ are still independent of the mass, and N has a leading three-dimensional delta function but a single Planckian factor. With the Rindler (i.e. exponential) mapping for (x', t') a global thermal equilibrium is again obtained.

With respect to applications for curved space–time, it should be noted that a large number of the exact solutions of General Relativity which possess event horizons (removable singularities) have their maximal extension defined in terms of exponential mappings for a pair of suitable chosen co-ordinates, from which the determination of their 'natural' temperature immediately follows (Sánchez 1982). In this connection, Bernard (1985) has recently included the effects of rotation and electric fields with the exponential mapping, and has specified precisely how the definition of temperature must be altered for such generalizations and how to define, in terms of the mapping, the relevant chemical potentials which enter into the Planckian factor, to give the flat space analogues of the Kerr, Reissner–Nordström, and even Kerr–Newman black holes.

Although the exponential mapping applies to a 'sub-manifold' with a finite boundary, which is then an event horizon, one result not commonly appreciated is that thermal properties are not uniquely associated with the existence of global events horizons, but can also be directly associated to the infinite boundaries of a 'sub-manifold' (in fact, as shown in Page (1982), even for Schwarzschild, truly thermal properties for $T^{\mu\nu}$ really only apply at infinity). This fact can be most clearly demonstrated with a mapping which has no global event horizons and for which the sub-manifold is the whole of Minkowski space, but for which the asymptotic acceleration is everywhere uniform, i.e. the mapping is asymptotically exponential, implying asymptotic thermal equilibrium (Sánchez 1979).

24.5. Normalization and renormalization

In the course of constructing the stress-energy operator for a quantum field one is naturally led to consider the need for either renormalization (such as by geodetic point separation (Christensen 1976)) or, as Brown *et al.* (1982) and Brown and Ottewill (1983) have recently argued, the definition of a reasonable 'normalization' prescription. In this regard, the practical advantage of renormalization is that it requires only one complete set of modes (in some sense, it is meant to be 'absolute'), but the technical disadvantage is that it will usually involve mode-sums subject to a careful cancellation of infinities. For normalization (Christensen and Fulling 1977), a locally constructed 'normalizing' Green function is also required, for which a unique characterization is, to date, problematical. Nevertheless, Brown and Ottewill (1983) still argue that renormalization, even in curved space–time, is inappropriate until interactions are explicitly introduced.

In flat Minkowski space one is prepared simply to 'normalize' for free quantum fields since we know that there is a good ground state (which might be regarded as a *true* vacuum) with satisfactory properties both locally and globally; but one would have to resort to 'renormalization' for interacting fields. This 'normalization' is actually equivalent to defining that a particular ground state (the Minkowski vacuum) has zero energy. The difficulties which arise in the case of curved space–time are:

(i) we do not know that there is any global ground state which we should regard as empty;

(ii) in general, locally defined ground states (vacua, Green's functions, . . .) need not be unique and may have unphysical singularities in their global extension (Brown and Ottewill 1983);

(iii) renormalization, effected simply by adjusting the mass, charge (or other couplings) and fields appearing in the Lagrangian, is insufficient, since new counter-terms must also be introduced (at each level, say, of a loop expansion about some classically-valued geometry).

Now, normalization certainly has the advantage of preserving more of the properties of the classical problem, but it hardly seems valid to regard even a 'free' (scalar) field in a curved space–time as thought it were non-interacting with the geometry: surely it *is* interacting in a way which cannot be undone. On the other hand, renormalizing a theory which is perturbatively 'un-renormalizable' presents its own (grave) difficulties, and the choice of how to proceed needs some physical guidance currently lacking. However, at least on those 'sub-manifolds' for which there is (a 'mapping' to) a global (or pseudo-inertial) extension, we have seen above that there is another naturally defined normalizing procedure which can be expressed in terms of the mapping, i.e.

$$\langle T^{\mu\nu} \rangle_{\text{'GLOBAL'}} - \langle T^{\mu\nu} \rangle_{\text{'CURVILINEAR'}} = + \mathcal{T}^{\mu\nu}(x'),$$

to which one might affix the interpretation that with respect to an observer at rest in a co-ordinate frame on the curvilinear 'sub-manifold', the full global ground state appears normalized as indicated. Note that Brown and Ottewill would prefer to consider

$$\langle T^{\mu\nu} \rangle_{\text{'CURVILINEAR'}} - \langle T^{\mu\nu} \rangle_{\text{'LOCAL'}} = \mathcal{T}^{\mu\nu}(x')$$

which would give the negative of the above for Rindler space, since their 'local' Green function is also the 'global' one for Minkowski space–time.

Our result properly reflects a truly thermal property (rather than it is negative) for Rindler space, as it does also for the Schwarzschild black hole (Christensen and Fulling 1977), where an example of a 'global' state would be the Krustal or Hartle–Hawking state, while for the exterior region, the 'curvilinear' state, at infinity (say), would be the Boulware state (which correctly reflects the difficulty of pursuing an $r = $ constant path too close to the horizon, by its singular nature there). Our point of view would seem to be indicative of the fact that an accelerated observer may not be able to impose those boundary conditions on his past horizon which would prevent a thermal flux there. It would be interesting to see if one could gain some physical insight into this question by experiment with, for example, particle accelerators.

Finally, with reference to a discussion of the back-reaction in curved space–time, it is difficult to argue that (covariant) renormalization might not be more appropriate, if any thing is. It is not clear that geometry will respond to (quantum) matter in the same way as does an observer in a laboratory which is accelerating.

24.6. Conclusion

The general result which these investigations exemplify is that the 'mapping' from a 'sub-manifold' to its global extension is remarkably useful in the full description of the interesting physical properties which one might investigate on the 'sub-manifold'. In fact, for infinite volumes, it is often only the asymptotic properties of the mapping near the boundaries which are most relevant.

Acknowledgement

This work has been supported by the CNRS, France.

References

Bernard, D. and Sánchez, N. (1985). Paper in preparation, G.A.R., Meudon.
Brown, M. R. and Ottewill, A. C. (1983). *Proc. R. Soc. Lond.* A**389,** 379.
—— —— and Siklos, S. T. C. (1982). *Phys. Rev.* D**26,** 1881.
Christensen, S. M. (1976). *Phys. Rev.* D**14,** 2490.
—— and Fulling, S. A. (1977). *Phys. Rev.* D**15,** 2088.
Gibbons, G. W. and Perry, M. J. (1978). *Proc. R. Soc. Lond.* A**358,** 467.
Page, D. N. (1982). *Phys. Rev.* D**25,** 1499.
Sánchez, N. (1979). *Phys. Lett.* **87B,** 212.
—— (1981). *Phys. Rev.* D**24,** 2100.
—— (1982). Analytic mappings: a new approach to particle production by accelerated observers. In *Proceedings of the second Marcel Grossman meeting on general relativity, ICTP Trieste, 5–11 July 1979* (ed. R. Ruffini). North-Holland, Amsterdam, p. 501.
—— and Whiting, B. F. (1985). Quantization of fields for accelerated frames in flat and curved space–time. G.A.R., Meudon.

25

Effective action for expectation values

Bryce DeWitt

25.1. Introduction

This is a brief report on some work that I have begun with my student, Richard Jordan, concerning the effective action.† The effective action in quantum field theory summarizes, in a single functional, all the quantum properties of the fields under consideration. It has become increasingly useful in recent years in discussing such disparate topics as renormalization theory, the S-matrix, and quantum cosmology. Its functional derivative yields the so-called effective field equations, which replace the classical field equations as descriptors of the dynamical behaviour of the quantized fields.

Solutions of the effective field equations, as commonly derived, are 'in–out' matrix elements of the field operators and, when substituted back into the effective action itself, yield logarithms of the corresponding 'in–out' transition amplitudes. These solutions, called effective fields, are generally complex-valued even when the original classical fields are real, and hence their immediate physical significance is somewhat obscure. In order to obtain a more intuitive understanding of the effective field idea it is desirable to replace the conventional effective fields by 'in–in' expectation values, which are necessarily real. What follows is an outline of how to define a new effective action, the field equations of which yield directly the 'in–in' expectation values.‡

25.2. Representation of 'in–in' expectation values

It will be assumed that no gauge groups are present. Generalization to the case in which the classical action is invariant under local symmetry

---

† The effective action was first introduced into physics by Julian Schwinger in 1954, in some unpublished lectures given at the Institute for Advanced Study. It is sometimes attributed (incorrectly) to Jona Lasinio.

‡ Other approaches to this problem will be found in Schwinger (1961), Buchbinder *et al.* (1981), Fradkin and Gitman (1981), and Hajicek (1982).

transformations is relatively straightforward with techniques that are, by now, standard. Let $A[\varphi]$ be any Hermitian operator built out of Hermitian fields $\varphi^i(t, x)$ in some compact region of space–time, and let T be a time lying to the future of this region. Let $|\text{in}\rangle$ be the vector corresponding to a relative-vacuum 'in' state, i.e. a coherent state determined by the positive frequency components of a weak (and hence linear) background in the remote past. Then the 'in–in' expectation value of $A[\varphi]$ may be expressed in the form

$$\langle \text{in}| A[\varphi] |\text{in}\rangle = \int \langle \text{in}| A[\varphi] |\varphi, T\rangle \langle \varphi, T | \text{in}\rangle \, d\varphi, \tag{25.1}$$

$$d\varphi = \prod_{i,x} d\varphi^i(x), \tag{25.2}$$

where $|\varphi, T\rangle$ is either the eigenvector of the operators $\varphi^i(T, x)$ corresponding to the c-number eigenvalues $\varphi^i(x)$ when the fields are bosonic, or, when the fields are fermionic, satisfies the equation

$$\langle \varphi, T| \varphi^i(T, x) = [\tfrac{1}{2}\varphi^i(x) + \delta/\delta\varphi^i(x)] \langle \varphi, T|, \tag{25.3}$$

where the $\varphi^i(x)$ are real a-numbers (i.e. anticommuting Grassmann variables).

At this point one may introduce functional integrals. The two factors appearing in the integrand of eqn (25.1) can be expressed in the form

$$\langle \text{in}| A[\varphi] |\varphi, T\rangle = N \int A[\varphi] \exp(iS[\varphi]_{\text{in}}^T) \mu[\varphi] \, d\varphi, \tag{25.4}$$

$$\langle \varphi, T | \text{in}\rangle = N \int \exp(-iS[\varphi]_{\text{in}}^T) \mu[\varphi] \, d\varphi, \tag{25.5}$$

$$d\varphi = \prod_{-\infty < t < T} \prod_{i,x} d\varphi^i(t, x), \tag{25.6}$$

where $S[\varphi]_{\text{in}}^T$ is the classical action appropriate to the boundary conditions defined by the vectors $|\text{in}\rangle$ and $|\varphi, T\rangle$, N is a normalizing constant, and $\mu[\varphi]$ is an appropriate functional measure. Expressions (25.4) and (25.5) can be inserted into eqn (25.1) provided the dummy variables of integration are separately labelled:

$$\langle \text{in}| A[\varphi] |\text{in}\rangle$$

$$= N^2 \int d\varphi \int \mu[\varphi_+] \, d\varphi_+ \int \mu[\varphi_-] \, d\varphi_- A[\varphi_+] \exp(iS[\varphi_+]_{\text{in}}^T - iS[\varphi_-]_{\text{in}}^T). \tag{25.7}$$

It is easy to see that the expectation value is equally well obtained by replacing $A[\varphi_+]$ by $A[\varphi_-]$ in the integrand.

25.3. Alternative representation

An alternative representation of $\langle \text{in}| A[\varphi] |\text{in}\rangle$ is obtained by extending the time interval $(-\infty, T)$ formally to the whole t-axis through reflection in the point T. Thus

$$\langle \text{in}| A[\varphi] |\text{in}\rangle = N^2 \int A[\varphi]\exp(i\mathcal{S}[\varphi])\mu[\varphi]\,d\varphi, \qquad (25.8)$$

where now

$$\varphi^i(t, \mathbf{x}) = \begin{cases} \varphi^i_+(t, \mathbf{x}) & t < T \\ \varphi^i(\mathbf{x}) & t = T \\ \varphi^i_-(2T - t, \mathbf{x}) & t > T \end{cases} \qquad (25.9)$$

$$d\varphi = \prod_{i,t,x} d\varphi^i(t, \mathbf{x}) \qquad (25.10)$$

$$\mu[\varphi] = \mu[\varphi_+]\mu[\varphi_-] \qquad (25.11)$$

and (suppressing the spatial coordinates)

$$\mathcal{S}[\varphi] = \int_{-\infty}^{\infty} \mathcal{L}[\varphi, \dot{\varphi}, t]\,dt + \text{boundary terms}, \qquad (25.12)$$

$$\mathcal{L}[\varphi, \dot{\varphi}, t] = \theta(T - t)L[\varphi, \dot{\varphi}, t] - \theta(t - T)L[\varphi, -\dot{\varphi}, 2T - t], \quad (25.13)$$

L being the Lagrangian functional for the original classical action $S[\varphi]$. It is not difficult to verify that because of the step functions θ appearing in expression (25.13), a delta function $\delta(t - T)$ appears explicitly in the field equations generated by the new action functional $\mathcal{S}[\varphi]$. Through use of the rule

$$\dot{\varphi}(t)\delta(t - T) = \lim_{\varepsilon \to 0} \tfrac{1}{2}[\dot{\varphi}(T - \varepsilon) + \dot{\varphi}(T + \varepsilon)]\delta(t - T) \qquad (25.14)$$

one finds that the solutions of these equations are forced to satisfy

$$\varphi^i(t, \mathbf{x}) = \varphi^i(2T - t, \mathbf{x}), \qquad (25.15)$$

where, for $t < T$, $\varphi^i(t, \mathbf{x})$ is a solution of the field equations generated by $S[\varphi]$.

In the notation of eqn (25.9), relation (25.15) implies that φ_+ and φ_- are equal when the exponent in the functional integral (25.8) is stationary. It also implies that the 'in' boundary conditions suffice completely to determine the physics and that the precise value of the time T is irrelevant. In some applications it is actually convenient at the end to shift T to $+\infty$.

25.4. Equation for small disturbances. Mode functions

The functional integral (25.8) can be treated in exactly the same way as the conventional functional integral for 'in–out' matrix elements, and a corresponding effective action can be constructed. However, in this construction some new features arise. It is convenient to begin with the equation for small disturbances $\delta\varphi$ around an arbitrary background φ satisfying eqn (25.15). This takes the form

$$\left[-\frac{\partial}{\partial t}\mathscr{A}(t)\frac{\partial}{\partial t} + \frac{1}{2}\left\{\mathscr{B}(t), \frac{\partial}{\partial t}\right\} - \mathscr{C}(t)\right]\delta\varphi = 0, \qquad (25.16)$$

where $\mathscr{A}, \mathscr{B}$, and $\mathscr{C}$ are matrix-operators (involving derivatives with respect to the spatial co-ordinates x) given by

$$\mathscr{A}(t) = \theta(T-t)A(t) - \theta(t-T)A(2T-t),$$
$$\mathscr{B}(t) = \theta(T-t)B(t) + \theta(t-T)B(2T-t), \qquad (25.17)$$
$$\mathscr{C}(t) = \theta(T-t)C(t) - \theta(t-T)C(2T-t),$$

A, B, and C being the corresponding operators for $S[\varphi]$. If one introduces the abbreviation

$$\bar{t} = 2T - t, \qquad (25.18)$$

one observes that $\mathscr{A}, \mathscr{B}$, and $\mathscr{C}$ possess the symmetries

$$\mathscr{A}(t) = -\mathscr{A}(\bar{t}),$$
$$\mathscr{B}(t) = \mathscr{B}(\bar{t}), \qquad (25.19)$$
$$\mathscr{C}(t) = -\mathscr{C}(\bar{t}).$$

The Green's functions for eqn (25.16) are the Green's functions that appear in the Gaussian expansion of the functional integral (25.8). In order to construct these Green's functions it is convenient to introduce mode functions in terms of which the small disturbances may be expanded. These have the form

$$u^i(t, x, A) = \theta(T-t)u^i_{\text{in}}(t, x, A) + \theta(t-T)u^i_{\text{in}}(2T-t, x, A), \quad (25.20)$$

where the u^i_{in} are 'in' mode functions for the original dynamical system $S[\varphi]$ and A is a mode label. In the conventional theory one also introduces 'out' mode functions and relates them to the u_{in} by Bogoliubov coefficients. The 'out' mode functions will play no role here.

The functions $u^i(t, x, A)$ satisfy the usual Wronskian orthonormality relations (see, for example, DeWitt 1984) associated with the differential operator appearing in eqn (25.16). From this it follows that the so-called supercommutator function, which solves the Cauchy problem for this operator, is given by

$$\tilde{\mathcal{G}} = \mathcal{G}^{(+)} + \mathcal{G}^{(-)}, \tag{25.21}$$

where

$$\begin{aligned} \mathcal{G}^{(+)} &= -iuu^\dagger, \\ \mathcal{G}^{(-)} &= \pm iu^*u^\sim, \end{aligned} \tag{25.22}$$

the upper sign holding for bosons and the lower sign for fermions. Here the labels i, t, $\boldsymbol{x}$, and A have been suppressed and u is regarded as a continuous matrix. If we reinsert the label t we may distinguish four regimes.

$$\begin{aligned} t, t' < T: \quad & \mathcal{G}^{(+)}(t, t') = -iu_{in}(t)u_{in}^\dagger(t') = G_{in}^{(+)}(t, t'), \\ t < T, t' > T: \quad & \mathcal{G}^{(+)}(t, t') = -iu_{in}(t)u_{in}^\dagger(\bar{t}') = G_{in}^{(+)}(t, \bar{t}'), \\ t > T, t' < T: \quad & \mathcal{G}^{(+)}(t, t') = -iu_{in}(\bar{t})u_{in}^\dagger(t') = G_{in}^{(+)}(\bar{t}, t'), \\ t, t' > T: \quad & \mathcal{G}^{(+)}(t, t') = -iu_{in}(\bar{t})u_{in}^\dagger(\bar{t}') = G_{in}^{(+)}(\bar{t}, \bar{t}'). \end{aligned} \tag{25.23}$$

Returning to the earlier notation, given in eqn (25.9), we may alternatively express $\mathcal{G}^{(+)}$ in the block form

$$\mathcal{G}^{(+)} = \begin{pmatrix} G_{in}^{(+)} & G_{in}^{(+)} \\ G_{in}^{(+)} & G_{in}^{(+)} \end{pmatrix}, \tag{25.24}$$

where the blocks are labelled by $++$, $+-$, $-+$, and $--$ respectively, and the parameter t now ranges only from $-\infty$ to T. In a similar manner one finds

$$\mathcal{G}^{(-)} = \begin{pmatrix} G_{in}^{(-)} & G_{in}^{(-)} \\ G_{in}^{(-)} & G_{in}^{(-)} \end{pmatrix} \tag{25.25}$$

and

$$\tilde{\mathcal{G}} = \begin{pmatrix} \tilde{G} & \tilde{G} \\ \tilde{G} & \tilde{G} \end{pmatrix}, \tag{25.26}$$

where

$$G_{in}^{(-)} = \pm iu_{in}^* u_{in}^\sim = \pm G_{in}^{(+)*} = \mp G_{in}^{(+)\sim} \tag{25.27}$$

and

$$\tilde{G} = G_{in}^{(+)} + G_{in}^{(-)} = \pm \tilde{G}^* = \mp \tilde{G}^\sim = -\tilde{G}^\dagger. \tag{25.28}$$

No suffix is placed on $\tilde{G}$ because it remains unchanged if the u_{in}s are replaced by u_{out}s, or by any other mode functions.

25.5. Retarded and advanced Green's functions

If the differential operator appearing in eqn (25.16) is denoted by $\mathscr{F}$, it too may be expressed in block form:

$$\mathscr{F} = \begin{pmatrix} F & 0 \\ 0 & -F \end{pmatrix} \tag{25.29}$$

where

$$F = -\frac{\partial}{\partial t}A(t)\frac{\partial}{\partial t} + \frac{1}{2}\left\{B(t),\frac{\partial}{\partial t}\right\} - C(t). \tag{25.30}$$

All the important Green's functions of the operator $\mathscr{F}$ may be constructed from $\tilde{\mathscr{G}}$, $\mathscr{G}^{(+)}$ and $\mathscr{G}^{(-)}$. The block forms that these Green's functions take are less trivial than those appearing in eqns (25.24)–(25.26). For example, the retarded Green's function of $\mathscr{F}$ is given by

$$\mathscr{G}^-(t,t') = -\theta(t-t')\tilde{\mathscr{G}}(t,t'), \tag{25.31}$$

whence

$$\mathscr{G}^- = \begin{pmatrix} G^- & 0 \\ -\tilde{G} & -G^+ \end{pmatrix}, \tag{25.32}$$

where G^- and G^+ are respectively the retarded and advanced Green's functions of F, being related to one another by

$$G^+ = \pm G^{-\sim}, \tag{25.33}$$

the upper sign, as usual, referring to the boson sector and the lower sign to the fermion sector. In a similar manner the advanced Green's function is given by

$$\mathscr{G}^+(t,t') = \theta(t'-t)\tilde{\mathscr{G}}(t,t'), \tag{25.34}$$

which yields

$$\mathscr{G}^+ = \begin{pmatrix} G^+ & \tilde{G} \\ 0 & -G^- \end{pmatrix}. \tag{25.35}$$

The following symmetries may be noted:

$$\tilde{G} = G^+ - G^- = \mp\tilde{G}^\sim, \tag{25.36}$$

$$\mathscr{G}^+ = \pm\mathscr{G}^{-\sim}, \tag{25.37}$$

$$\tilde{\mathscr{G}} = \mathscr{G}^+ - \mathscr{G}^- = \mp\tilde{\mathscr{G}}^\sim. \tag{25.38}$$

25.6. Variational law and superdeterminants

The functions $\mathscr{G}^\pm$ satisfy an important variational law which follows from the well-known law (see, for example, DeWitt 1984)

$$\delta G^\pm = G^\pm\,\delta F G^\pm. \tag{25.39}$$

Thus

$$\delta\mathscr{G}^- = \begin{pmatrix} \delta G^- & 0 \\ -\delta\tilde{G} & -\delta G^+ \end{pmatrix} = \begin{pmatrix} G^-\,\delta F G^- & 0 \\ -G^+\,\delta F G^+ + G^-\,\delta F G^- & -G^+\,\delta F G^+ \end{pmatrix}$$

$$= \begin{pmatrix} G^-\,\delta F G^- & 0 \\ -\tilde{G}\,\delta F G^- - G^+\,\delta F\tilde{G} & -G^+\,\delta F G^+ \end{pmatrix}$$

$$= \begin{pmatrix} G^- & 0 \\ -\tilde{G} & -G^+ \end{pmatrix}\begin{pmatrix} \delta F & 0 \\ 0 & -\delta F \end{pmatrix}\begin{pmatrix} G^- & 0 \\ -\tilde{G} & -G^+ \end{pmatrix} = \mathscr{G}^-\,\delta\mathscr{F}\mathscr{G}^-, \tag{25.40}$$

and, similarly,

$$\delta \mathscr{G}^+ = \mathscr{G}^+ \, \delta \mathscr{F} \mathscr{G}^+. \tag{25.41}$$

This variational law is important because it allows one to define without ambiguity (up to a constant factor) the formal superdeterminant of $\mathscr{G}^+$. Because of the triangular block structure of $\mathscr{G}^+$ one has

$$\text{sdet } \mathscr{G}^+[\varphi] = (\text{sdet } G^+[\varphi_+])(\text{sdet } G^-[\varphi_-]), \tag{25.42}$$

where sdet G^+ and sdet G^- are defined appropriately to the interval $(-\infty, T)$.

Consider now the measure $\mu[\phi]$ for the functional integrals (25.4) and (25.5). It may be shown (DeWitt 1984) to be given by

$$\mu[\varphi] = (\text{sdet } G^+[\varphi])^{-1/2}. \tag{25.43}$$

Because of the relation (25.33) and the (super)transposition invariance of the superdeterminant one may equally well write

$$\mu[\varphi] = (\text{sdet } G^-[\varphi])^{-1/2}. \tag{25.44}$$

It follows from eqns (25.11) and (25.42) that the measure for the functional integral (25.8) is simply

$$\mu[\varphi] = (\text{sdet } \mathscr{G}^+[\varphi])^{-1/2}. \tag{25.45}$$

That is, the advanced Green's function for the 'in–in' theory enters in exactly the same way that it does in the conventional 'in–out' theory.

Feynman propagator analogue

In the Gaussian expansion of the functional integral (25.8) the Green's function that arises from the Gaussian integrations is the following analogue of the Feynman propagator:

$$\mathscr{G}(t, t') = -\theta(t - t')\mathscr{G}^{(+)}(t, t') + \theta(t' - t)\mathscr{G}^{(-)}(t, t'). \tag{25.46}$$

In block form this becomes

$$\mathscr{G} = \begin{pmatrix} G_{\text{in}} & G_{\text{in}}^{(-)} \\ -G_{\text{in}}^{(+)} & \mp G_{\text{in}}^* \end{pmatrix}, \tag{25.27}$$

where

$$G_{\text{in}}(t, t') = -\theta(t - t')G_{\text{in}}^{(+)}(t, t') + \theta(t' - t)G_{\text{in}}^{(-)}(t, t'). \tag{25.48}$$

The suffix 'in' is placed on G_{in} to distinguish it from the conventional Feynman propagator for the operator F, which satisfies 'in–out' boundary conditions. This distinction is necessary because G_{in} does *not* satisfy the simple variational law that the conventional propagator does. To obtain the variational law for G_{in}, begin with

$$Fu_{\text{in}} = 0. \tag{25.49}$$

Variation of this equation yields

$$F \, \delta u_{\text{in}} = -\delta F u_{\text{in}},\qquad(25.50)$$

of which the solution incorporating the 'in' boundary condition is

$$\delta u_{\text{in}} = G^- \, \delta F u_{\text{in}}.\qquad(25.51)$$

The adjoint of eqn (25.51), in both the boson and fermion sectors, is

$$\delta u_{\text{in}}^{\dagger} = u_{\text{in}}^{\dagger} \, \delta F G^+,\qquad(25.52)$$

whence

$$\delta G_{\text{in}}^{(+)} = -\mathrm{i} \, \delta u_{\text{in}} u_{\text{in}}^{\dagger} - \mathrm{i} u_{\text{in}} \, \delta u_{\text{in}}^{\dagger} = G^- \, \delta F G_{\text{in}}^{(+)} + G_{\text{in}}^{(+)} \, \delta F G^+.\quad(25.53)$$

In a similar manner one finds

$$\delta G_{\text{in}}^{(-)} = G^- \, \delta F G_{\text{in}}^{(-)} + G_{\text{in}}^{(-)} \, \delta F G^+.\qquad(25.54)$$

Combining eqns (25.39), (25.53), and (25.54) with the identities

$$G_{\text{in}} = G^+ - G_{\text{in}}^{(+)} = G^- + G_{\text{in}}^{(-)},\qquad(25.55)$$

$$\pm G_{\text{in}}^* = G^+ - G_{\text{in}}^{(-)} = G^- + G_{\text{in}}^{(+)},\qquad(25.56)$$

one easily obtains the laws

$$\delta G_{\text{in}} = G_{\text{in}} \, \delta F G_{\text{in}} + G_{\text{in}}^{(-)} \, \delta F G_{\text{in}}^{(+)},\qquad(25.57)$$

$$\delta G_{\text{in}}^{(-)} = G_{\text{in}} \, \delta F G_{\text{in}}^{(-)} \pm G_{\text{in}}^{(-)} \, \delta F G_{\text{in}}^*,\qquad(25.58)$$

$$\delta G_{\text{in}}^{(+)} = \pm G_{\text{in}}^* \, \delta F G_{\text{in}}^{(+)} + G_{\text{in}}^{(+)} \, \delta F G_{\text{in}},\qquad(25.59)$$

$$\delta G_{\text{in}}^* = G_{\text{in}}^* \, \delta F G_{\text{in}}^* + G_{\text{in}}^{(+)} \, \delta F G_{\text{in}}^{(-)},\qquad(25.60)$$

whence

$$\begin{aligned}
\delta \mathcal{G} &= \begin{pmatrix} \delta G_{\text{in}} & \delta G_{\text{in}}^{(-)} \\ -\delta G_{\text{in}}^{(+)} & \mp \delta G_{\text{in}}^* \end{pmatrix}\\
&= \begin{pmatrix} G_{\text{in}} & G_{\text{in}}^{(-)} \\ -G_{\text{in}}^{(+)} & \mp G_{\text{in}}^* \end{pmatrix}\begin{pmatrix} \delta F & 0 \\ 0 & -\delta F \end{pmatrix}\begin{pmatrix} G_{\text{in}} & G_{\text{in}}^{(-)} \\ -G_{\text{in}}^{(+)} & \mp G_{\text{in}}^* \end{pmatrix}\\
&= \mathcal{G} \, \delta \mathcal{F} \mathcal{G}.
\end{aligned}\qquad(25.61)$$

25.8. The effective action and the effective field equations

Equation (25.61) shows that $\mathcal{G}$, unlike G_{in}, *does* satisfy the standard variational law for a Feynman propagator. This means that its superdeterminant can be unambiguously defined. The logarithm of this superdeterminant, in fact, is part of the one-loop contribution to the new effective action. In terms of the fields φ_+ and φ_- the new action (denoted here by $\Gamma[\varphi_+, \varphi_-]$) has a perturbation expansion of the form

$$\Gamma[\varphi_+, \varphi_-] = \mathcal{S}[\varphi_+, \varphi_-] - \frac{i}{2} \ln \mathrm{sdet}\, \mathcal{G}[\varphi_+, \varphi_-] + \frac{i}{2} \ln \mathrm{sdet}\, \mathcal{G}^+[\varphi_+, \varphi_-]$$

$$+ \text{higher loop terms.} \tag{25.62}$$

The first and third terms of this series may be decomposed into parts that depend solely on the field φ_+ or the field φ_-:

$$\mathcal{S}[\varphi_+, \varphi_-] = S[\varphi_+] - S[\varphi_-], \tag{25.63}$$

$$\ln \mathrm{sdet}\, \mathcal{G}^+[\varphi_+, \varphi_-] = \ln \mathrm{sdet}\, G^+[\varphi_+] + \ln \mathrm{sdet}\, G^+[\varphi_-], \tag{25.64}$$

but none of the other terms can be decomposed in this way. All terms can be expressed as standard one-particle-irreducible vacuum diagrams just as in the conventional theory, but most of these diagrams now involve both off-diagonal and diagonal blocks of the propagator $\mathcal{G}$, as well as vertices coming from both $S[\varphi_+]$ and $-S[\varphi_-]$. The diagonal blocks of $\mathcal{G}^+$ also make important contributions to the multiloop graphs.

If one makes the abbreviation

$$\bar{\varphi}^i = \langle \mathrm{in} | \, \boldsymbol{\varphi}^i \, | \mathrm{in} \rangle, \tag{25.65}$$

then the expectation value $\bar{\varphi}^i$ is a solution of the equations

$$\left(\frac{\delta}{\delta \varphi_+^i} \Gamma[\varphi_+, \varphi_-] \right)_{\varphi_+ = \varphi_-} = 0, \tag{25.66}$$

or, equivalently, of the equations

$$\left(\frac{\delta}{\delta \varphi_-^i} \Gamma[\varphi_+, \varphi_-] \right)_{\varphi_+ = \varphi_-} = 0. \tag{25.67}$$

These are the new effective field equations. In obtaining them it is important to let φ_+ and φ_- become equal only after the functional differentiation has been performed. $\Gamma[\varphi_+, \varphi_-]$ in fact vanishes when $\varphi_+ = \varphi_-$.

25.9. Reality, causality, and the role of the measure

From eqn (25.62) it is easy to obtain the effective field equations correct to one-loop order. Setting φ_+ and φ_- both equal to $\bar{\varphi}$ and using a well-known notation (DeWitt 1984), one finds that eqn (25.66) takes the form

$$0 = S_{,i}[\bar{\varphi}] - \frac{i}{2} S_{,ijk}[\bar{\varphi}](G_{\mathrm{in}}^{kj}[\bar{\varphi}] - G^{+kj}[\bar{\varphi}]) + \ldots$$

$$= S_{,i}[\bar{\varphi}] + \frac{i}{2} S_{,ijk}[\bar{\varphi}] G_{\mathrm{in}}^{(+)kj}[\bar{\varphi}] + \ldots$$

$$= S_{,i}[\bar{\varphi}] + \tfrac{1}{2} S_{,ijk}[\bar{\varphi}] u_{\mathrm{in}}^k u_{\mathrm{in}}^{j+} + \ldots, \tag{25.68}$$

where the u_{in} are the 'in' mode functions appropriate to a background equal to $\bar{\varphi}$. The form of the second term on the right is seen to confirm the validity of a well known *Ansatz* that is often used in studying the back reaction problem for nonlinear fields to one-loop order: one replaces the lowest-order nonlinear term in the operator field equations by its expectation value, evaluated as if the field operators satisfied linear equations governed by the operator $S_{,ij}[\bar{\varphi}]$.

Both terms on the right of eqn (25.68) are manifestly real or manifestly imaginary, according as the index i is bosonic or fermionic, and this ensures the reality of $\bar{\varphi}$. The reality (imaginarity) of the second term follows from the fact that, for each bosonic (fermionic) value of the index i, the matrix $S_{,ijk}[\bar{\varphi}]$, in j and k, is Hermitian (anti-Hermitian).

Both terms are also manifestly causal: i.e. $\bar{\varphi}$ at any time depends only on what $\bar{\varphi}$ was doing at earlier times. The causality of the first term follows from the (assumed) locality of the classical theory. The causality of the second term follows from the locality of the vertex function $S_{,ijk}$ and from the fact that the 'in' mode functions at any time are completely determined by conditions to the past of that time, including the 'in' boundary conditions.

Reality (imaginarity) and causality of the higher terms in the series (25.68) can also be shown by straightforward computation, although the steps involved in the demonstration, even for only two loops, are both lengthy and highly non-trivial. Contributions from the measure functional (25.45), which introduces advanced Green's functions $G^+[\bar{\varphi}]$ into many diagram lines, prove to be essential. This is in sharp contrast to the situation that holds in the case of the conventional effective action. There, although the measure functional is known formally to guarantee unitarity (DeWitt 1984), it is invariably neglected, because its effect is already contained in the standard regularization procedures, particularly dimensional regularization. In the case of the effective action $\Gamma[\varphi_+, \varphi_-]$ the measure functional cannot be disposed of so easily.

25.10. The role of the heat kernel

With a formal definition of $\Gamma[\varphi_+, \varphi_-]$ at hand the question immediately arises how to use it. In the case of the conventional effective action Vilkovisky (1985) has shown that it is possible to calculate not merely one- and two-body propagators in momentum space on a flat empty background, but the full effective action itself, in multi-loop order, directly in real space–time. The chief tool for this is the heat kernel representation of the Feynman propagator, which can be used, for example, to obtain the anomalous magnetic moment of the electron in multi-loop order without ever performing a momentum integration (Vilkovisky 1985).

The propagator $\mathcal{G}$, out of which $\Gamma[\varphi_+, \varphi_-]$ is built, unfortunately does not have a heat kernel representation. It can, however, be obtained from a propagator that does, by a simple process of analytic continuation. The latter propagator arises in the evaluation of the functional integral

$$\langle \text{in}| A[\varphi] |\text{in}\rangle_{\text{E}} = N^2 \int A[\varphi] \exp(-\mathcal{S}_{\text{E}}[\varphi]) \mu[\varphi] \, d\varphi, \qquad (25.69)$$

where

$$\mathcal{S}_{\text{E}}[\varphi] = \int_{-\infty}^{\infty} \mathcal{L}_{\text{E}}[\varphi, \dot{\varphi}, \tau] \, d\tau, \qquad (25.70)$$

$$\mathcal{L}_{\text{E}}[\varphi, \dot{\varphi}, \tau] = \theta(T - \tau) L[\varphi, -i\dot{\varphi}, i\tau] + \theta(\tau - T) L[\varphi, -i\dot{\varphi}, -i(2T - \tau)]. \qquad (25.71)$$

The functional $\mathcal{S}_{\text{E}}$ is a 'Euclideanized' form of $\mathcal{S}$, obtained by making the following complex-plane rotations of the variables t and $\bar{t}$:

$$t = i\tau, \qquad \bar{t} = -i\bar{\tau}. \qquad (25.72)$$

The Green's function of the small-disturbance operator $\mathcal{F}_{\text{E}}$ associated with $\mathcal{S}_{\text{E}}$ has the heat-kernel representation

$$\mathcal{G}_{\text{E}} = \int_0^{\infty} e^{\mathcal{F}_{\text{E}} s} \, ds, \qquad (25.73)$$

or, more explicitly,

$$\mathcal{G}_{\text{E}}(\tau, \tau') = \int_0^{\infty} \mathcal{K}(\tau, \tau', s) \, ds, \qquad (25.74)$$

where

$$\frac{\partial}{\partial s} \mathcal{K}(\tau, \tau', s) = \mathcal{F}_{\text{E}} \mathcal{K}(\tau, \tau', s), \qquad (25.75)$$

$$\mathcal{K}(\tau, \tau', 0) = \mathbf{1} \, \delta(\tau - \tau'), \qquad (25.76)$$

$\mathbf{1}$ being the unit operator. The four blocks of which the propagator $\mathcal{G}(t, t')$ is composed are obtained from $\mathcal{G}_{\text{E}}(\tau, \tau')$ by making appropriate complex-plane rotations inverse to those eqns (25.72).

Use of the heat kernel $\mathcal{K}$ has not yet been implemented in any explicit calculations. Nor has a demonstration been constructed showing that $\mathcal{K}$, in the relevant τ and τ' domains, is independent of how far forward, in imaginary time, the constant T is placed. These remain problems for the future.

References

Buchbinder, I. L., Fradkin, E. S., and Gitman, D. M. (1981). *Fortschr. Phys.* **29,** 187.
DeWitt, B. (1984). In *Relativity, groups and topology II* (ed. B. DeWitt and Stora). North-Holland, Amsterdam.
Fradkin, E. S. and Gitman, D. M. (1981). *Fortschr. Phys.* **29,** 381.
Hajicek, P. (1982). In *Proceedings of the 2nd Marcel Grossmann meeting on general relativity* (ed. R. Rufini). North-Holland, Amsterdam.
Schwinger, J. (1961). *J. Math. Phys.* **2,** 407.
Vilkovisky, G. A. (1985) (to be published).

26

Charged matter from a Kaluza–Klein-like theory

Tsou Sheung Tsun

First let me apologize for both the title, which does not contain the word 'quantum', and the writer, who is a particle physicist—and all relativists know how naive some particle physicists can be when talking about relativity! Notwithstanding this apology, I shall indeed describe to you an extremely naive geometric model which, however, seems to give quite a number of interesting results.

In particle physics it is generally believed that gauge theories describe the fundamental interactions. However, there are at least three immediate problems: no theory tells us which gauge group to use, matter and charges have to be introduced by hand as sources, and space–time is assumed to be flat. As a very crude first step towards meeting these problems, I shall present a highly unrealistic toy model, which, starting from an Einstein theory in $4 + 1$ dimensions, gives charged matter in $3 + 1$ dimensions.

Kaluza (1921) and Klein (1926) postulated that the vacuum state for five-dimensional gravity is four-dimensional Minkowski space M^4 times a (very small) circle S^1. If we put in a magnetic monopole, space–time can no longer be a global product, because the S^1 describing electromagnetism must be twisted. One sees then that such a configuration has actually the simple topology of $\mathbb{R}^5$. This suggests that one may look at things the other way round, i.e. in a five-dimensional Einstein theory one can find a solution which has topology $\mathbb{R}^5$ and which, when suitably interpreted, represents a Kaluza–Klein world populated with a single magnetic monopole. In other words, if one starts with the five-dimensional Einstein–Hilbert action

$$\mathcal{A} = \int \mathrm{d}^5 x \sqrt{-\mathcal{G}} \, \mathcal{R}, \qquad (26.1)$$

where $\mathcal{G}$ and $\mathcal{R}$ refer to five-dimensional metric and curvature, makes the Kaluza–Klein approximation but with topology $\mathbb{R}^5$, puts in the condition that there exists a magnetic monopole, then one might hope to obtain equations of motion appropriate to the interaction of a heavy magnetic monopole with its electromagnetic and gravitational fields (Chan and Tsou 1985). Notice that because of electromagnetic duality the above procedure works equally for an electric monopole, i.e. an electric charge such as an electron. In view of the fact that the actual existence of magnetic monopoles has not been experimentally confirmed despite continued efforts, the second alternative seems to be actually more practical. It is made even more attractive for some of us who are dissatisfied with the current treatment of matter and charges which are just added by hand to the theory with a coupling that is rarely theoretically justified, whereas here the charge (and matter) occurs naturally as a topological entity. However, we stress that the mathematics is the same for either case, and whether we call the charged matter a magnetic monopole or an electric charge is entirely a matter of interpretation (and preference). In order to retain this freedom we can refer to the charged matter simply as a monopole.

To obtain the equations of motion under the above conditions we shall use the method of Lagrange multipliers. These conditions then translate into the following three:

(1) Low energy approximation in Kaluza–Klein theories—we assume that all distances are much larger than the size of the compactified fifth dimension, so that all dependence on x^5 can be ignored. The action then becomes:

$$\mathcal{A} = 2\pi \int d^4x \sqrt{-g} \ \{R + \tfrac{1}{4}\phi F_{ab}F^{ab} + \tfrac{1}{6}\phi^{-2}g^{ab}\phi_{,a}\phi_{,b}\}, \qquad (26.1')$$

where the 5-dimensional metric is written in the form

$$\mathcal{G}_{AB} = \begin{pmatrix} g_{ab}\phi^{1/2} + \phi^{2/3}A_aA_b & A_a\phi^{2/3} \\ A_a\phi^{2/3} & \phi^{2/3} \end{pmatrix},$$

A, B running from 1 to 5, a, b from 1 to 4, with the metric convention $(+++-+)$, and where $F_{ab} = A_{b,a} - A_{a,b}$.

(2) The constraint that at each time t, space has topology $\mathbb{R}^4$. $\mathbb{R}^4$ is considered as a series of S^3 together with a point, the S^3 being circle bundles over nesting S^2 of first Chern class 1, with the fibre shrinking to a point over the distinguished point.

In terms of the variables introduced in (1), we have:

(a) $\phi(Y(s)) = 0$ for a time-like world-line $Y(s)$, $\qquad (26.2a)$

(b) $^*F^{ab}{}_{;b}(x) = 2\pi \int ds \dfrac{1}{\sqrt{-g}} \dfrac{dY^a}{ds} \delta^4(x - Y(s)), \qquad (26.2b)$

where $^*F^{ab} = \tfrac{1}{2}\varepsilon^{abcd}F_{cd}$.

(3) Boundary condition: the size of the circle remains small, i.e. $\phi \rightarrow \phi_0$ at spatial infinity.

Then the action principle says that $\mathscr{A}$ is stationary against variations of g_{ab}, A_a, ϕ and Y^a under the constraints (26.2). The variation is straightforward except for the potential A_a, which has either to be patched or attached to a Dirac string. However, in our particular case we can replace A_a by F_{ab} as variables because of constraint (26.2b). In fact, eqn (26.2b) implies that

$$*F^{ab}{}_{;b} = 0 \qquad (26.3)$$

except at the monopole position. Now by the Poincaré lemma, $*F_{ab}$ divergence-free is equivalent to the existence of a potential A_a of which F_{ab} is the curl. In the usual formulation of Maxwell's theory the reason for using A_a as variable is precisely that F_{ab} is not free. If, however, we put in the divergence-free condition (26.3) as a constraint we can obtain the same equations by varying F_{ab} instead of A_a. Hence in our problem, since we have already put in the constraint (26.2b) we should really use F_{ab} as variable rather than A_a. This then resolves the difficulty because F_{ab} is not patched, i.e. well defined everywhere as a single-valued function.

The equations of motion thus obtained are:

$$R^{ab} - \tfrac{1}{2}Rg^{ab} + \tfrac{1}{2}\{F^{ca}F_c{}^b - \tfrac{1}{4}F_{cd}F^{cd}g^{ab}\}\phi$$
$$+ \tfrac{1}{6}\phi^{-2}\{g^{ca}g^{db}\phi_{,c}\phi_{,d} - \tfrac{1}{2}g^{cd}g^{ab}\phi_{,c}\phi_{,d}\} = 0, \qquad (26.4)$$

$(\phi F^{ab})_{;b} = 0$, together with the two constraint equations.

To facilitate physical interpretation, let us effect the following change of variables:

$$\sigma = \phi/\phi_0$$

$$f_{ab} = \frac{e}{2\pi}\sigma * F_{ab} \qquad (26.5)$$

$$\phi_0 = \frac{4}{\pi}Ge^2,$$

where f_{ab} is now the electromagnetic field due to an (electric) charge e, and G is Newton's constant. After some simplification and elimination, the equations of motion now become:

$$R^{ab} - \tfrac{1}{2}Rg^{ab} + 8\pi G\sigma^{-1}E^{ab} + T^{ab} = 0,$$

$$(\sigma^{-1}f^{ab})_{;b}(x) = e\int ds \frac{1}{\sqrt{-g}}\frac{dY^a}{ds}\delta^4(x - Y(s)), \qquad (26.6)$$

$$\sigma(Y(s)) = 0,$$

where f_{ab} is a gauge field,

$$E^{ab} = g^{ac}f_c{}^b - \tfrac{1}{4}f_{cd}f^{cd}g^{ab},$$
$$T^{ab} = \tfrac{1}{6}(g^{ca}g^{db} - \tfrac{1}{2}g^{cd}g^{ab})(\ln\sigma)_{,c}(\ln\sigma)_{,d}.$$

We notice that (i) σ vanishes at $Y(s)$ and goes to 1 at spatial infinity, which implies that the material tensor T^{ab} is localized around the world-line $Y(s)$, and (ii) at sufficiently large distance from $Y(s)$, f^{ab} looks like a gauge field with a source of strength e also localized around $Y(s)$. In other words, eqns (26.6) describe the motion of what looks like at large distances a lump of charged matter. Furthermore, one can generalize eqns (26.6) to multi-lump equations. The topology then may be more complicated than just $\mathbb{R}^5$.

So, if solutions to eqns (26.6) do exist, we shall have obtained charged lumps of matter (particles?) *together with* their dynamics. Furthermore, at finite temperatures, such solutions will dominate over the Kaluza–Klein vacuum ($M^4 \times S^1$). These solutions are labelled by world-lines $Y^a(s)$ and therefore are statistically more favoured than the vacuum which has no such label, similar to the occurrence of vortices in a large bucket of liquid. Hence the world is likely to be populated by these particle-like charges.

The question remains whether solutions do exist. Sorkin (1983) and Gross and Perry (1983) have exhibited explicit static solutions obtained from the Taub-NUT and multi-Taub-NUT metrics. But because these are static, all interactions exactly cancel and there is no dynamics. The only hope is that after quantization these cancellations would not be exact. Also one does not know if there are other solutions.

To summarize, the model is undoubtedly highly unrealistic, and possibly intrinsically sick. But it does show that it is possible to obtain from pure geometry alone the gauge group, matter, charge and their dynamics, together with two 'bonuses':

(a) only singly-charged 'particles' occur, otherwise space–time will not be a manifold (Hitchin 1979);

(b) the model works only for special cases (e.g. Hopf fibrations), and may thus furnish some element of choice in the correct gauge group.

Acknowledgements

This work is done in collaboration with Chan Hong-Mo. I thank Mike Eastwood for conversations.

References

Chan, Hong-Mo and Tsou, Sheung Tsun (1986). *Acta Phys. Pol.* (to appear).
Gross, D. J. and Perry, M. J. (1983). *Nucl. Phys.* B226, 29.
Hitchin, N. J. (1979). *Math. Proc. Camb. Phil. Soc.* 85, 465.
Kaluza, Th. (1921). *Sitz. Peuss. Akad. Wiss.* K1, 966.
Klein, O. (1926). *Z. Phys.* 37, 895.
Sorkin, R. D. (1983). *Phys. Rev. Lett.* 51, 87.

27

Quantum supergravity via canonical quantization

P. D. D'Eath

27.1. Introduction

In quantum field theory one usually works with the amplitude to go from an initial to a final state, given by a Feynman path integral. However, one can instead describe the theory using a canonical formulation which works with functional differential equations, to which the path integral provides the solution. While the path integral provides a powerful method of evaluating amplitudes, the canonical approach can give considerable insight into the theory, helping with its interpretation and in attacking a variety of problems. The canonical approach to quantizing general relativity has had a long and fruitful history, growing in particular from the work of Dirac (1958*a*,*b*, 1959) and that of DeWitt (1967), in which the quantum state of the gravitational field is described by a wave functional $\varLambda(h_{ij})$, where $h_{ij}(x)$ is the intrinsic metric on a space-like hypersurface. A recent review is given by Kuchař (1981). One would expect similarly to learn a great deal from the canonical quantization of supergravity.

One important application of the canonical approach to supergravity is to the study of the familiar field-theoretic divergences in quantum amplitudes. It is already known that the supersymmetry invariance of the theory severely limits the allowed divergences in the S-matrix (Duff 1982). However, there is at present no guarantee that quantum supergravity is a finite theory beyond two loops in perturbation theory. But when one works with the amplitude to go from data on an initial boundary surface to data on a final surface, rather than the S-matrix, one finds from the canonical approach that the amplitude obeys additional conditions which arise from the freedom to make certain supersymmetry transformations at the boundaries, as will be described in this lecture. These conditions may possibly impose further non-trivial restrictions on the counter-terms allowed in supergravity, helping to show whether or not the theory is indeed finite—a major goal of investigations in quantum gravity.

Another way of attacking the question of divergences in the theory is to study alternatives to the standard semi-classical perturbation approach to computing quantum amplitudes. Canonical quantization leads naturally to an alternative perturbation theory, the multiple-scattering expansion (D'Eath 1981, 1984), which provides a re-summation of the standard theory and may help in limiting quantum divergences in supergravity.

A further natural application of the canonical quantization of supergravity is to quantum cosmology. Considerable progress has been made in understanding the quantum behaviour of the universe as a whole by applying canonical quantization to the interaction of gravity and matter and restricting attention only to highly symmetrical configurations of the gravitational and matter fields. The resulting quantization of cosmological geometries with only a small number of degrees of freedom may then be tractable analytically. Recently, interest in this field has been renewed by the work of Hartle and Hawking (Hartle and Hawking 1983; Hawking 1984), who have suggested a candidate path-integral expression for the 'ground state' wave function of the universe. In all this work, the matter is put in by hand in a rather *ad hoc* way. It would be far more natural to start with a unified theory of matter and gravity such as simple $N = 1$ supergravity, considered here, or one of the higher-N supergravity theories. Interesting results should follow from applying the approach described in this lecture to cosmological models.

In Section 27.2 the representation of quantum states and operators is outlined, leading to a discussion in Section 27.3 of the quantum version of the constraints occurring in the classical Hamiltonian theory. The quantum amplitude to go from data on an initial surface to data on a final surface is treated in Section 27.4. The quantum constraints impose conditions on this amplitude which may limit the allowed counter-terms in the theory, as discussed in Section 27.5. A more detailed exposition of this material may be found in D'Eath (1984).

27.2. Hamiltonian formulation, quantum states and operators

The canonical quantization of any classical theory starts from its Hamiltonian formulation. Classical supergravity may be described in terms of the tetrad $e_\mu^{AA'}$, the spin-$\frac{3}{2}$ field ψ_μ^A and its hermitian conjugate $\bar\psi_\mu^{A'}$. Here $e_\mu^{AA'}$ is a hermitian spinor-valued form which is even, commuting with all other variables, while the spinor-valued forms ψ_μ^A and $\bar\psi_\mu^{A'}$ are odd, anti-commuting among themselves. Indices $A, \dots, A', \dots$ are two-component spinor indices, while $\mu, \nu, \dots$ are space–time indices. The auxiliary fields of supergravity (van Nieuwenhuizen 1981) do not need to be included, since they vanish for classical solutions and are set to zero in the Hamiltonian formulation.

In the Hamiltonian treatment, as presented by Pilati (1978), the basic dynamical variables on space-like hypersurfaces $t = x^0 = $ constant are the spatial tetrad components $e_i^{AA'}$ and their conjugate momenta $p^i_{AA'}$, where $i, j, \ldots$ are spatial indices, together with ψ_i^A and $\bar{\psi}_i^{A'}$. Since the Lagrangian of supergravity involves only first-order derivatives of the spin-$\frac{3}{2}$ fields, their conjugate momenta π_A^i and $\bar{\pi}_{A'}^i$ can be written in terms of the other variables. Using these conditions, which are second-class constraints in the sense of Dirac (1965), to eliminate π_A^i and $\bar{\pi}_{A'}^i$ from the theory, one finds that the usual Poisson brackets suitable for a classical theory containing both bosons and fermions (Casalbuoni 1976) are replaced by more complicated Dirac brackets $[,]^*$ among the basic variables.

The Hamiltonian takes the form

$$H = \int d^3x \left(\begin{array}{c} N\mathcal{H}_\perp + N^i \mathcal{H}_i + \psi_0^A S_A + \bar{S}_{A'} \bar{\psi}_0^{A'} \\ - \omega_{AB0} J^{AB} - \bar{\omega}_{A'B'0} \bar{J}^{A'B'} \end{array} \right) \tag{27.1}$$

plus terms at spatial infinity, where relevant. The quantities $\mathcal{H}_\perp$ and $\mathcal{H}_i$ are modified generators of normal and tangential deformations of the hypersurface, S_A and $\bar{S}_{A'}$ are the generators of supersymmetry transformations, and $J^{AB}, \bar{J}^{A'B'}$ are the generators of local Lorentz transformations acting on the basic dynamical variables (Teitelboim 1977). All of these quantities are formed from the basic variables. The remaining quantities in eqn (27.1) are the lapse function N and shift vector N^i (which involve the components $e_0^{AA'}$), describing the amount by which the hypersurface is deformed normally and tangentially per unit co-ordinate time, together with ψ_0^A, $\bar{\psi}_0^{A'}$ and the components ω_{AB0} and $\bar{\omega}_{A'B'0}$ of the connection forms $\omega_{AB\mu}$, $\bar{\omega}_{A'B'\mu}$. Here ψ_0^A and $\bar{\psi}_0^{A'}$ give the amount of supersymmetry transformation and ω_{AB0}, $\bar{\omega}_{A'B'0}$ the amount of Lorentz transformation applied to the basic variables per unit time.

Classically the dynamical variables obey the constraint equations

$$\mathcal{H}_{AA'} = 0,$$
$$S_A = 0, \qquad \bar{S}_{A'} = 0, \tag{27.2}$$
$$J^{AB} = 0, \qquad \bar{J}^{A'B'} = 0.$$

Here

$$\mathcal{H}_{AA'} = -n_{AA'} \mathcal{H}_\perp + e^i_{AA'} \mathcal{H}_i, \tag{27.3}$$

where $n_{AA'}$ is the spinor version of the unit time-like future-directed normal n^μ to the surface, and $e^i_{AA'} = h^{ij} e_{AA'j}$ where h^{ij} is the inverse of the spatial metric $h_{ij} = -e_{AA'i} e_j^{AA'}$.

In quantizing the system, Dirac brackets must be replaced by (anti-) commutators according to

$$[E_1, E_2] = i\hbar[E_1, E_2]^*,$$
$$[O, E] = i\hbar[O, E]^*, \qquad (27.4)$$
$$\{O_1, O_2\} = i\hbar[O_1, O_2]^*,$$

where E denotes an even element and O an odd element (Dirac 1965; Casalbuoni 1976). One has to find a description of quantum states and a representation of the basic dynamical variables in terms of operators acting on those states, such that their (anti-) commutators provide a quantum version of the classical Dirac bracket relations. There are many alternative bases which can be chosen here, but it would be convenient to have a generalization of the description of states in canonically quantized general relativity by functionals $f(h_{ij}(x))$ of the spatial metric, and so to work with eigenstates of the spatial tetrad variables $e_i^{AA'}(x)$. One cannot in addition work with simultaneous eigenstates of both $\psi_i^A(x)$ and $\bar{\psi}_i^{A'}(x)$, since their Dirac bracket is non-zero; e.g. $\bar{\psi}_i^{A'}(x)$ is related to the momentum conjugate to $\psi_i^A(x)$. One must rather choose to work with eigenstates of $e_i^{AA'}(x)$ and (say) the right-handed part $\psi_i^A(x)$ of the spatial spin-$\frac{3}{2}$ field, describing quantum states by wave-functionals $f(e_i^{AA'}(x), \psi_i^A(x))$. This provides a way of treating bosonic and fermionic variables on a roughly equal footing.

The variables $\psi_i^{A'}(x)$ and $p^i_{AA'}(x)$ can then be represented by the momentum-like operators

$$\bar{\psi}_i^{A'}(x) = -i\hbar D_{ji}^{AA'}(x)\frac{\delta}{\delta\psi_j^A(x)}, \qquad (27.5)$$

$$p^i_{AA'}(x) = -i\hbar\frac{\delta}{\delta e_i^{AA'}(x)} + \tfrac{1}{2}\varepsilon^{ijk}\psi_{Aj}(x)\bar{\psi}_{A'k}(x), \qquad (27.6)$$

where

$$D_{ji}^{AA'} = -2ih^{-1/2}e_i^{AB'}e_{BB'j}n^{BA'} \qquad (27.7)$$

and $h \equiv \det h_{ij}$. Note that there is some freedom in the factor ordering of $p^i_{AA'}(x)$: alternative choices could have been made in the second term in (27.6). A formal inner product can be found in which $\psi_i^A(x)$ and $\bar{\psi}_i^{A'}(x)$ are hermitian conjugates, and $p^i_{AA'}(x)$ is hermitian.

27.3. The quantum constraints

Following the work of Dirac (1965), the first-class constraints (27.2) in the classical theory become constraints on physically allowed states in the quantum theory. For example, to take the simplest of these, the operator versions of the classical quantities J^{AB} and $\bar{J}^{A'B'}$ must annihilate physical wave functions $f(e_i^{AA'}(x), \psi_i^A(x))$:

$$J^{AB}f = 0, \qquad \bar{J}^{A'B'}f = 0. \tag{27.8}$$

These conditions just imply that f is invariant under right- and left-handed local Lorentz transformations applied to $e_i^{AA'}(x)$ and $\psi_i^A(x)$, acting on their spinor indices.

Next consider the operator $\bar{S}_{A'}$. Classically, $\bar{S}_{A'}$ is given by

$$\bar{S}_{A'} = \varepsilon^{ijk}e_{AA'i}{}^{3s}D_j\psi_k^A + \tfrac{1}{2}i\kappa^2\psi_i^A p_{AA'}^i, \tag{27.9}$$

where ${}^{3s}D_j$ is a torsion-free spatial covariant derivative (D'Eath 1984), and $\kappa^2 = 8\pi$ in geometrical units. Quantum-mechanically it is natural to choose the same factor-ordering for $\bar{S}_{A'}$, which in our representation is given by the first-order differential operator

$$\bar{S}_{A'} = \varepsilon^{ijk}e_{AA'i}{}^{3s}D_j\psi_k^A - \tfrac{1}{2}\hbar\kappa^2\psi_i^A \frac{\delta}{\delta e_i^{AA'}}. \tag{27.10}$$

With this choice, the quantum constraint

$$\bar{S}_{A'}f = 0 \tag{27.11}$$

simply implies that under a left-handed supersymmetry transformation

$$\delta e_i^{AA'} = -i\kappa\bar{\varepsilon}^{A'}\psi_i^A, \qquad \delta\psi_i^A = 0 \tag{27.12}$$

specified by the infinitesimal odd spinor parameter $\bar{\varepsilon}^{A'}(x)$, a physical wave functional $f(e_i^{AA'}(x), \psi_i^A(x))$ changes by

$$\delta f = \frac{-2i}{\hbar\kappa} f \int d^3x\, \varepsilon^{ijk}e_{AA'i}({}^{3s}D_j\psi_k^A)\bar{\varepsilon}^{A'}. \tag{27.13}$$

The quantum operator S_A can then be found as the Hermitian conjugate of $\bar{S}_{A'}$. In this representation it is a second-order differential operator, and the constraint $S_A f = 0$ appears complicated. Of course, when one changes over to a basis of eigenstates of $e_i^{AA'}(x)$ and $\bar{\psi}_i^{A'}(x)$ (using a suitable functional Fourier transform), this constraint also gives a simple transformation property analogous to eqns (27.12) and (27.13) under right-handed supersymmetry transformations.

It remains to determine the form of the quantum operator $\mathcal{H}_{AA'}$ and the corresponding quantum constraint. In this one is helped by the form of the classical Dirac bracket between $S_A(x)$ and $\bar{S}_{A'}(x')$, which equals $\tfrac{1}{2}i\kappa^2\mathcal{H}_{AA'}(x)\,\delta(x,x')$ plus terms proportional to the Lorentz generators J and $\bar{J}$. [Originally it was thought that the extra J, $\bar{J}$ terms were absent (Teitelboim 1977), but their presence was pointed out by Henneaux

(1983) and can be verified by direct computation.] Quantum-mechanically the anti-commutators among S_A and $\bar{S}_{A'}$ are

$$\{S_A(x), S_B(x')\} = 0, \qquad \{\bar{S}_{A'}(x), \bar{S}_{B'}(x')\} = 0,$$
$$\{s_A(x), \bar{S}_{A'}(x')\} = -\tfrac{1}{2}\hbar\kappa^2 \,_2\mathscr{H}_{AA'}(x)\,\delta(x, x'),$$

(27.14)

where the operator $_2\mathscr{H}_{AA'}(x)$ is given in D'Eath (1984). Thus a state obeying the constraints $\bar{S}_{A'}f = 0$ and $S_A f = 0$ will automatically also obey the condition $_2\mathscr{H}_{AA'}f = 0$. One now needs to check whether the operator $_2\mathscr{H}_{AA'}$ can be written as a quantum version of $\mathscr{H}_{AA'}$ plus terms containing factors of J^{BC} or $\bar{J}^{B'C'}$ appearing on the right. This check on factor ordering has not yet been carried out. However, if it holds, one can use this property to define the quantum operator $\mathscr{H}_{AA'}$, and the quantum constraint $\mathscr{H}_{AA'}f = 0$ will then follow from the other quantum constraints. If it doesn't hold, then this attempt at factor-ordering is inconsistent and one must try a different one. Similarly, one should check that all the remaining commutator relations among the operators S_A, $\bar{S}_{A'}$ and $\mathscr{H}_{AA'}$, following the classical pattern given by Teitelboim (1977), again apart from suitable extra terms containing factors of J^{BC} or $\bar{J}^{B'C'}$, lead to no further possible inconsistencies.

Provided that $\mathscr{H}_{AA'}$ can be defined, the tangential projection $e_i^{AA'}\mathscr{H}_{AA'}f = 0$ of the corresponding constraint onto a surface $t = \text{const.}$ should imply that f is invariant under spatial co-ordinate transformations applied to its arguments $e_i^{AA'}(x)$ and $\psi_i^A(x)$, subject to the other constraints. Again this will require a non-trivial calculation, since one already knows classically that $\mathscr{H}_{AA'}$ generates modified co-ordinate transformations (Teitelboim 1977). The normal projection $n^{AA'}\mathscr{H}_{AA'}f = 0$ will give a many-time evolution equation for f, analogous to the Wheeler–DeWitt equation of canonically quantized general relativity (Wheeler 1968; DeWitt 1967).

There is no guarantee that the factor ordering suggested here will turn out to be consistent. However, one may take some hope at this point, since the form of the operator $\bar{S}_{A'}$ associated with the simple transformation property (eqns (27.12) and (27.13)), together with the bracket relations among S_A and $\bar{S}_{A'}$, do offer some guidance in the canonical quantization of supergravity which is lacking in the canonical treatment of ordinary gravity.

27.4. The quantum amplitude

The most fundamental quantity in the theory is the amplitude to go from data given on an initial hypersurface to data on a final hypersurface. In our approach, the most natural way to pose such data turns out to be to specify the spatial tetrad $e_i^{AA'}(x)$ and left-handed part $\psi_i^{A'}(x)$ of the spatial spin-$\frac{3}{2}$ field (say) on the initial surface (I), and to give the tetrad $e_i^{AA'}(x)$ and right-handed part $\psi_i^A(x)$ on the final surface (F). This will provide a generalization of the amplitude $K(h_{ijF}; h_{ijI})$ in quantum gravity to go from a 3-metric h_{ijI} on an initial surface to a 3-metric h_{ijF} on a final surface. Here the amplitude is given formally by a path integral

$$K(e_F, \psi_F; e_I, \tilde{\psi}_I) = \int \exp(iS/\hbar)\mathcal{D}e\mathcal{D}\psi\mathcal{D}\tilde{\psi} \qquad (27.15)$$

over all infilling fields $e_\mu^{AA'}$, ψ_μ^A, $\tilde{\psi}_\mu^{A'}$, where $(e_\mu^{AA'}, \tilde{\psi}_\mu^{A'})$ restricted to the initial surface give the correct initial data $(e_i^{AA'}, \psi_i^{A'})_I$, and similarly $(e_\mu^{AA'}, \psi_\mu^A)$ give the correct final data. Berezin integration (Berezin 1966) is used for the fermionic variables. The integral can be regarded as a contour integral, where the contour can be deformed such that $e_\mu^{AA'}$ is no longer hermitian; similarly ψ_μ^A and $\tilde{\psi}_\mu^{A'}$ are to be regarded as independent quantities, not necessarily hermitian conjugates.

The action S contains the usual spin-2 and spin-$\frac{3}{2}$ volume terms S_2 and $S_{3/2}$ of supergravity (van Nieuwenhuizen 1981) as well as boundary terms S_{2B} and $S_{3/2B}$ at the initial and final surfaces, together with a contribution S_∞ at spatial infinity where relevant:

$$S = S_2 + S_{3/2} + S_{2B} + S_{3/2B} + S_\infty. \qquad (27.16)$$

Strictly one should also include gauge-fixing and ghost terms to make the path integral (27.15) well-defined (van Nieuwenhuizen 1981). The boundary terms are necessary in order that when one composes the amplitude to go from data on an initial surface I to an intermediate surface J with the amplitude to go from J to a final surface F, one correctly recovers the amplitude to go from I to F, just as in quantum gravity (Gibbons and Hawking 1977). They are given for our choice of data by

$$S_{2B} = \kappa^2\left(\int_F - \int_I\right) d^3x h^{1/2} \operatorname{Tr} K,$$

$$\qquad (2.17)$$

$$S_{3/2B} = \tfrac{1}{2}\left(\int_I + \int_F\right) d^3x \varepsilon^{ijk}\psi_i^A e_{AA'j}\tilde{\psi}_k^{A'},$$

where $\operatorname{Tr} K$ is the trace of the second fundamental form K_{ij} on the surface. These terms also ensure that if one imposes the condition $\delta S = 0$ that S be stationary under small variations of $e_\mu^{AA'}$, ψ_μ^A and $\tilde{\psi}_\mu^{A'}$, subject to our boundary data, one correctly recovers the classical solution joining the initial to the final data.

The amplitude K will obey the quantum constraints of Section 27.3 with respect to its arguments at both the initial and final surfaces. A general physical wave functional $f(e_i^{AA'}, \psi_i^A)$ can be built up by super-position from amplitudes K corresponding to different initial data.

One way of evaluating the amplitude is to take the standard semi-classical expansion

$$K(e_F, \psi_F; e_I, \bar{\psi}_I) \sim (A + \hbar A_1 + \hbar^2 A_2 + \ldots) \exp\left[\frac{i}{\hbar} S_c(e_F, \psi_F; e_I, \bar{\psi}_I)\right]$$

(27.18)

about the classical solution $(e_\mu^{AA'}, \psi_\mu^A, \bar{\psi}_\mu^{A'})$ obeying $\delta S = 0$ with the prescribed boundary data. Here S_c is the action of the classical solution, while A, A_1, A_2, ... are one-, two-, three- and higher-order loop terms found by expanding around it. Note that in the classical solution $e_\mu^{AA'}$ will not generally be Hermitian, even though the boundary data $e_i^{AA'}$ may be, nor will ψ_μ^A and $\bar{\psi}_\mu^{A'}$ in general be Hermitian conjugates.

The quantum constraints have important implications for the terms in this expansion. The J^{AB} and $\bar{J}^{A'B'}$ constraints imply that K is invariant under local Lorentz transformations acting on the initial and final data, and the tangential projection of the $\mathcal{H}_{AA'}$ constraint should imply that K is also invariant under spatial co-ordinate transformations of the boundary data. Each of the terms S_c, A, A_1, A_2, ... in the expansion will share these invariance properties. The situation with regard to the supersymmetry constraints is somewhat different. For example, let us regard the amplitude as a function of the data $e_i^{AA'}(x)$, $\psi_i^A(x)$ on the final surface, for given initial data $e_i^{AA'}(x)$, $\bar{\psi}_i^A(x)$, and consider the constraint

$$\bar{S}_{A'}K = 0$$

(27.19)

at the final surface. Following eqns (27.12) and (27.13), this shows that under the left-handed supersymmetry transformation

$$\delta e_i^{AA'} = -i\kappa \bar{\varepsilon}^{A'} \psi_i^A, \qquad \delta \psi_i^A = 0$$

(27.20)

at the final surface, the action of the classical solution *changes* by the amount

$$\delta S_c = \frac{-2}{\kappa} \int d^3x \varepsilon^{ijk} e_{AA'i} (^{3s}D_j \psi_k^A) \bar{\varepsilon}^{A'},$$

(27.21)

while *the one- and higher-loop terms are invariant*:

$$\delta A = \delta A_1 = \delta A_2 = \ldots = 0.$$

(27.22)

A similar property holds for right-handed supersymmetry transformations at the initial surface. These transformation properties may help in limiting the quantum divergences in the theory (Section 27.5).

The constraint $S_A K = 0$ at the final surface is more complicated, since S_A is a second-order differential operator with respect to the variables $e_i^{AA'}(x)$ and $\psi_i^A(x)$. The second-order nature of S_A enforces the presence of the higher-loop terms A_1, A_2, . . . in the amplitude; e.g. one cannot solve the constraint $S_A K = 0$ exactly with an amplitude K of the one-loop form $A \exp(iS_c/\hbar)$. Following the discussion of Section 27.3, one can regard this constraint as effectively governing the dynamics of the amplitude as a functional of the final data, subject to the transformation properties above, since its anti-commutator with $\bar{S}_{A'}$ involves $\mathcal{H}_{AA'}$.

27.5. Divergences in the quantum amplitude in the presence of boundaries

Typically, in a semi-classical expansion of the form (27.18) in a quantum field theory, the one- and higher-loop terms $A, A_1, \ldots$ will involve divergent integrals, requiring the inclusion of counter-terms in the action. For example, in quantum gravity, when one considers the amplitude to go from a given 3-metric h_{ijI} on an initial surface to a 3-metric h_{ijF} on a final surface, the counter-terms should involve volume integrals of invariants formed from the Riemann tensor and its derivatives, together with surface integrals of invariants formed from the spatial metric h_{ij}, its curvature, the second fundamental form K_{ij} and their spatial derivatives. At one loop, evaluated at a classical solution joining the initial and final data, one of the counter-terms is proportional to the Euler number of the space–time, which here involves both volume and surface contributions (Eguchi *et al.* 1980). There are also four possible counter-terms at one loop formed purely from surface quantities, namely $\int d^3x h^{1/2} \, {}^3R_{ij}K^{ij}$, where ${}^3R_{ij}$ is the spatial Ricci scalar, and integrals of the three invariants formed from the cube of K_{ij}. Possible volume and surface counter-terms in quantum gravity proliferate at higher loop order.

In quantum supergravity at one loop there is again a counter-term proportional to the Euler number (Duff 1982), which being a topological invariant automatically yields $\delta A = 0$ [condition (27.22)] under a supersymmetry transformation (27.20) at the final surface (and similarly at the initial surface). However, the invariance property (27.22) does forbid the purely surface counter-terms (D'Eath 1984). There are no surface terms corresponding either to the $\int d^3x h^{1/2} \, {}^3h_{ij}K^{ij}$ to the $\int d^3x h^{1/2}KKK$ terms of quantum gravity, as no partners involving fermionic terms can be found for them to make a supersymmetric invariant.

At higher-loop order in quantum supergravity there will again be two types of counter-term to be considered. First, there may again be purely surface counter-terms at the final surface invariant under left-handed supersymmetry (and corresponding terms at the initial surface). These can also be shown to be forbidden at least up to two-loop order. Second, the invariance property (eqns (27.20) and (22.22)) requires that a surface contribution be added to each of the usual volume counter-terms, just as the Euler number is given by a sum of volume and surface terms. It is conceivable that this might place further restrictions on the usual volume terms, eliminating some of them. In this case the formulation of quantum supergravity in terms of the amplitude to go from data on an initial surface to data on a final surface would have interesting implications for the alternative S-matrix description. The existence of counter-terms will be most easily investigated using superfield methods. It would also be valuable to investigate the corresponding question for the higher-N extended supergravity theories, particularly for $N = 8$, which is the theory most likely to be finite.

References

Berezin, F. A. (1966). *The method of second quantization*. Academic Press, New York.

Casalbuoni, R. (1976). *Nuovo Cim.* **33A,** 115.

D'Eath, P. D. (1981). *Phys. Rev.* D **24,** 811.

—— (1984). *Phys. Rev.* D**29,** 2199; erratum, to appear.

DeWitt, B. S. (1967). *Phys. Rev.* **160,** 1113.

Dirac, P. A. M. (1958*a*). *Proc. R. Soc. Lond.* A **246,** 326.

—— (1958*b*). *Proc. R. Soc. Lond.* A **246,** 333.

—— (1959). *Phys. Rev.* **114,** 924.

—— (1965). *Lectures on quantum mechanics*. Academic Press, New York.

Duff, M. J. (1982). In *Supergravity '81* (ed. S. Ferrara and J. G. Taylor). Cambridge University Press, Cambridge.

Eguchi, T., Gilkey, P. B., and Hanson, A. J. (1980). *Phys. Rep.* **66,** 213.

Gibbons, G. W. and Hawking, S. W. (1977). *Phys. Rev.* D **15,** 2752.

Hartle, J. B. and Hawking, S. W. (1983). *Phys. Rev.* D **28,** 2960.

Hawking, S. W. (1984). In *Relativity, groups and topology II* (ed. B. S. DeWitt and R. Stora). North-Holland, Amsterdam.

Henneaux, M. (1983). *Phys. Rev.* D**27,** 986.

Kuchař, K. (1981). In *Quantum Gravity 2* (ed. C. J. Isham, R. Penrose, and D. W. Sciama). Oxford University Press, Oxford.

van Nieuwenhuizen, P. (1981). *Phys. Rep.* **68,** 189.

Pilati, M. (1978). *Nucl. Phys.* B **132,** 138.

Teitelboim, C. 1977. *Phys. Rev. Lett.* **38,** 1106.

Wheeler, J. A. (1968). In *Battelle Rencontres* (ed. C. M. DeWitt and J. A. Wheeler). Benjamin, New York.

Index